Mathematical Engineering

Today, the development of high-tech systems is unthinkable without mathematical modeling and analysis of system behavior. As such, many fields in the modern engineering sciences (e.g. control engineering, communications engineering, mechanical engineering, and robotics) call for sophisticated mathematical methods in order to solve the tasks at hand.

The series Mathematical Engineering presents new or heretofore little-known methods to support engineers in finding suitable answers to their questions, presenting those methods in such manner as to make them ideally comprehensible and applicable in practice.

Therefore, the primary focus is—without neglecting mathematical accuracy—on comprehensibility and real-world applicability.

To submit a proposal or request further information, please use the PDF Proposal Form or contact directly: Dr. Thomas Ditzinger (thomas.ditzinger@springer.com)

Indexed by SCOPUS, zbMATH, SCImago.

Jorge Angeles · Shaoping Bai

Kinematics of Mechanical Systems

Fundamentals, Analysis and Synthesis

Springer

Jorge Angeles
Department of Mechanical Engineering,
Centre for Intelligent Machines
McGill University
Montreal, QC, Canada

Shaoping Bai
Department of Materials and Production
Aalborg University
Aalborg, Denmark

ISSN 2192-4732 ISSN 2192-4740 (electronic)
Mathematical Engineering
ISBN 978-3-031-09543-6 ISBN 978-3-031-09544-3 (eBook)
https://doi.org/10.1007/978-3-031-09544-3

This Springer imprint is published by the registered company Springer Nature Switzerland AG
The registered company address is: Gewerbestrasse 11, 6330 Cham, Switzerland

Preface

The *kinematics of mechanisms* is a classical discipline in the broader domain of *mechanical systems.* The latter include systems of rigid and deformable solids, while the former has mainly focused on systems of rigid bodies. Given that the mechanics of deformable solids depends on the forces applied to the body under study, their motion is not *determined* solely by the motion undergone by some of the system bodies. Indeed, the interplay of forces and moments (wrenches) with the displacements of the system bodies must be taken into account via the principles of *continuum mechanics*, which is a significantly broad multidiscipline, involving fluids and deformable solids. In this multidiscipline, the relations between *wrenches* and *displacements* are included via the *constitutive equations* relating them.

On the contrary, the motion of a system of coupled *rigid bodies* depends only on the motion undergone by a given subset of the bodies. In this case, knowing the motion of that subset, the motion of all the bodies of the system is *determined*—modulo the possible *finite* number of postures adopted by a subset of the system bodies, a concept termed *assembly modes.* The mechanical systems of interest to the book involve only *rigid bodies.*

Broadly speaking, the systems of rigid bodies studied here can form *open* and *closed* chains. The paradigm mechanical system studied in the realm of kinematics is the planar *four-bar linkage*, composed of four rigid bodies (the links) coupled by means of what is known as *lower kinematic pairs* (LKP). The book focuses on systems of rigid bodies forming closed loops, mainly intended for one-degree-of-freedom (one-dof) tasks. The four-bar linkage is given due attention, in its three domains: *planar*, *spherical*, and *spatial.* Multi-degree-of-freedom mechanisms of either open or closed kinematic chains are of utmost importance in robotics. These systems are extensively and intensively studied in the realm of robotic mechanical systems, whereby a rich literature is available.

The book contains six chapters. Chapter 1 provides the fundamentals for kinematic synthesis, covering two basic concepts, synthesis-equation solving, and numerical methods. Chapter 2 is devoted to the *qualitative synthesis* of linkages, in which the layout of kinematic chains and pairs is considered, to determine the linkage degree of freedom. Chapters 3–5 cover in full detail single-dof linkages of three major types, namely *planar, spherical, and spatial* four-bar linkages. Their functional, motion and path syntheses are described in sequence in the three chapters. In the last chapter, the synthesis of single-dof complex linkages, including six-bar and ten-bar linkages, is introduced as cases of application of the methodology studied in the book.

Our aim is to provide comprehensive and systematic knowledge of kinematic-synthesis theory and methods. The book includes not only the classical foundations of kinematic synthesis, such as the *Burmester theorem*, the *Roberts–Chebyshev theorem*, and the *principle of transference*, but also the latest advances in this area that were made by the authors, particularly on coupler-curve synthesis. Moreover, synthesis examples are included in the book to illustrate the synthesis theory and methods. The book is suitable for graduate students of mechanical engineering, researchers of mechanism and robot design, and machine design engineers. We trust that the book will guide and inspire engineers, researchers, and students in the design, analysis, and development of innovative and high-performance machines in all relevant fields of applications.

Jorge Angeles acknowledges the financial support received from Canada's *Natural Sciences and Engineering Research Council* (NSERC) throughout his academic career in Canada. Indeed, NSERC's *Discovery Grant* program allowed him to hire top talent at the *Robotic Mechanical Systems Laboratory* (RMSLab), McGill University, since the mid-1980s. Various graduate students and visiting fellows contributed, with their work, to the development of ideas which, eventually, found their way into the book. A complete list of these trainees is available in the RMSLab web page.[1]

Shaoping Bai acknowledges Danish national and EU funding bodies, including *Innovationsfonden*, *Independent Research Fund Denmark*, and *EU AAL Programme*, among others, for providing financial support for research in innovative mechanisms and exoskeletons at Aalborg University, Denmark. He acknowledges also collaboration with local industry and international universities, such as *PN Beslag Fabric* and *Dalian University of Technology*, among others, which motivated research topics on linkage synthesis and applications.

Jorge Angeles, Ph.D.
Professor Emeritus, McGill University
Montreal, QC, Canada

Shaoping Bai, Ph.D.
Professor, Aalborg University
Aalborg, Denmark

[1] http://www.cim.mcgill.ca/:rmsl/Index/index.htm.

Contents

List of Figures

List of Tables

Chapter 1
Introduction to Kinematic Synthesis

1.1 The Role of Kinematic Synthesis in Mechanical Design

When designing a *mechanical system*, whether a *structure* or a *machine*, the first step is to produce a *conceptual design* that will meet the *design specifications*. Broadly speaking, the main *function* of a structure is *to be capable of withstanding the anticipated loads without exhibiting major deformations that would hamper the integrity of the structure, and hence, the safety of its occupants*. Likewise, the main function of a machine is *to be able to perform the intended task, involving* **finite displacements** *of its parts, without major deformations that would hamper the integrity of the machine or the safety of its users*.

To be true, mechanical systems exhibit, more often than not, features of both structures and mechanisms. Such examples occur in transportation machinery, such as landing gears in aircraft. Figure 1.1 illustrates the structure of the deployed landing gear of the *Airbus A300-600* aircraft. In this *posture*, the landing gear works as a structure, to withstand not only the static load of the airplane on the tarmac, but also the *dynamic load* exerted by the unavoidable impact upon landing.

Transportation machinery is a domain in which mechanisms, especially linkages—the focus of this book—play a major role. Shown in Fig. 1.2 is a depiction of the powertrain of a Class-C Mercedes Benz. A key subsystem of the system in question—the powertrain—is the steering linkage, whose main components are visible in the figure. Again, this mechanism plays the dual role of a structure and a machine, as its function is not only to properly orient the planes of the front wheels upon turning, but also to support the wheels and the loads transmitted by the ground onto the vehicle frame.

In the above paragraphs, we have introduced concepts of *engineering design* as pertaining to mechanical systems at large. Of these, we have focused on structures and machines. In fact, design, together with *manufacturing*, is the *raison d'être* of engineers, all disciplines known as *engineering science*, namely, mechanics, thermofluids and numerics, to name but just the major branches, playing a supporting

J. Angeles and S. Bai, *Kinematics of Mechanical Systems*, Mathematical Engineering,
https://doi.org/10.1007/978-3-031-09544-3_1

Fig. 1.1 The landing gear of an *Airbus A300–600*

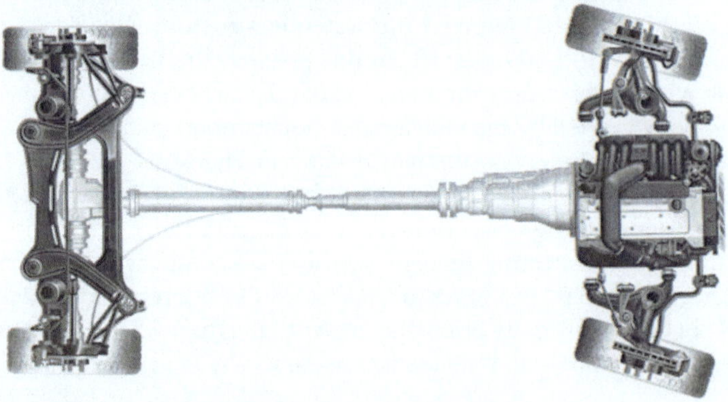

Fig. 1.2 A view of the powertrain of a C-Class Mercedes Benz

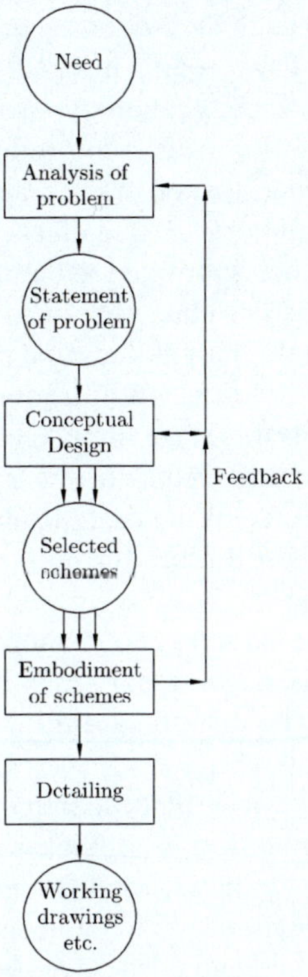

Fig. 1.3 French's model of the design process

role in the production process. For this reason, it is necessary to dwell on this concept. Because of its importance, the *engineering design process* has been the subject of study over the centuries, starting with Marcus Vitruvius Polio(ca. 75 BCE–ca. 15 CE) and his 10-volume work under the title *De Architectura* (Vitruvius, 28 B.C.E.). Modern engineering design theory owes its origins, largely, to Franz Reuleaux (1829–1905), who first proposed a *grammar* to describe the kinematic chain of a machine (Reuleaux, 1875).

A modern model of the design process, due to French (1992), is depicted in Fig. 1.3. In this model, four stages are distinguished: (a) analysis of the problem; (b) conceptual design; (c) embodiment design; and (d) detailing, or detailed design.

In the first stage, analysis of the problem, the *functions* required from the object under design, in our case, a machine, are clearly defined, in general, but precise terms. At this stage, the task of the design engineer is to produce (a) design requirements, in terms as general as possible, in order to avoid biasing the design team towards a specific layout of the solution, and (b) *design specifications*, so as to satisfy the rather vaguely spelled-out needs of the *client*.

In the second stage, the design team produces a set of *design variants*, as rich as possible, after several sessions, structured or unstructured, which are part of the *creative* aspect of the design process.

In the third stage, the design team focuses on a reduced set of design variants, those having the highest likelihood of succeeding in satisfying the client's demands within the resources—budget, milestones, technology—set at the disposal of the design team. In this stage, the task of the team is to produce a preliminary model of the design solution, with a clear identification of the main parameters defining a specific design variant. Further, a *parametric model* of each of the short-listed candidate variants is produced, which is then submitted to optimization with the aim of finding the specific fundamental dimensions that either maximize a profit or minimize a cost of the design solution, or even do both at the same time, in a process that is known as *multiobjective optimization*. Design optimization is thus a key activity in the embodiment stage, which makes this stage *iterative*, as optimization requires several rounds of assignment of numerical values to the parameters of the mathematical model; evaluating the performance of the design solution thus resulting; improving this performance, when there is still room for improvement; and stopping when no further improvement is possible.

The final stage involves material selection, manufacturing issues, and production-cost analysis. As a result of this stage, a set of manufacturing drawings is produced that is then sent out for prototype manufacturing, when the design job so requires, or directly to production.

Kinematic synthesis plays a key role in the first three stages of the foregoing design process, as pertaining to machine design. In fact, in the first stage, analysis of the problem, more than kinematic synthesis, what is required is *kinematics knowledge*, as design functions and specifications are to be understood by any engineer trained in the discipline. It is in the second and third stages where kinematic synthesis plays a fundamental role, as explained below.

As pertaining to machine design or, more specifically, to mechanism design, Hartenberg and Denavit proposed three phases of kinematic synthesis: (a) *type synthesis*; (b) *number synthesis*; and (c) *dimensional synthesis* (Hartenberg and Denavit, 1964a). Both type and number syntheses pertain to the *conceptual design* phase, as the former refers to choosing the type of mechanism to perform the required function, namely, a linkage, a cam-follower mechanism, a belt-pulley transmission, or a gear train, for example. Number synthesis refers to the numbers of links and joints in a linkage, along with the type of joints to be used—kinematic joints, or lower kinematic pairs, are studied in Chap. 2. In this book, only mechanical systems coupled by *lower kinematic pairs* are studied. *Higher kinematic pairs* (HKP) occur in gear trains and cam mechanisms, which lie outside of the scope of the book. An extensive discussion of HKP, with pertinent references, is available in the literature (Angeles and López-Cajún, 1991).

The conceptual phase of the design process is fundamental. Moreover, this phase is the one that has posed the major challenges to those attempting to automate the design process. In the realm of kinematic synthesis, we introduce a methodology, termed *qualitative synthesis*, in Chap. 2, in an attempt to provide a structure to the rather unstructured stage of conceptual mechanism design. Qualitative synthesis focuses on the synthesis of linkages.

Chapters 3–6 are devoted to what Hartenberg and Denavit called dimensional synthesis, as the main objective here is to find the dimensions defining the geometry of the various links and joints of the kinematic chain underlying the mechanism under design. In these chapters, we assume that a *preliminary layout* of the conceptual design—obtained as a result of the type and number syntheses of the kinematic chain at hand—is available, our main job being to contribute to the production of the embodiment of this design. The design embodiment is the realization of a kinematic chain in the form of a table of what is known as the *Denavit–Hartenberg parameters*—to be introduced in Chap. 3—that define uniquely the kinematic chain at hand.

Going back to the more general machine-design process, *dimensioning* involves two phases: *functional dimensioning* and *mechanical dimensioning*. The former is previous to the latter, and includes the determination of the *fundamental dimensions* of the machine, prior to the shaping of all its parts. It is the functional dimensioning where kinematic synthesis plays a major role. Mechanical dimensioning pertains to the dimensioning of the machine elements for stress, strength, heat capacity, and dynamic-response requirements, which lie outside of the scope of the book.

Before we embark on the details of the book, a review of the glossary is in order.

1.2 Glossary

Some general definitions are first recalled:

- Kinematics: The branch of mechanics that studies motion, independent of its relation with forces (Norton, 2004; Wittenburg, 2016).
- Statics: The branch of mechanics that studies the equilibrium of forces and moments acting on particles, rigid bodies, and flexible bodies.
- Kinematic constraint: The physical prevention of relative motion—rotation and translation—between two bodies in one or more directions. The term also denotes the algebraic or differential relations representing the physical constraint.
- Kinetostatics: The branch of mechanics that studies the interplay of forces and moments with motion variables under static, conservative conditions.

The concepts of *machine* and *mechanism* are frequently interchanged as if they were equivalent, but they are not. We give below some definitions from various sources, with added comments:

Machine

- Below we give an account of the definitions of machine, as given by Dudiţă and Diaconescu (1987).
  Different definitions of machine have been given by scholars for more than two millennia, starting with Vitruvius in 28 BCE, namely,

- *A machine is a combination (system, assemblage) of moving material bodies* (Borgnis, 1818; Beck, 1859; Hachette, 1811; Koenigs, 1901; Reuleaux, 1875; Vitruvius, 28 B.C.E.)
- *A machine is generally composed of three parts: a motor part, a transmission part, and an execution part* (Bogolyubov, 1976; Euler, 1753)
- *A machine produces mechanical work, or performs productive operations, actions, or effects* (Bogolyubov, 1976; Koenigs, 1901; Poncelet, 1824; Reuleaux, 1900; Vitruvius, 28 B.C.E.)
- *A machine transforms or transmits forces* (Bogolyubov, 1976; Euler, 1753; Leupold, 1725; Reuleaux, 1900; Vitruvius, 28 B.C.E.)
- *A machine is characterized by deterministic motions* (Borgnis, 1818; Hachette, 1811; Leupold, 1725; Reuleaux, 1900)
- *A machine is an artifact* (Leupold, 1725).

• Some dictionary definitions:

- Merriam-Webster (2003):

  (archaic): a constructed thing whether material or immaterial;
  an assemblage of parts that transmit forces, motion, and energy one to another in a predetermined manner;
  an instrument (as a lever) designed to transmit or modify the application of power, force, or motion;
  a mechanically, electrically, or electronically operated device for performing a task (a calculating machine, a card-sorting machine.)

  Comment: *comprehensive definitions when considered as a whole.*
- Fowler et al. (1995):
  An *apparatus* for applying mechanical power, having several parts, each with a definite function.
  Comment: *leaves computers out.*
- *The Random House College Dictionary* (Stein, 1979):
  An *apparatus* consisting of interrelated parts with separate functions, used in the performance of some kind of work.
  Comment: *ditto.*
- Le Petit Robert (Robert, 1994):
  Any *system* in which a specific correspondence exists between an input form of energy or information and the corresponding ones at the output (loosely translated from French).
  Comment: *a comprehensive definition, that includes computers.*

• *An apparatus* for transformation of power, materials, and information to substitute or simplify physical or intellectual work (Frolov, 1987).
Comment: *a comprehensive definition, that includes computers.*

- *Mechanical system* that performs a specific task, such as the forming of material, and the transference and transformation of *motion* and *force* (IFToMM PC SoT, 2003).
  Comment: *leaves computers out.*

Mechanism

- A piece of machinery (Merriam-Webster, 2003).
  Comment: *too vague.*
- Definitions in IFToMM PC SoT (2003):
  - *System* of bodies designed to convert *motions* of, and *forces* on, one or several bodies into constrained motions of, and *forces* on, other bodies.
    Comment: *Could be much terser and more informative.*
  - *Kinematic chain* with one of its components (*link* or *joint*) connected to the *frame*.
    Comment: *confuses mechanism with its kinematic chain.*
- Structure, adaptation of parts of machine; system of mutually adapted parts working together (as) in machine (Fowler et al., 1995).
- An assembly of moving parts performing a complete functional motion (Stein, 1979).
- A combination layout of pieces or elements, assembled with the goal of (producing) an operation as a unit—loosely translated from (Robert, 1994).

Comment: *In all above definitions, the concept of goal or task is present.*

Linkage

- Definitions in Merriam-Webster (2003):
  - A system of links.
    Comment: *concise and comprehensive.*
  - a system of links or bars which are jointed together and more or less constrained by having a link or links fixed and by means of which straight or nearly straight lines or other point paths may be traced.
    Comment: *unnecessarily cumbersome and limited to path-generating linkages.*
- *Kinematic chain* whose *joints* are equivalent to *lower pairs* only (IFToMM PC SoT, 2003).
  Comment: *confuses linkage with its kinematic chain.*

Rigid Body
A continuum whose points remain equidistant under any possible motion.

Rigid-body Pose
The position of one landmark point of the body, together with the orientation of a coordinate frame fixed to the body with respect to a reference frame.

Rigid-body Twist
The velocity of one landmark point of the body and the angular velocity of the body.

Mechanical-System Posture
The set of link poses allowed by the kinematic constraints imposed by the various kinematic pairs. The concept also applies to humans and other living organisms.

Mechanical-System Gesture
The set of link twists allowed by the kinematic constraints imposed by the kinematic pairs. The concept also applies to humans—e.g., the *surgical gesture*—and other living organisms.

1.3 Kinematic Analysis Versus Kinematic Synthesis

The fundamental problems in mechanism kinematics can be broadly classified into

(a) **Analysis**: Given a linkage, find the motion of its links, for a prescribed motion of its input joint(s).
(b) **Synthesis**: Given a *task*, to be produced by a linkage, find a linkage[1] that performs the task.

The task at hand can be one of three, in this context:

(a) **Function generation**: the motion of the output joint(s) is prescribed as a function of the motion variable(s) of the input joint(s);
(b) **Motion generation** (a.k.a. rigid-body guidance): the motion of the output link(s)—translation and rotation—is prescribed in terms of the motion variable(s) of the input link(s) or joint(s);
(c) **Path generation**: the path traced by a point on a *floating link*—a link not anchored to the mechanism frame—is prescribed as a curve, possibly timed with the motion of the input joint(s).

Kinematic synthesis being a quite broad concept, it involves various aspects (Hartenberg and Denavit, 1964a):

- *Type synthesis*: Given a task to be produced by a mechanism, find the type that will best perform it, e.g., a linkage, a cam mechanism, a gear train, or a combination thereof.
- *Number synthesis*: Given a task to be produced by a mechanism of a given type, find the number of links and joints that are needed to execute the task.
- *Dimensional synthesis* Given a task to be produced by a mechanism, find its relevant geometric parameters.

[1] The multiplicity of solutions is noticed.

In our case, we treat both type and number synthesis under one single umbrella, *qualitative synthesis*; for consistency, dimensional synthesis will be termed *quantitative synthesis*. There are, moreover, two types of quantitative synthesis:

1. **Exact synthesis:** The number of linkage parameters available is sufficient to produce *exactly* the prescribed motion. The problem leads to linear or, most frequently, nonlinear equation solving.
2. **Approximate synthesis:** The number of linkage parameters available is **not** sufficient to produce exactly the prescribed motion. Optimum dimensions are sought that *approximate* the prescribed motion with the *minimum* error. The problem leads to *mathematical programming* (*optimization*).

Furthermore, quantitative synthesis can be achieved, to a variable degree of success, via one of three types of methods:

- Graphical: Under this type, the geometric relations of the prescribed task are manipulated *directly* as such. In the pre-computer era, this was done by means of drafting instruments alone. Nowadays, the drafting instruments have been replaced by CAD software. Although this tool has expanded significantly the capabilities of geometric methods, these are still limited to the *primitives* in the menu of CAD software packages.
- Algebraic: In these methods, the geometric relations in question are manipulated by algebraic means of computer-algebra software, to produce the desired linkage parameters as the solutions to the underlying *synthesis equations*. It is noteworthy that, by virtue of the Theorems of (Kempe, 1875) and (Koenigs, 1897), the geometric relations of any linkage containing any combination of five of the six lower kinematic pairs,[2] the screw pair excluded, lead to systems of multivariate polynomial equations. This is excellent news because the best-developed computer-algebra software available caters to systems of multivariate polynomial equations.
- Semigraphical: Purely algebraic methods entail some drawbacks, like *algebraic singularities*, which are conditions under which some solutions cannot be found for reasons other than kinematic. Semigraphical methods reduce the system of algebraic equations to a subsystem of *bivariate equations*, i.e., equations involving only two variables. The bivariate equations defining a set of contours in the plane of those two variables, the *real* solutions to the problem at hand are found as the intersections of all those contours.

Caveat: Sometimes algebraic methods are billed as "analytic methods," which is to be avoided. Reason: according to common usage—see, for example, *The Random House College Dictionary*—"analytic" is anything "pertaining to or proceeding by analysis (as opposed to *synthetic*)". Careless usage of the qualifier leads to awkward phrases like "analytic synthesis."

---

[2] To be introduced in Chap. 2.

1.3.1 A Summary of Systems of Algebraic Equations

The relevance of systems of algebraic equations is to be highlighted. The synthesis of linkages, as pointed out above, *usually* leads to *algebraic* systems of equations. Non-algebraic equations are termed *transcendental*. We expand first on the former.

Systems of algebraic equations are sets of *multivariate polynomial equations*. That is, each equation, written in *homogeneous form* $P_k(\mathbf{x}) = 0,\ k = 1, \ldots, n$, where we have assumed that the system has as many equations as unknowns, is a linear combination of products of *integer powers* of its n unknown variables, e.g., $x_1^{p_1} x_2^{p_2} \cdots x_n^{p_n}$. The sum of the exponents of each product, $\sum_1^n p_j$, is known as the *degree* of the product; the highest product-degree of the system of the ith equation is termed the *degree* d_i of the same equation. Using a suitable elimination procedure, it is conceptually possible, although not really possible all the time, to reduce the system of n equations in n unknowns to one single *univariate polynomial equation*. According to a result due to Bezout (Salmon, 1964), the degree of the resulting univariate polynomial, termed *the resolvent polynomial*, the *resolvent*, a.k.a. the *eliminant*, of the algebraic system at hand, cannot be greater than the product $P = d_1 d_2 \cdots d_n$. The number of possible solutions, thus, can be as high as P. The degree of the resolvent polynomial is known as the *Bezout number* of the given system of equations.

While some simple kinematic synthesis problems lead to either linear or quadratic equations, some not so complicated problems can lead to polynomials of a degree of the order of 10, which in this case means that the degree of the system can be as high as 10^n—the case in which all equations of the system are of the 10th degree. For example, the synthesis of a six-bar *function generator*—see Chap. 3—for eight accuracy points that approximates a parabolic function led to a resolvent polynomial of degree 64,858 (Plecnik and McCarthy, 2013). Some synthesis problems of planar four-bar linkages lead to resolvent polynomials of a degree lying in the billions! (Chen and Angeles, 2008). Because of the above reasons, we stress in this book *semigraphical methods* of kinematic synthesis.

One major difference between algebraic and transcendental systems of equations is recalled: an algebraic system admits a finite—predictably bounded by the system Bezout number; systems of transcendental equations can admit infinitely many. Examples of the latter occur in the characteristic equation of continuous bodies, e.g., a taut string with its ends fixed to an inertial frame. The roots of the characteristic equation, which involves trigonometric functions associated with a variable λ that has units of length-inverse, determine the infinitely many vibration modes of the string, one mode per root.

1.4 Algebraic and Computational Tools

In deriving the kinematic relations that lead to the various synthesis equations, we shall resort to the *two-dimensional* representation of the cross product. To do this, we introduce below a 2×2 orthogonal matrix $\mathbf{E}$ that will prove to be extremely useful.

An alternative to the use of this matrix for the same purpose is the use of *complex numbers*. While complex numbers are quite useful to represent two-dimensional vectors, their application to three and higher dimensions remains unknown. On the contrary, the two-dimensional representation of the cross product is just a particular case of three-dimensional vector algebra. Moreover, the cross product matrix of two three-dimensional vectors, an extremely useful tool in mechanical analysis, is introduced in Sect. 2.2.

We will also need some quick computations with 2×2 matrices, which will be revised in this section. Methods for the numerical solution of linear systems of equations are included in Sects. 1.4.4 and 1.4.5.

1.4.1 The Two-Dimensional Representation of the Cross Product

The cross product occurs frequently in planar kinematics and statics, and hence, in planar *kinetostatics*. However, planar problems involve only two-dimensional vectors and 2×2 matrices, while the cross product is limited to three-dimensional spaces. Here we describe how to represent, in two dimensions, the cross product, without resorting to three-dimensional vectors. Let: $\mathbf{r}$ be the position vector of a point of a rigid body under planar motion; ω be the angular-velocity vector of the rigid body and assumed normal to the plane of motion.

Without loss of generality, assume, moreover, that $\mathbf{r}$ lies entirely in the plane of motion, which is normal to $\boldsymbol{\omega}$. Below we compute $\boldsymbol{\omega} \times \mathbf{r}$ using only two-dimensional vectors.

Let $\mathbf{E}$ be an *orthogonal matrix* that rotates vectors in the plane through an angle of 90° counterclockwise (ccw):

$$\mathbf{E} \equiv \begin{bmatrix} 0 & -1 \\ 1 & 0 \end{bmatrix} \tag{1.1a}$$

Notice that

$$\mathbf{E}^T\mathbf{E} = \mathbf{E}\mathbf{E}^T = \mathbf{1} \tag{1.1b}$$

with $\mathbf{1}$ denoting the 2×2 *identity matrix*. Also notice that $\mathbf{E}$ is *skew-symmetric*:

$$\mathbf{E} = -\mathbf{E}^T \quad \Rightarrow \quad \mathbf{E}^2 = -\mathbf{1}, \quad \mathbf{E}^{-1} = -\mathbf{E} \tag{1.1c}$$

Therefore, $\mathbf{E}$ rotates vectors $\mathbf{r}$ in the plane through an angle of 90° ccw, as depicted in Fig. 1.4, i.e.,

$$\mathbf{r} = \begin{bmatrix} x \\ y \end{bmatrix} \quad \Rightarrow \quad \mathbf{E}\mathbf{r} = \begin{bmatrix} -y \\ x \end{bmatrix} \tag{1.1d}$$

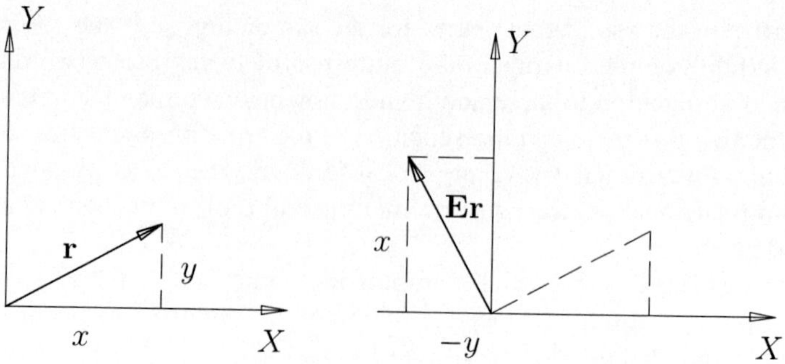

Fig. 1.4 Vector **r** and its image under **E**

Now, for the purpose at hand, we start with the usual three-dimensional vectors **r** and $\boldsymbol{\omega}$ and assume an *orthonormal basis* for the three-dimensional space, {**i**, **j**, **k**}, with **k** defined normal to the plane of motion and pointing toward the viewer. Thus,

$$\boldsymbol{\omega} = \omega \mathbf{k} = \begin{bmatrix} 0 \\ 0 \\ \omega \end{bmatrix}, \quad \mathbf{r} = \begin{bmatrix} x \\ y \\ 0 \end{bmatrix} \tag{1.2}$$

where $\omega > 0$ if the angular velocity is ccw; if cw, then $\omega < 0$. Therefore,

$$\boldsymbol{\omega} \times \mathbf{r} = \det \begin{bmatrix} \mathbf{i} & \mathbf{j} & \mathbf{k} \\ 0 & 0 & \omega \\ x & y & 0 \end{bmatrix} = -\omega y \mathbf{i} + \omega x \mathbf{j}$$

Then, the two-dimensional form of the foregoing product is

$$[\boldsymbol{\omega} \times \mathbf{r}]_{2\mathrm{D}} = \omega \begin{bmatrix} -y \\ x \end{bmatrix} \equiv \omega \mathbf{E} \mathbf{r} \tag{1.3}$$

As a second use of matrix **E**, we derive the two-dimensional form of the cross product $\mathbf{r} \times \mathbf{f}$ yielding the moment of the force **f** about the origin. We assume that **r** and **f** both lie in a plane normal to the unit vector **k**.

First, we start with the usual three-dimensional representation of vectors **r** and **f**, and hence,

$$\mathbf{r} \times \mathbf{f} = \det \begin{bmatrix} \mathbf{i} & \mathbf{j} & \mathbf{k} \\ x & y & 0 \\ f_x & f_y & 0 \end{bmatrix} = (x f_y - y f_x)\mathbf{k}$$

Now, let

$$x f_y - y f_x \equiv n \tag{1.4}$$

which can be readily recognized as the dot product of the two-dimensional vectors **Er** and **f**, i.e.,

$$n = \mathbf{f}^T\mathbf{E}\mathbf{r} \equiv (\mathbf{E}\mathbf{r})^T\mathbf{f} = -\mathbf{r}^T\mathbf{E}\mathbf{f} \tag{1.5}$$

If $n > 0$, then the moment is ccw; otherwise, cw. We have thus shown that the cross product of two two-dimensional vectors reduces to a scalar, i.e., n.

Matrix **E** also appears in the representation of the rotation of a rigid body in planar motion through an angle θ. This rotation is represented algebraically by means of a *proper orthogonal matrix* **Q**. This matrix is proper orthogonal because it is orthogonal and its determinant is $+1$. Matrix **Q** is given by

$$\mathbf{Q} = \begin{bmatrix} \cos\theta & -\sin\theta \\ \sin\theta & \cos\theta \end{bmatrix} \equiv (\cos\theta)\mathbf{1} + (\sin\theta)\mathbf{E} \tag{1.6}$$

Thus, if a vector $\mathbf{r}_0$ is "fixed" to a rigid body rotating about the origin through an angle θ, after the rotation, $\mathbf{r}_0$ becomes $\mathbf{r}$, which is given by

$$\mathbf{r} = \mathbf{Q}\mathbf{r}_0 = (\cos\theta)\mathbf{r}_0 + (\sin\theta)\mathbf{E}\mathbf{r}_0 \tag{1.7}$$

1.4.2 Algebra of 2 × 2 Matrices

A 2×2 matrix **A** can be partitioned either columnwise or rowwise:

$$\mathbf{A} \equiv \begin{bmatrix} \mathbf{a} & \mathbf{b} \end{bmatrix} \equiv \begin{bmatrix} \mathbf{c}^T \\ \mathbf{d}^T \end{bmatrix}$$

where **a**, **b**, **c**, and **d** are all two-dimensional column vectors. We have

Fact 1.4.1

$$\det(\mathbf{A}) = -\mathbf{a}^T\mathbf{E}\mathbf{b} = \mathbf{b}^T\mathbf{E}\mathbf{a} = -\mathbf{c}^T\mathbf{E}\mathbf{d} = \mathbf{d}^T\mathbf{E}\mathbf{c}$$

and

Fact 1.4.2

$$\mathbf{A}^{-1} = \frac{1}{\det(\mathbf{A})}\begin{bmatrix} \mathbf{b}^T \\ -\mathbf{a}^T \end{bmatrix}\mathbf{E} = \frac{1}{\det(\mathbf{A})}\mathbf{E}\begin{bmatrix} -\mathbf{d} & \mathbf{c} \end{bmatrix} \tag{1.8}$$

Componentwise, if a_{ij} denotes the ith entry of the jth column of **A**,

$$\mathbf{A}^{-1} = \frac{1}{\det(\mathbf{A})}\begin{bmatrix} a_{22} & -a_{12} \\ -a_{21} & a_{11} \end{bmatrix}$$

That is, the inverse of a 2×2 nonsingular matrix is obtained upon

(a) exchanging the diagonal entries of the given matrix;
(b) reversing the sign of its off-diagonal entries; and
(c) dividing the matrix thus resulting by the determinant of the given matrix.

Notice that the two 2×2 matrix factors multiplying the reciprocal of $\det(\mathbf{A})$ in Eq. (1.8) are both, in fact, nothing but $\mathrm{Adj}(\mathbf{A})$, i.e., the *adjoint* of **A**.

1.4.3 Algebra of 3 × 3 Matrices

A 3×3 matrix **A** can be partitioned columnwise into three columns, each having as entries the components of a three-dimensional vector, namely,

$$\mathbf{A} = \begin{bmatrix} \mathbf{a}_1 & \mathbf{a}_2 & \mathbf{a}_3 \end{bmatrix}$$

its determinant being readily computed as the *mixed* vector-scalar product of its column vectors:

$$\det(\mathbf{A}) = \mathbf{a}_1 \times \mathbf{a}_2 \cdot \mathbf{a}_3$$

The inverse of **A** can also be readily computed symbolically if we resort to the concept of *reciprocal bases*:

$$\mathbf{A}^{-1} = \frac{1}{\Delta} \begin{bmatrix} (\mathbf{a}_2 \times \mathbf{a}_3)^T \\ (\mathbf{a}_3 \times \mathbf{a}_1)^T \\ (\mathbf{a}_1 \times \mathbf{a}_2)^T \end{bmatrix} \tag{1.9a}$$

where

$$\Delta \equiv \mathbf{a}_1 \times \mathbf{a}_2 \cdot \mathbf{a}_3 \tag{1.9b}$$

The reader can verify the validity of the foregoing formula by straightforward computation of the product $\mathbf{A}\mathbf{A}^{-1}$ or, equivalently, of $\mathbf{A}^{-1}\mathbf{A}$, which should yield the 3×3 identity matrix.

1.4.4 Linear-Equation Solving: Determined Systems

Consider solving for **x** the system below

$$\mathbf{A}\mathbf{x} = \mathbf{b} \tag{1.10}$$

where **A** is a $n \times n$ matrix of *known* coefficients; **b** is the n-dimensional right-hand side *known* vector; and **x** is the n-dimensional vector of *unknowns*.
Definition: **A** is said to be *singular* if

$$\det(\mathbf{A}) = 0 \tag{1.11}$$

Otherwise, **A** is *nonsingular*

Fact 1.4.3 If **A** is nonsingular, then Eq. (1.10) has a unique solution, namely,

$$\mathbf{x} = \mathbf{A}^{-1}\mathbf{b} \tag{1.12}$$

Caveat: Never compute $\mathbf{A}^{-1}$ explicitly, unless instructed to do so! A matrix inverse is seldom needed and incurs a waste of precious CPU time! Instead, find a *good* numerical approximation to the solution, while taking into account that **A** and **b** are usually known only up to a certain roundoff error.

In computing the solution of system (1.10) for **x**, we must take into account the unavoidable *roundoff error* of the data, **A** and **b**. Let

- $\delta\mathbf{A}$ be the matrix roundoff error in **A**
- $\delta\mathbf{b}$ be the vector roundoff-error in **b**
- $\delta\mathbf{x}$ be the vector roundoff-error incurred when solving Eq. (1.10) for **x**, by virtue of $\delta\mathbf{A}$ and $\delta\mathbf{b}$

The *relative* roundoff errors in the data, $\epsilon_{\mathbf{A}}$ and $\epsilon_{\mathbf{b}}$, as well as in the computed solution, $\epsilon_{\mathbf{x}}$, are defined as

$$\epsilon_{\mathbf{A}} \equiv \frac{\|\delta\mathbf{A}\|}{\|\mathbf{A}\|}, \quad \epsilon_{\mathbf{b}} \equiv \frac{\|\delta\mathbf{b}\|}{\|\mathbf{b}\|}, \quad \epsilon_{\mathbf{x}} \equiv \frac{\|\delta\mathbf{x}\|}{\|\mathbf{x}\|} \tag{1.13}$$

where $\|\cdot\|$ denotes *any* vector or matrix norm,[3] as the case may be.

The relative roundoff error in the computed solution is known to be related to the relative roundoff error in the data via the relation (Golub and Van Loan, 1983)

$$\epsilon_{\mathbf{x}} \leq \kappa(\mathbf{A})(\epsilon_{\mathbf{A}} + \epsilon_{\mathbf{b}}) \tag{1.14}$$

where $\kappa(\mathbf{A})$ is the *condition number* of matrix **A** of Eq. (1.10), defined as

$$\kappa(\mathbf{A}) \equiv \|\mathbf{A}\|\|\mathbf{A}^{-1}\| \geq \|\mathbf{A}\mathbf{A}^{-1}\| = 1 \tag{1.15}$$

Various matrix norms are at our disposal, such as the *Euclidean norm*, a.k.a. the *2-norm*, the *Frobenius norm* and the *infinity norm*, a.k.a. the *Chebyshev norm*, denoted, respectively, by $\|\mathbf{A}\|_2$, $\|\mathbf{A}\|_F$ and $\|\mathbf{A}\|_\infty$. The definitions of these norms are given below:

[3] The matrix norm is a generalization of the vector norm, the latter being, in turn, a generalization of the module of complex numbers or the absolute value of real numbers.

$$\|\mathbf{A}\|_2 \equiv \max_i \{ \sqrt{\lambda_i} \}_1^n \tag{1.16a}$$

$$\|\mathbf{A}\|_F \equiv \sqrt{\mathrm{tr}(\mathbf{AWA}^T)} \tag{1.16b}$$

$$\|\mathbf{A}\|_\infty \equiv \max_j \{\max_i \{ |a_{ij}| \}_{i,\, j=1}^n\} \tag{1.16c}$$

where $\{ \lambda_i \}_1^n$ denotes the set of eigenvalues of $\mathbf{AA}^T$, and $\mathbf{W}$ is a *weighting* positive-definite matrix, that is defined according to the user's needs. For example, if $\mathbf{W} = (1/n)\mathbf{1}$, with $\mathbf{1}$ defined as the $n \times n$ identity matrix, then the Frobenius norm of the identity matrix is unity, regardless of the value of n, which is convenient. Not only this; with the foregoing value of $\mathbf{W}$, $\|\mathbf{A}\|_F$ is the rms value of the *singular values*[4] of $\mathbf{A}$. Moreover, $\mathrm{tr}(\cdot)$ denotes the *trace* of its $n \times n$ matrix argument $(\cdot)$, i.e., the sum of the diagonal entries of the matrix. Also notice that the eigenvalues of $\mathbf{A}$ can be real or complex, $|\lambda_i|$ denoting the *module* of λ_i. Computing the eigenvalues of arbitrary matrices is cumbersome because of the complex nature of the eigenvalues, in general. Computing the eigenvalues of symmetric matrices, on the contrary, is much simpler, because these are known to be real. In fact, the set $\{ \lambda_i \}_1^n$ is most conveniently computed as the eigenvalues of $\mathbf{AA}^T$, which is not only symmetric, but also positive-definite, and hence, its eigenvalues are all positive. Moreover, if $\mathbf{A}$ is singular, then $\mathbf{AA}^T$ is only positive-semidefinite, meaning that some of its eigenvalues vanish, but none is negative. We thus have

$$\lambda_i \geq 0, \quad i = 1, \ldots, n \tag{1.17}$$

Whenever we have chosen one specific norm to define the condition number, we indicate the condition number as κ_2, κ_F or κ_∞. In particular,

$$\kappa_2 = \frac{\sqrt{\lambda_l}}{\sqrt{\lambda_s}} \tag{1.18}$$

where λ_l and λ_s denote the largest and the smallest eigenvalues of $\mathbf{AA}^T$, respectively. In fact, $\{ \sqrt{\lambda_i} \}_1^n$ is known as the set of *singular values* of matrix $\mathbf{A}$. Moreover,

$$\kappa_F = \sqrt{\frac{1}{n}\mathrm{tr}(\mathbf{AA}^T)}\sqrt{\frac{1}{n}\mathrm{tr}(\mathbf{A}^{-1}\mathbf{A}^{-T})} \tag{1.19}$$

where $\mathbf{A}^{-T}$ denotes the inverse of the transpose of $\mathbf{A}$ or, equivalently, the transpose of the inverse of the same matrix.

It is now apparent that κ, regardless of the matrix norm used to compute it, is bounded from below, but unbounded from above:

$$1 \leq \kappa < \infty \tag{1.20}$$

[4] The singular values of a $m \times n$ matrix $\mathbf{A}$, with $m \leq n$ are the (nonnegative) eigenvalues of the product $\mathbf{AA}^T$.

Remark 1.4.1 The condition number of a singular matrix is infinitely large.

Remark 1.4.2 If a matrix $\mathbf{A}\mathbf{A}^T$ has all its eigenvalues identical, then $\mathbf{A}$ is said to be *isotropic*. Isotropic matrices have a $\kappa = 1$, *regardless of the matrix norm used to compute* κ. Isotropic matrices are, thus, *optimally conditioned*.

Methods for computing a *good numerical approximation* to the solution (1.12):

- Gaussian elimination, a.k.a. LU-decomposition: Based on the observation that a *triangular system* is readily solved by either *backward* or *forward substitution*, $\mathbf{A}$ is decomposed into a *lower-* and an *upper-triangular* factor, $\mathbf{L}$ and $\mathbf{U}$, respectively.
- Iteratively: Various types of methods of this kind are available, by the names Gauss-Jordan, Gauss-Seidel, successive-overrelaxation (SOR), etc., used mainly for "large" systems (hundreds or thousands of unknowns, which we will not be handling
- Symbolically: Only possible for certain classes of $\mathbf{A}$ matrices, like tridiagonal, and for arbitrary matrices of modest size (n is below 5 or so).

We focus here on Gaussian elimination, or LU-decomposition. We start by *decomposing* the $n \times n$ matrix $\mathbf{A}$ into the form

$$\mathbf{A} = \mathbf{L}\mathbf{U} \tag{1.21}$$

with $\mathbf{L}$ and $\mathbf{U}$ given below

$$\mathbf{L} = \begin{bmatrix} 1 & 0 & \cdots & 0 \\ l_{21} & 1 & \cdots & 0 \\ \vdots & \vdots & \ddots & \vdots \\ l_{n1} & l_{n2} & \cdots & 1 \end{bmatrix}, \quad \mathbf{U} = \begin{bmatrix} u_{11} & u_{12} & \cdots & u_{1n} \\ 0 & u_{22} & \cdots & u_{2n} \\ \vdots & \vdots & \ddots & \vdots \\ 0 & 0 & \cdots & u_{nn} \end{bmatrix} \tag{1.22}$$

Now Eq. (1.10) is rewritten as

$$\mathbf{L}\mathbf{U}\mathbf{x} = \mathbf{b} \Rightarrow \begin{cases} \mathbf{L}\mathbf{y} = \mathbf{b} \\ \mathbf{U}\mathbf{x} = \mathbf{y} \end{cases} \tag{1.23}$$

and hence, $\mathbf{x}$ is computed in two stages: First, $\mathbf{y}$ is computed from a lower-triangular system; then, $\mathbf{x}$ is computed from an upper-triangular system, whose right-hand side is $\mathbf{y}$. The lower-triangular system is solved for $\mathbf{y}$ by *forward substitution*; the upper-triangular system is solved for $\mathbf{x}$ by *backward substitution*.

Note that

$$\det(\mathbf{A}) = \det(\mathbf{L})\det(\mathbf{U}) \tag{1.24a}$$

But, apparently,

$$\det(\mathbf{L}) = 1, \quad \det(\mathbf{U}) = \Pi_1^n u_{ii} \tag{1.24b}$$

Hence,

$$\det(\mathbf{A}) = \det(\mathbf{U}) = \Pi_1^n u_{ii} \tag{1.24c}$$

Therefore, **A** is singular if any of the diagonal entries of **U** vanishes.

The Case of a Positive-Definite Matrix

A $n \times n$ matrix **A** is said to be *positive-definite* if (i) it is symmetric and (ii) all its eigenvalues are positive. Under these conditions, the quadratic form $\mathbf{x}^T\mathbf{A}\mathbf{x} > 0$, for any $\mathbf{x} \in \mathbb{R}^n$. If the same matrix verifies property (i) but verifies property (ii) in the alternative form $(ii$-alt) *none of its eigenvalues is negative*, then the matrix is said to be *positive-semidefinite* and the foregoing quadratic form becomes $\mathbf{x}^T\mathbf{A}\mathbf{x} \geq 0$.

If **A** is either *positive-definite* or *positive-semidefinite*, then it admits the **Cholesky decomposition**:

$$\mathbf{A} = \mathbf{C}^T\mathbf{C} \tag{1.25a}$$

where **C** is a *real, lower-triangular* matrix, namely,

$$\mathbf{C} = \begin{bmatrix} c_{11} & 0 & \cdots & 0 \\ c_{21} & c_{22} & \cdots & 0 \\ \vdots & \vdots & \ddots & \vdots \\ c_{n1} & c_{n2} & \cdots & c_{nn} \end{bmatrix} \tag{1.25b}$$

According to *Sylvester's Theorem*,[5] **C** is invertible if **A** is positive-definite; if positive-semidefinite, then **C** is singular, of the same rank as **A**.

The solution of system (1.10), in the absence of singularities, proceeds as in the general case, in two steps:

$$\mathbf{C}^T\mathbf{y} = \mathbf{b} \tag{1.26}$$

$$\mathbf{C}\mathbf{x} = \mathbf{y} \tag{1.27}$$

1.4.5 *Linear-Equation Solving: Overdetermined Systems*

We are now confronted with solving a system of linear equations formally identical to that given in Eq. (1.10). The difference now is that matrix **A** is no longer square, but rectangular, of the "tall" type, i.e., with n columns of dimension m, namely,

[5] See Theorem 1.4.3.

$$\mathbf{A}\mathbf{x} = \mathbf{b}, \quad \mathbf{A} : m \times n, \ m > n \tag{1.28}$$

where $\mathbf{b}$ is, obviously, m-dimensional. Now, given that we have a surplus of equations over the number of unknowns, *it is not possible*, in general, to find a vector $\mathbf{x}$ that will verify all m equations, and hence, an error will be incurred, the purpose here being to find the vector $\mathbf{x}$ that renders the error of minimum norm. That is, we cannot *actually solve* system (1.28); all we can do is find an acceptable approximation $\mathbf{x}$ to the system. The *error vector* $\mathbf{e}$ in this approximation is defined as

$$\mathbf{e} \equiv \mathbf{b} - \mathbf{A}\mathbf{x} \tag{1.29}$$

Again, we have various norms at our disposal, that we can choose to minimize. All norms of $\mathbf{e}$ can be expressed as

$$\|\mathbf{e}\|_p \equiv \left(\sum_1^m |e_k|^p \right)^{1/p} \tag{1.30}$$

with e_k being understood as the kth component of the m-dimensional vector $\mathbf{e}$. When $p = 2$, the foregoing norm is known as the *Euclidean norm*, which is used most frequently in mechanics, because of its association with *energy*.[6] When $p \to \infty$, the *infinity norm*, also known as the *Chebyshev norm*, is obtained. This norm is, in fact, nothing but the largest absolute value of the components of the vector at hand; finding this norm, thus, *incurs no computational cost*. It turns out that upon seeking the value of $\mathbf{x}$ that minimizes a norm of $\mathbf{e}$, the simplest is the Euclidean norm, for the minimization of its square leads to a linear system of equations whose solution can be obtained *directly*, as opposed to *iteratively*. Indeed, let us set up the minimization problem below

$$z(\mathbf{x}) \equiv \frac{1}{2}\|\mathbf{e}\|_2^2 \quad \to \quad \min_{\mathbf{x}} \tag{1.31}$$

The *normality condition* of the minimization problem at hand is derived upon setting the *gradient* of z with respect to $\mathbf{x}$ equal to zero, i.e.,

$$\frac{dz}{d\mathbf{x}} = \mathbf{0} \tag{1.32a}$$

which is *shorthand notation* for

$$\frac{dz}{dx_1} = 0, \quad \frac{dz}{dx_2} = 0, \quad \ldots \quad \frac{dz}{dx_n} = 0 \tag{1.32b}$$

However, z is not an explicit function of $\mathbf{x}$, but of $\mathbf{e}$, while $\mathbf{e}$ is an explicit function of $\mathbf{x}$. Hence, in order to find $dz/d\mathbf{x}$, the "chain rule" has to be applied. As this "rule"

---

[6] The kinematic energy of a particle of mass m traveling at a velocity $\mathbf{v}$ is given by $m||\mathbf{v}||^2/2$.

is to be applied over vectors, an explanation is in order: first, $d\mathbf{e}$ is derived upon a differential increment dx_i of the ith component of $\mathbf{x}$, for $i = 1, 2, \ldots, n$:

$$d\mathbf{e} = \frac{d\mathbf{e}}{dx_1}dx_1 + \frac{d\mathbf{e}}{dx_2}dx_2 + \ldots + \frac{d\mathbf{e}}{dx_n}dx_n$$

or, in array form,

$$d\mathbf{e} = \underbrace{\begin{bmatrix} \frac{d\mathbf{e}}{dx_1} & \frac{d\mathbf{e}}{dx_2} & \cdots & \frac{d\mathbf{e}}{dx_n} \end{bmatrix}}_{\equiv \frac{d\mathbf{e}}{d\mathbf{x}}} \underbrace{\begin{bmatrix} dx_1 \\ dx_2 \\ \vdots \\ dx_n \end{bmatrix}}_{d\mathbf{x}}$$

where $d\mathbf{e}/d\mathbf{x}$ is, apparently, a $m \times n$ array, namely,

$$\frac{d\mathbf{e}}{d\mathbf{x}} = \begin{bmatrix} \frac{de_1}{dx_1} & \frac{de_1}{dx_2} & \cdots & \frac{de_1}{dx_n} \\ \frac{de_2}{dx_1} & \frac{de_2}{dx_2} & \cdots & \frac{de_2}{dx_n} \\ \vdots & \vdots & \ddots & \vdots \\ \frac{de_m}{dx_1} & \frac{de_m}{dx_2} & \cdots & \frac{de_m}{dx_n} \end{bmatrix} \in \mathbb{R}^{m \times n} \tag{1.33}$$

On the other hand, upon application of the "chain rule" to z in terms of the components of $\mathbf{e}$,

$$\frac{dz}{d\mathbf{x}} = \frac{de_1}{d\mathbf{x}}\frac{dz}{de_1} + \frac{de_2}{d\mathbf{x}}\frac{dz}{de_2} + \ldots + \frac{de_m}{d\mathbf{x}}\frac{dz}{de_m}$$

or, in array form,

$$\frac{dz}{d\mathbf{x}} = \begin{bmatrix} \frac{de_1}{d\mathbf{x}} & \frac{de_2}{d\mathbf{x}} & \cdots & \frac{de_m}{d\mathbf{x}} \end{bmatrix} \begin{bmatrix} \frac{dz}{de_1} \\ \frac{dz}{de_2} \\ \vdots \\ \frac{dz}{de_m} \end{bmatrix}$$

If the first factor in the foregoing product is expressed in component form, then

$$\frac{dz}{d\mathbf{x}} = \begin{bmatrix} \frac{de_1}{dx_1} & \frac{de_2}{dx_1} & \cdots & \frac{de_m}{dx_1} \\ \frac{de_1}{dx_2} & \frac{de_2}{dx_2} & \cdots & \frac{de_m}{dx_2} \\ \vdots & \vdots & \ddots & \vdots \\ \frac{de_1}{dx_n} & \frac{de_2}{dx_n} & \cdots & \frac{de_m}{dx_n} \end{bmatrix} \begin{bmatrix} \frac{dz}{de_1} \\ \frac{dz}{de_2} \\ \vdots \\ \frac{dz}{de_m} \end{bmatrix} \tag{1.34}$$

whose first factor can be readily identified, in light of Eq. (1.33), as $(d\mathbf{e}/d\mathbf{x})^T$, and hence,

$$\frac{dz}{d\mathbf{x}} = \left(\frac{d\mathbf{e}}{d\mathbf{x}}\right)^T \frac{dz}{d\mathbf{e}} \tag{1.35}$$

an expression that can be fairly termed the *chain rule* for vector arrays.

Now, from Eq. (1.29),

$$\frac{d\mathbf{e}}{d\mathbf{x}} = -\mathbf{A} \tag{1.36}$$

Hence, the final expression for the gradient of z, $dz/d\mathbf{x}$, is

$$\frac{dz}{d\mathbf{x}} = -\mathbf{A}^T\mathbf{e} \tag{1.37}$$

Therefore, the error vector of minimum Euclidean norm, or *least-square error* for brevity, represented henceforth by $\mathbf{e}_0$, satisfies the *normality condition*

$$\mathbf{A}^T\mathbf{e}_0 = \mathbf{0}_n \tag{1.38}$$

with $\mathbf{0}_n$ denoting the n-dimensional zero vector. Now we have the first result:

Theorem 1.4.1 *The least-square error* $\mathbf{e}_0$ *of the overdetermined system of linear equations (1.10) lies in the null space of the transpose of the full-rank* $m \times n$ *matrix* $\mathbf{A}$, *with* $m > n$.

In order to gain insight into the above result, let $\{\mathbf{a}_i\}_1^n$ represent the n m-dimensional columns of matrix $\mathbf{A}$. Hence, $\mathbf{A}^T$ can be expressed as a column array of vectors $\mathbf{a}_i^T$, for $i = 1, \ldots, n$, Eq. (1.38) thus leading to

$$\mathbf{a}_i^T\mathbf{e}_0 = 0, \quad i = 1, \ldots, n \tag{1.39}$$

Furthermore, if Eq. (1.29) is substituted into Eq. (1.37), and the product thus resulting is substituted, in turn, into the normality condition (1.32a), we obtain

$$\mathbf{A}^T\mathbf{A}\mathbf{x} = \mathbf{A}^T\mathbf{b} \tag{1.40}$$

which is known as the *normal equations* of the minimization problem at hand. By virtue of the assumption on the rank of $\mathbf{A}$, the product $\mathbf{A}^T\mathbf{A}$ is *positive-definite* and hence, invertible. As a consequence, the value $\mathbf{x}_0$ of $\mathbf{x}$ that minimizes the Euclidean norm of the approximation error of the given system is

$$\mathbf{x}_0 = (\mathbf{A}^T\mathbf{A})^{-1}\mathbf{A}^T\mathbf{b} \tag{1.41}$$

the matrix coefficient of $\mathbf{b}$ being known as a *generalized inverse* of $\mathbf{A}$; we shall denote this generalized inverse as $\mathbf{A}^I$, i.e.,

$$\mathbf{A}^I \equiv (\mathbf{A}^T\mathbf{A})^{-1}\mathbf{A}^T \tag{1.42}$$

More specifically, $\mathbf{A}^I$ is known as the *left Moore–Penrose generalized inverse*[7] of $\mathbf{A}$, because, when $\mathbf{A}$ is multiplied by $\mathbf{A}^I$ *from the left*, the product becomes

$$\mathbf{A}^I\mathbf{A} = \mathbf{1}_n \tag{1.43}$$

in which $\mathbf{1}_n$ denotes the $n \times n$ identity matrix. The error obtained with this value is known as the *least-square error* of the approximation, i.e.,

$$\mathbf{e}_0 \equiv \mathbf{b} - \mathbf{A}\mathbf{x}_0 \tag{1.44}$$

Now we have one more result:

Theorem 1.4.2 (Projection Theorem) *The least-square error is orthogonal to* $\mathbf{A}\mathbf{x}_0$, *i.e.,*

$$\mathbf{e}_0^T\mathbf{A}\mathbf{x}_0 \equiv \mathbf{x}_0^T\mathbf{A}^T\mathbf{e}_0 = 0 \tag{1.45}$$

Proof Readily follows from Eq. (1.38), and hence, from Theorem 1.4.1.
The Projection Theorem is illustrated in Fig. 1.5.

While the formula yielding the foregoing generalized inverse is quite simple to implement, the number of floating-point operations (flops) it takes to evaluate, along with the ever-present roundoff errors in both the data and the results, renders it not only *inefficient*, but also *unreliable*, if applied verbatim. What is at stake here is the concept of condition number, introduced in Sect. 1.4.4 for square matrices. The same concept is applicable to rectangular matrices, if the matrix inverse is replaced by its left Moore–Penrose generalized inverse. In fact, the singular values of rectangular $\mathbf{A}$ are the nonnegative square roots of the nonnegative eigenvalues of the $n \times n$ positive-semidefinite matrix $\mathbf{A}^T\mathbf{A}$, exactly as in the case of square matrices. If $\mathbf{A}$ is of full rank, i.e., if its n m-dimensional columns are linearly independent, then $\mathbf{A}^T\mathbf{A}$ is positive-definite. However, notice that $\mathbf{A}\mathbf{A}^T$ is singular, whether $\mathbf{A}$ is of full rank or not. The foregoing statement is a result of Sylvester's Theorem (Strang, 1988):

---

[7] The right counterpart of this inverse applies to the same matrix $\mathbf{A}$ when $m < n$, which occurs in the presence of underdetermined systems of linear equations.

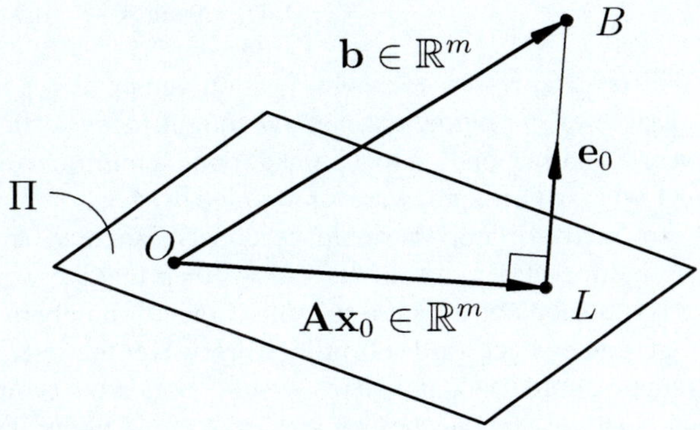

Fig. 1.5 The projection theorem

Theorem 1.4.3 (Sylvester's Theorem) *Let* $p \times q$ **A** *and* $q \times r$ **B** *be two arbitrary matrices, which are thus compatible under multiplication. Then,*

$$\text{rank}(\mathbf{AB}) \leq \min\{\text{rank}(\mathbf{A}),\ \text{rank}(\mathbf{B})\} \tag{1.46}$$

Therefore, if **A** is of full rank, then $\text{rank}(\mathbf{A}) = n$, and hence, $\text{rank}(\mathbf{AA}^T) = n < m$, which means that the product $\mathbf{AA}^T \in \mathbb{R}^{m\times m}$ is rank-deficient, i.e., singular in this case.

Remark 1.4.3 If the working condition number is either κ_2 or κ_F, then the condition number of $\mathbf{A}^T\mathbf{A}$ is *exactly* the square of the condition number of **A**.

As a consequence, then, even if **A** is only slightly ill-conditioned, the product $\mathbf{A}^T\mathbf{A}$ can be *catastrophically* ill-conditioned, the moral of the story being that the normal equations (1.40) are much more sensitive to data roundoff error than the original equations (1.28). Therefore, **solving the normal equations should be avoided**.

Below we outline two procedures to calculate efficiently the least-square approximation of the overdetermined system (1.10) that do not resort to the normal equations, and hence, preserve the condition number of **A**, and do this with a low number of flops.

In figuring out a numerical method suitable to finding the *least-square approximation* of the overdetermined system of linear equations (1.28), it is convenient to resort to the geometric interpretation of the problem at hand: Let us assume that **A** is of full rank, and hence, its n m-dimensional columns $\{\, \mathbf{a}_i \,\}_1^n$, introduced in Eq. (1.39), are linearly independent. However, notice that this set cannot constitute a basis of the m-dimensional space of these vectors, or of vector **b** for that matter, because of a deficit of $m - n$ vectors in the set. Hence, there is no guarantee that, given an arbitrary m-dimensional vector **b**, we can find n real numbers $\{\, x_k \,\}_1^n$ that will produce **b** as a linear combination of the given set of vectors—the columns of **A**. Now, let us regard **b** as the position vector of a point B in m-dimensional space, with **l** denoting the vector spanned by the linear combination

$$\mathbf{l} \equiv \mathbf{a}_1 x_1 + \mathbf{a}_2 x_2 + \cdots + \mathbf{a}_n x_n \tag{1.47}$$

We can also regard $\mathbf{l}$[8] as the position vector of a point L in the same space, the purpose of the numerical method sought being to find the set $\{ x_i \}_1^n$ that yields a vector $\mathbf{l}$ corresponding to a point L lying a minimum distance from B, as illustrated in Fig. 1.5. If vector $\mathbf{a}_i$ were represented in a basis in which only its first i components were nonzero, then $\mathbf{A}$ would be upper-triangular, and the task at hand would be straightforward: it would be obvious then that we would be able to match the first n components of $\mathbf{b}$ with a suitable choice of numbers $\{ x_i \}_1^n$—these numbers could be found by backward substitution! However the last $m - n$ components of $\mathbf{b}$ would remain unmatched, and hence, would contain the error in the approximation.

Now, in general, the columns of $\mathbf{A}$ most likely will be full. Nevertheless, it is always possible to find a suitable coordinate system, i.e., a suitable basis, under which the columns of $\mathbf{A}$ will have the special structure described above.

In seeking the new coordinate system, we aim at a transformation of both all columns of $\mathbf{A}$ that will render $\mathbf{A}$ in upper-triangular form, similar to the effect of the LU-decomposition applied to the solution of system (1.10). However, in seeking the suitable transformation in the case at hand, we should preserve the distances between points in m-dimensional space; else, the Euclidean norm will not be preserved and the approximation obtained will not yield the minimum distance between points B and L. A safe numerical procedure should thus preserve the Euclidean norm of the columns of $\mathbf{A}$ and, hence, the inner product between any two columns of this matrix. Therefore, a triangularization procedure like LU-decomposition **would not work**, because this does not preserve inner products. Obviously, the transformations that do preserve these inner products are orthogonal, either rotations or reflections. Examples of these methods are (a) the Gram–Schmidt orthogonalization procedure and (b) *Householder reflections*, both described below.[9]

The Gram–Schmidt Orthogonalization Procedure

This procedure consists in regarding each column of $\mathbf{A}$ as a set of n m-dimensional vectors $\{ \mathbf{a}_k \}_1^n$. From this set, a new set $\{ \mathbf{c}_k \}_1^n$ is obtained that is *orthonormal*. The first step consists in defining a *normal* vector—i.e., of *unit norm*—$\mathbf{c}_1$ as[10]

$$\mathbf{c}_1 = \frac{\mathbf{a}_1}{\|\mathbf{a}_1\|} \tag{1.48}$$

[8] $\mathbf{l}$ should not be confused with $\mathbf{1}$, which is used in this book to denote the $n \times n$ identity matrix for any finite integer n.

[9] These methods are implemented in Maple, a language for computer algebra, under the command `LeastSquares(A, B, ...)`. Similar commands are available in other similar languages, like *Mathematica*.

[10] In the balance of this section only the Euclidian norm is used; for the sake of brevity, this norm is simply referred to as "the norm".

Further, we define $\mathbf{c}_2$ as the *normal component* of $\mathbf{a}_2$ onto $\mathbf{c}_1$, namely,

$$\mathbf{b}_2 \equiv (\mathbf{1} - \mathbf{c}_1\mathbf{c}_1^T)\mathbf{a}_2 \tag{1.49a}$$

$$\mathbf{c}_2 \equiv \frac{\mathbf{b}_2}{\|\mathbf{b}_2\|} \tag{1.49b}$$

In the next step, we define $\mathbf{c}_3$ as the unit vector normal to the plane defined by $\mathbf{c}_1$ and $\mathbf{c}_2$ and in the direction in which the inner product $\mathbf{c}_3^T\mathbf{a}_3$ is positive, which is possible because all vectors of the set $\{\,\mathbf{a}_k\,\}_1^m$ have been assumed to be linearly independent—remember that $\mathbf{A}$ has been assumed to be of full rank. To this end, we subtract from $\mathbf{a}_3$ its projection onto the plane mentioned above, i.e.,

$$\mathbf{b}_3 = (\mathbf{1} - \mathbf{c}_1\mathbf{c}_1^T - \mathbf{c}_2\mathbf{c}_2^T)\mathbf{a}_3 \tag{1.50a}$$

$$\mathbf{c}_3 \equiv \frac{\mathbf{b}_3}{\|\mathbf{b}_3\|} \tag{1.50b}$$

and so on, until we obtain $\mathbf{c}_{n-1}$, the last unit vector of the orthogonal set, $\mathbf{c}_n$, being obtained as

$$\mathbf{b}_n = (\mathbf{1} - \mathbf{c}_1\mathbf{c}_1^T - \mathbf{c}_2\mathbf{c}_2^T - \cdots - \mathbf{c}_{n-1}\mathbf{c}_{n-1}^T)\mathbf{a}_n \tag{1.51a}$$

Finally,

$$\mathbf{c}_n \equiv \frac{\mathbf{b}_n}{\|\mathbf{b}_n\|} \tag{1.51b}$$

In the next stage, we represent all vectors of the set $\{\,\mathbf{a}_k\,\}_1^n$ in *orthogonal coordinates*, i.e., in the base $\mathcal{O} = \{\,\mathbf{c}_k\,\}_1^n$, which are then arranged in a $m \times n$ array $\mathbf{A}_o$. By virtue of the form in which the set $\{\,\mathbf{c}_k\,\}_1^n$ was defined, the last $m - k$ components of vector $\mathbf{a}_k$ vanish. We thus have, in the said orthonormal basis,

$$[\mathbf{a}_k]_{\mathcal{O}} = \begin{bmatrix} \alpha_{1k} \\ \alpha_{2k} \\ \vdots \\ \alpha_{kk} \\ 0 \\ \vdots \\ 0 \end{bmatrix}, \qquad [\mathbf{b}]_{\mathcal{O}} = \begin{bmatrix} \beta_1 \\ \beta_2 \\ \vdots \\ \beta_m \end{bmatrix}$$

Therefore, Eq. (1.28), when expressed in $\mathcal{O}$, becomes

$$
\begin{bmatrix}
\alpha_{11} & \alpha_{12} & \cdots & \alpha_{1n} \\
0 & \alpha_{22} & \cdots & \alpha_{2n} \\
\vdots & \vdots & \ddots & \vdots \\
0 & 0 & \cdots & \alpha_{nn} \\
0 & 0 & \cdots & 0 \\
\vdots & \vdots & \ddots & \vdots \\
0 & 0 & \cdots & 0
\end{bmatrix}
\begin{bmatrix} x_1 \\ x_2 \\ \vdots \\ x_n \end{bmatrix}
=
\begin{bmatrix} \beta_1 \\ \beta_2 \\ \vdots \\ \beta_n \\ \beta_{n+1} \\ \vdots \\ \beta_m \end{bmatrix}
\tag{1.52}
$$

Hence, $\mathbf{x}$ can be computed by back-substitution. It is apparent, then, that the last $m-n$ equations of the foregoing system are incompatible: their left-hand counterparts are zero, while their right-hand counterparts are not necessarily so. What the right-hand sides of these equations represent is then the *approximation error* in orthogonal coordinates, its Euclidian norm being

$$
\|\mathbf{e}_0\| \equiv \sqrt{\beta_{n+1}^2 + \ldots + \beta_m^2} \tag{1.53}
$$

Householder Reflections

The familiar concept of reflection about a plane—that of a flat mirror, for example—in 3D space can be generalized to n dimensions. Indeed, a reflection onto a plane Π of unit normal $\mathbf{n}$ is illustrated in Fig. 1.6. In that figure, the projection of point P, of position vector $\mathbf{p}$, from an origin lying in the plane—this is necessary in order for Π to be a *vector space*—onto Π is labeled P', of position vector $\mathbf{p}'$. Obviously, point P' is the intersection of the normal to Π from P with Π itself. If now a third point P'' is found on the extension of the normal from P lying exactly the same distance d between P and P', then P'' becomes the *reflection* of P with respect to Π.

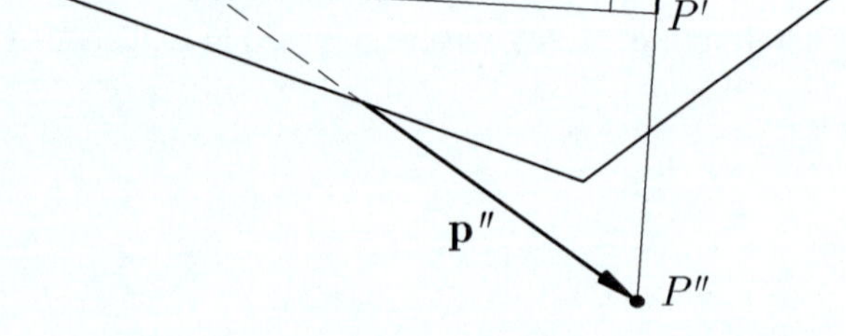

Fig. 1.6 A reflection onto a plane Π of unit normal $\mathbf{n}$

Now it is apparent that the position vector $\mathbf{p}''$ of P'' is given by

$$\mathbf{p}'' = \mathbf{p} - 2(\mathbf{p}^T\mathbf{n})\mathbf{n} \tag{1.54}$$

or

$$\mathbf{p}'' = (\mathbf{1} - 2\mathbf{n}\mathbf{n}^T)\mathbf{p} \tag{1.55}$$

with $\mathbf{1}$ denoting the 3×3 *identity matrix*.[11] Hence, the reflection is represented algebraically by a matrix $\mathbf{R}$ given by

$$\mathbf{R} = \mathbf{1} - 2\mathbf{n}\mathbf{n}^T \tag{1.56}$$

The extension of the 3D concept of reflection is now straightforward: vector $\mathbf{n}$ in Eq. (1.56) can be thought of as being n-dimensional.

With the above background, the second method discussed here, based on the application of a chain of n transformations $\{\mathbf{H}_k\}_1^n$, known as *Householder reflections*, to both sides of Eq. (1.10), can now be introduced. The purpose of these reflections is, again, to obtain a representation of matrix $\mathbf{A}$ in upper-triangular form (Golub and Van Loan, 1983).

The algorithm proceeds as follows: We assume that we have applied reflections $\mathbf{H}_1, \mathbf{H}_2, \ldots, \mathbf{H}_{i-1}$, in this order, to $\mathbf{A}$, that have rendered it in *upper-trapezoidal form*:

$$\mathbf{A}_{i-1} \equiv \mathbf{H}_{i-1} \ldots \mathbf{H}_2\mathbf{H}_1\mathbf{A}$$

or

$$\mathbf{A}_{i-1} = \begin{bmatrix} a_{11}^* & a_{12}^* & \cdots & a_{1,i-1}^* & a_{1i}^* & \cdots & a_{1n}^* \\ 0 & a_{22}^* & \cdots & a_{2,i-1}^* & a_{2i}^* & \cdots & a_{2n}^* \\ 0 & 0 & \cdots & a_{3,i-1}^* & a_{3i}^* & \cdots & a_{3n}^* \\ \vdots & \vdots & \ddots & \vdots & \vdots & \ddots & \vdots \\ 0 & 0 & \cdots & a_{i-1,i-1}^* & a_{i-1,i}^* & \cdots & a_{i-1,n}^* \\ 0 & 0 & \cdots & 0 & a_{i,i}^* & \cdots & a_{i,n}^* \\ \vdots & \vdots & \ddots & \vdots & \vdots & \ddots & \vdots \\ 0 & 0 & \cdots & 0 & a_{m,i}^* & \cdots & a_{mn}^* \end{bmatrix} \tag{1.57}$$

The next Householder reflection, $\mathbf{H}_i$, is determined so as to render the last $m - i$ components of the ith column of $\mathbf{H}_i\mathbf{A}_{i-1}$ equal to zero, while leaving its first $i - 1$ columns unchanged. We do this by setting

[11] **Caveat**: sometimes the identity matrix is referred to as "the unit matrix", but we do not follow this practice. The reason is that the quoted term can be confused with "unitary matrix", which is an orthogonal matrix defined over the *complex field*, a concept not needed in the book.

$$\alpha_i = \operatorname{sgn}(a_{ii}^*)\sqrt{(a_{ii}^*)^2 + (a_{i+1,i}^*)^2 + \cdots + (a_{mi}^*)^2} \tag{1.58a}$$

$$\mathbf{u}_i = \left[0\ 0 \cdots 0\ a_{ii}^* + \alpha_i\ a_{i+1,i}^* \cdots a_{mi}^*\right]^T \tag{1.58b}$$

$$\mathbf{H}_i = \mathbf{1} - \frac{\mathbf{u}_i\mathbf{u}_i^T}{\|\mathbf{u}_i\|^2/2} \tag{1.58c}$$

where $\operatorname{sgn}(x)$, the *signum function* of x, is defined as $+1$ if $x > 0$, as -1 if $x < 0$, and is left undefined when $x = 0$. As the reader can readily verify,

$$\frac{1}{2}\|\mathbf{u}_i\|^2 = \alpha_i(\mathbf{u}_i)_i = \alpha_i(a_{ii}^* + \alpha_i) \equiv \gamma_i \tag{1.59}$$

and hence, the denominator appearing in the expression for $\mathbf{H}_i$ is calculated with one single addition and one single multiplication. Notice that $\mathbf{H}_i$ *reflects* vectors in m-dimensional space onto a hyperplane of unit normal $\mathbf{u}_i/\|\mathbf{u}_i\|$.

It is important to realize that

(a) α_i is defined with the sign of a_{ii}^* because the denominator γ_i appearing in Eq. (1.59) is proportional to the sum of a_{ii}^* and α_i, thereby guaranteeing that the absolute value of this sum will always be greater than the absolute value of each of its terms. If this provision were not made, then the resulting sum could be of a negligibly small absolute value, which would thus render γ_i a small positive number. Such a small number would thus introduce unnecessarily an inadmissibly large roundoff-error amplification upon dividing the product $\mathbf{u}_i\mathbf{u}_i^T$ by γ_i;

(b) an arbitrary vector $\mathbf{v}$ is transformed by $\mathbf{H}_i$ with unusually few flops, namely,

$$\mathbf{H}_i\mathbf{v} = \mathbf{v} - \frac{1}{\gamma_i}(\mathbf{v}^T\mathbf{u}_i)\mathbf{u}_i$$

Upon application of the n Householder reflections thus defined, the system at hand becomes

$$\mathbf{HAx} = \mathbf{Hb} \tag{1.60}$$

with $\mathbf{H}$ defined as

$$\mathbf{H} \equiv \mathbf{H}_n \ldots \mathbf{H}_2\mathbf{H}_1 \tag{1.61}$$

Similar to that in equation (1.52), the matrix coefficient of $\mathbf{x}$ in Eq. (1.60), i.e., $\mathbf{HA}$, is in upper-triangular form. That is, we have

$$\mathbf{HA} = \begin{bmatrix} \mathbf{U} \\ \mathbf{O}_{m'n} \end{bmatrix}, \quad \mathbf{Hb} = \begin{bmatrix} \mathbf{b}_U \\ \mathbf{b}_L \end{bmatrix} \tag{1.62}$$

with $\mathbf{U}$ a $n \times n$ upper-triangular matrix identical to that appearing in Eq. (1.22), $\mathbf{O}_{m'n}$ denoting the $(m-n) \times n$ zero matrix, $m' \equiv m - n$, and $\mathbf{b}_U$ and $\mathbf{b}_L$ being n-

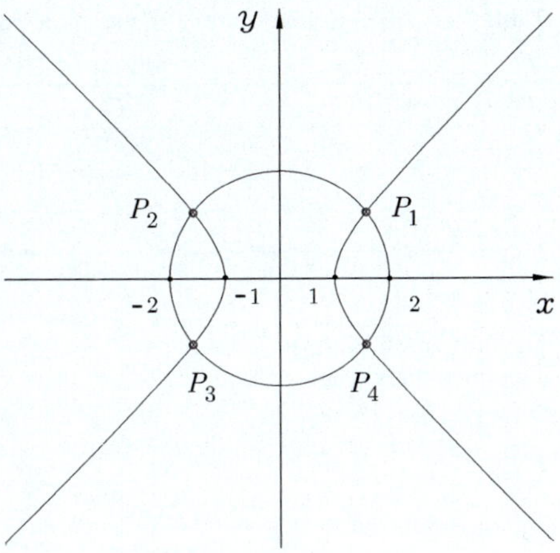

Fig. 1.7 Intersection of a circle and a hyperbola

and m'-dimensional vectors, respectively. The unknown $\mathbf{x}$ can thus be calculated from Eq. (1.60) by *back-substitution*.

Notice that the last m' components of the left-hand side of Eq. (1.60) are zero, while the corresponding components of the right-hand side of the same equation are not necessarily so. This apparent contradiction can be resolved by recalling that the overdetermined system at hand, in general, has no solution. The lower part of $\mathbf{b}$, $\mathbf{b}_L$, is then nothing but an m'-dimensional array containing the nonzero components of the approximation error in the new coordinates. That is, the least-square error, $\mathbf{e}_0$, in these coordinates takes the form

$$[\,\mathbf{e}_0\,]_{\mathcal{O}} = \begin{bmatrix} \mathbf{0}_n \\ \mathbf{b}_L \end{bmatrix} \tag{1.63a}$$

where subscripted brackets are used to denote a vector or matrix representation in a coordinate frame labeled with the subscript, usually in calligraphic fonts. In the above equation, $\mathcal{O}$ stands for *orthogonal frame*, which defines the new coordinates. Hence, the above expression readily leads to the norm (any norm) of the least-square error, namely,

$$\|\mathbf{e}_0\| = \|\mathbf{b}_L\| \tag{1.63b}$$

1.5 Nonlinear-Equation Solving: The Determined Case

Definition 1.5.1 A system of algebraic equations containing some that are not linear is termed nonlinear system. If the number of equations is identical to the number of unknowns, the system is, moreover, determined.

Example Find the intersection of the circle and the hyperbola depicted in Fig. 1.7.
Solution The equations of the circle and the hyperbola are, respectively,

Table 1.1 The four intersection points of the circle and the hyperbola of Fig. 1.7

P_i	x_i	y_i
1	$\sqrt{5/2}$	$\sqrt{3/2}$
2	$\sqrt{5/2}$	$-\sqrt{3/2}$
3	$-\sqrt{5/2}$	$\sqrt{3/2}$
4	$-\sqrt{5/2}$	$-\sqrt{3/2}$

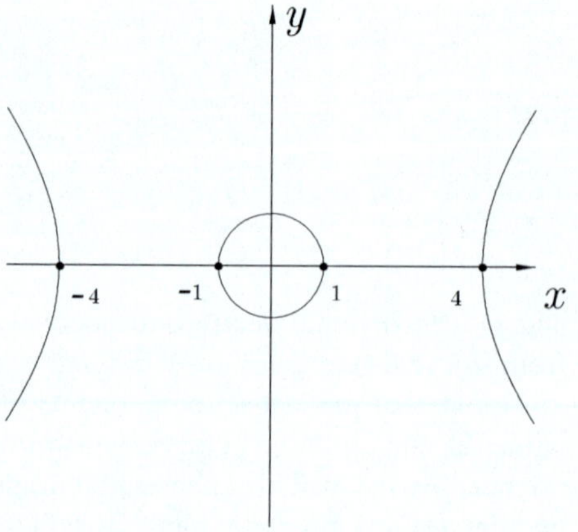

Fig. 1.8 A circle and a hyperbola that do not intersect

$$\phi_1(x, y) \equiv x^2 + y^2 - 4 = 0$$
$$\phi_2(x, y) \equiv x^2 - y^2 - 1 = 0$$

The solution to a nonlinear system of equations, when one exists at all, is usually *multiple*: The circle and the hyperbola of Fig. 1.7 intersect at four points $\{P_i\}_1^4$, of coordinates (x_i, y_i), as displayed in Table 1.1. The problem may have **no real solution**, e.g., the circle and the hyperbola of Fig. 1.8 do not intersect. For example, if the system of equations, from which the coordinates of the intersection points of the two curves of the foregoing figure, are to be computed is given below

$$\phi_1(x, y) \equiv x^2 + y^2 - 1 = 0$$
$$\phi_2(x, y) \equiv x^2 - y^2 - 16 = 0$$

Then this system of equations admits no real solution, as shown in Fig. 1.8.

In general, a determined nonlinear system of equations takes the form

$$\boldsymbol{\phi}(\mathbf{x}) = \mathbf{0} \tag{1.64}$$

where $\mathbf{x}$ and $\boldsymbol{\phi}$ are both n-dimensional vectors:

$$\mathbf{x} \equiv \begin{bmatrix} x_1 \\ x_2 \\ \vdots \\ x_n \end{bmatrix}, \quad \boldsymbol{\phi} \equiv \begin{bmatrix} \phi_1(x_1, x_2, \ldots, x_n) \\ \phi_2(x_1, x_2, \ldots, x_n) \\ \vdots \\ \phi_n(x_1, x_2, \ldots, x_n) \end{bmatrix} \tag{1.65}$$

1.5.1 The Newton–Raphson Method

We outline below the method of solution of determined nonlinear systems using the Newton–Raphson method. This is an *iterative method*, whereby a sequence of approximations is obtained that, if converging, it approaches the solution in a finite number of iterations *within a prescribed tolerance*.

A value $\mathbf{x}^0$ of $\mathbf{x}$ is given as an *initial guess*:

$$\mathbf{x}^0 \equiv \left[p_1 \; p_2 \; \ldots \; p_n\right]^T$$

and $\boldsymbol{\phi}$ is evaluated at $\mathbf{x}^0$:

$$\boldsymbol{\phi}^0 \equiv \boldsymbol{\phi}(\mathbf{x}^0)$$

If the value $\mathbf{x}^0$ was chosen arbitrarily, most likely it will not verify the given system of equations, i.e.,

$$\boldsymbol{\phi}^0 \neq \mathbf{0}$$

Next, we look for a "small" increment $\Delta\mathbf{x}$ of $\mathbf{x}$ (the increment is small if its norm—any norm—is small):

$$\Delta\mathbf{x} \equiv \left[\Delta x_1 \; \Delta x_2 \; \ldots \; \Delta x_n\right]^T$$

Now, $\boldsymbol{\phi}(\mathbf{x}^0 + \Delta\mathbf{x})$ is evaluated up to its *linear approximation* (all quadratic and higher-order terms are dropped from its series expansion):

$$\boldsymbol{\phi}(\mathbf{x}^0 + \Delta\mathbf{x}) \approx \boldsymbol{\phi}(\mathbf{x}^0) + \left.\frac{\partial\boldsymbol{\phi}}{\partial\mathbf{x}}\right|_{\mathbf{x}=\mathbf{x}^0} \Delta\mathbf{x} \tag{1.66}$$

The *Jacobian matrix* of $\boldsymbol{\phi}$ with respect to $\mathbf{x}$ is defined as the matrix of partial derivatives[12] of the components of $\boldsymbol{\phi}$ with respect to all the components of $\mathbf{x}$:

$$\boldsymbol{\Phi} \equiv \frac{\partial\boldsymbol{\phi}}{\partial\mathbf{x}} = \begin{bmatrix} \partial\phi_1/\partial x_1 & \partial\phi_1/\partial x_2 & \cdots & \partial\phi_1/\partial x_n \\ \partial\phi_2/\partial x_1 & \partial\phi_2/\partial x_2 & \cdots & \partial\phi_2/\partial x_n \\ \vdots & \vdots & \ddots & \vdots \\ \partial\phi_n/\partial x_1 & \partial\phi_n/\partial x_2 & \cdots & \partial\phi_n/\partial x_n \end{bmatrix} \tag{1.67}$$

---

[12] See Sect. 1.4.5 for a derivation of the (matrix) partial derivative of a vector with respect to another vector.

In the next step, we find $\Delta\mathbf{x}$ that renders zero the *linear approximation* of $\boldsymbol{\phi}(\mathbf{x}_0 + \Delta\mathbf{x})$:

$$\boldsymbol{\phi}_0 + \boldsymbol{\Phi}(\mathbf{x}^0)\Delta\mathbf{x} = \mathbf{0}$$

or

$$\boldsymbol{\Phi}(\mathbf{x}^0)\Delta\mathbf{x} = -\boldsymbol{\phi}^0 \tag{1.68}$$

whence $\Delta\mathbf{x}$ can be found using Gaussian elimination to yield

$$\Delta\mathbf{x} = -\boldsymbol{\Phi}_0^{-1}\boldsymbol{\phi}^0, \quad \boldsymbol{\Phi}_0 \equiv \boldsymbol{\Phi}(\mathbf{x}^0) \tag{1.69}$$

Next, $\mathbf{x}$ is updated:

$$\mathbf{x} \quad \leftarrow \quad \mathbf{x}^0 + \Delta\mathbf{x} \tag{1.70}$$

the procedure stopping when a prescribed tolerance ϵ_x is reached,

$$\|\Delta\mathbf{x}\| \leq \epsilon_x \tag{1.71}$$

Remarks

- Use the maximum norm to test convergence in Eq. (1.71), for it costs virtually nothing;
- no guarantee that the Newton–Raphson method will converge at all;
- whether the Newton–Raphson method converges is dependent upon the initial guess, $\mathbf{x}^0$;
- the boundary between regions of convergence and divergence is a *fractal* (Gleick, 1988; Mandelbrot, 1983);
- when the Newton–Raphson method converges, it does so *quadratically*: At every iteration, *two* decimal places of accuracy are gained (Björck and Dahlquist, 1974).

1.6 Overdetermined Nonlinear Systems of Equations

A system of nonlinear equations of the form

$$\boldsymbol{\phi}(\mathbf{x}) = \mathbf{0} \tag{1.72}$$

where $\mathbf{x}$ is a n-dimensional vector and $\boldsymbol{\phi}$ a q-dimensional vector is *overdetermined* if $q > n$. Just as in the linear case, in general, no vector $\mathbf{x}$ can be found that verifies *all* the q scalar equations of the system. However, approximations can be found that minimize the least-square error of the approximation, as described in the balance of this section. The method of solution adopted here is *Newton-Gauss*; this is the overdetermined counterpart of the Newton–Raphson method.

1.6.1 The Newton–Gauss Method

Problem Find an *approximate solution* to system (1.72) that verifies those equations with the *least-square error*:

$$f(\mathbf{x}) = \frac{1}{2}\phi^T \mathbf{W}\phi \quad \rightarrow \quad \min_{\mathbf{x}} \tag{1.73}$$

where $\mathbf{W}$ is a $q \times q$ positive-definite *weighting matrix*, which is essential if the physical units of ϕ are disparate. The weighting matrix can also be used to assign relative relevance of some equations over others.

Solution We follow a procedure similar to Newton-Raphson, which is known as the *Newton-Gauss method*, as described below

First, an initial guess $\mathbf{x}^0$ of $\mathbf{x}$ is given; then, we produce the sequence

$$\mathbf{x}^1, \mathbf{x}^2, \ldots, \tag{1.74}$$

such that

$$\mathbf{x}^{k+1} = \mathbf{x}^k + \Delta\mathbf{x}^k \tag{1.75}$$

Calculation of $\Delta\mathbf{x}^k$:

- Factor $\mathbf{W}$ into its two Cholesky factors:

$$\mathbf{W} = \mathbf{V}^T\mathbf{V} \tag{1.76}$$

which is possible because $\mathbf{W}$ is assumed positive-definite.[13]

- Compute $\Delta\mathbf{x}^k$ as the *least-square solution* of the overdetermined linear system

$$\mathbf{V}\Phi(\mathbf{x}^k)\Delta\mathbf{x}^k = -\mathbf{V}\phi(\mathbf{x}^k) \tag{1.77}$$

with $\Phi(\mathbf{x})$ defined as the $q \times n$ Jacobian matrix of the vector function $\phi(\mathbf{x})$, i.e.,

$$\Phi(\mathbf{x}) = \frac{\partial\phi(\mathbf{x})}{\partial\mathbf{x}} \tag{1.78}$$

Upon dropping superscripts for the sake of notation-simplicity, and recalling Eq. (1.41), $\Delta\mathbf{x}$ is obtained from Eq. (1.77) as in the linear case of Sect. 1.4.5, namely,

$$\Delta\mathbf{x} = -(\Phi^T\mathbf{W}\Phi)^{-1}\Phi^T\mathbf{W}\phi \tag{1.79}$$

This procedure is iterative, stopping when a *convergence criterion has been met.*

---

[13] In fact, the Cholesky decomposition is also possible for positive semidefinite matrices, as stated in Sect. 1.4.4, but these are of no interest in the context of the Newton–Gauss method.

The Damping Factor

When implementing the Newton–Gauss method, the objective function f may increase upon correcting $\mathbf{x}^k$ according to Eq. (1.75), i.e.

$$f(\mathbf{x}^{k+1}) > f(\mathbf{x}^k) \tag{1.80}$$

This increase gives rise to oscillations and sometimes even leads to *divergence*. One way to cope with this situation is by introducing *damping*. Instead of using the whole increment $\Delta\mathbf{x}^k$, we use a fraction of it, i.e.

$$\mathbf{x}^{k+1} = \mathbf{x}^k + \alpha\Delta\mathbf{x}^k, \quad 0 < \alpha < 1 \tag{1.81}$$

where α is known as the *damping factor*.

Convergence Criterion

Calculate first $\nabla f(\mathbf{x})$:

$$\nabla f(\mathbf{x}) \equiv \frac{\partial f}{\partial \mathbf{x}} = \left(\frac{\partial \phi}{\partial \mathbf{x}}\right)^T \frac{\partial f}{\partial \phi} \tag{1.82}$$

$$\frac{\partial \phi}{\partial \mathbf{x}} \equiv \Phi, \quad \frac{\partial f}{\partial \phi} = \mathbf{W}\phi \tag{1.83}$$

Hence, the condition for a stationary point is

$$\Phi^T\mathbf{W}\phi = \mathbf{0} \tag{1.84}$$

which is the *normality condition* of the problem stated in Eq. (1.73).

If now Eq. (1.79) is recalled, it is apparent that $\Delta\mathbf{x}$ vanishes when the normality condition (1.84) has been met.

It is thus apparent that, at a stationary point of f, $\phi(\mathbf{x})$ **need not vanish**, as is the case in overdetermined systems; however, $\phi(\mathbf{x}_0)$ **must lie in the null space of** $\Phi^T\mathbf{W}$. Moreover, from Eqs. (1.79) and (1.84) follows that, at a stationary point, $\Delta\mathbf{x}$ vanishes. Hence, the convergence criterion is

$$\|\Delta\mathbf{x}\| < \epsilon \tag{1.85}$$

where ϵ is a prescribed tolerance.

Remarks

- The normality condition (1.84) alone does not guarantee a minimum, but only a *stationary point*.

- However, as it turns out, if the procedure converges, then it does so, to a second-order approximation, to a minimum, and neither to a maximum nor to a saddle point, as we prove below.

The sequence $f(\mathbf{x}^0)$, $f(\mathbf{x}^1)$, ..., $f(\mathbf{x}^k)$, $f(\mathbf{x}^{k+1})$, ..., obtained from the sequence of $\mathbf{x}$ values, evolves, to a first order, as $\Delta f(\mathbf{x})$, given by

$$\Delta f = \left(\frac{\partial f}{\partial \mathbf{x}}\right)^T \Delta \mathbf{x} \tag{1.86}$$

i.e.,

$$\Delta f = \phi^T \mathbf{W} \Phi \Delta \mathbf{x} \tag{1.87}$$

Upon plugging expression (1.79) of $\Delta \mathbf{x}$ into Eq. (1.87), we obtain

$$\Delta f = -\phi^T \underbrace{\mathbf{W}\Phi(\Phi^T \mathbf{W}\Phi)^{-1}\Phi^T \mathbf{W}}_{\mathbf{M}} \phi = -\phi^T \mathbf{M} \phi \tag{1.88}$$

where, apparently, $\mathbf{M}$ is a $q \times q$ positive-definite matrix. As a consequence, $\phi^T \mathbf{M} \phi$ becomes a positive-definite quadratic expression of ϕ; hence, Δf is negative definite. Thus, the second-order approximation of $f(\mathbf{x})$ is negative-definite, and hence, the sequence of f values *decreases monotonically*. That is, in the neighborhood of a stationary point the first-order approximation of $\phi(\mathbf{x})$ is good enough, and hence, if the procedure **converges**, it does so to a **minimum**.

The reader may wonder whether the Newton–Raphson method can be used to solve nonlinear least-square problems. Although the answer is *yes*, the Newton–Raphson method is not advisable in this case, as made apparent below.

Recall ∇f from Eqs. (1.73) and (1.74):

$$\nabla f(\mathbf{x}) = \frac{\partial f}{\partial \mathbf{x}} = \underbrace{\mathbf{\Phi}^T(\mathbf{x})}_{n\times q} \underbrace{\mathbf{W}}_{q\times q} \underbrace{\phi(\mathbf{x})}_{q-\text{dim}}$$

$$\nabla f(\mathbf{x}) = \mathbf{0} \Rightarrow \underbrace{\mathbf{\Phi}^T(\mathbf{x})\mathbf{W}\phi(\mathbf{x})}_{\equiv \psi(\mathbf{x}) \in \mathbb{R}^n} = \mathbf{0}$$

thereby obtaining a determined system of n equations in n unknowns. This system can be solved using Newton–Raphson method, which requires $\nabla \psi(\mathbf{x})$:

$$\nabla \psi(\mathbf{x}) = \frac{\partial \psi}{\partial \mathbf{x}} = \frac{\partial}{\partial \mathbf{x}}[\underbrace{\mathbf{\Phi}^T(\mathbf{x})}_{(\partial\phi/\partial\mathbf{x})^T} \mathbf{W}\phi(\mathbf{x})]$$

That is, $\nabla \psi(\mathbf{x})$ involves second-order derivatives of ϕ with respect to $\mathbf{x}$:

$$\frac{\partial \psi_i}{\partial x_j} \equiv \frac{\partial^2 \phi_i}{\partial x_j \partial x_i}, \quad i, \; j = 1, \ldots, n$$

In summary, the Newton–Raphson method is too cumbersome and prone to ill-conditioning, for it is based on the normality conditions of the least-square problem at hand. It is strongly recommended to avoid the Newton–Raphson method when confronted with overdetermined systems of nonlinear equations.

1.7 Packages Relevant to Linkage Synthesis

Several software packages of interest to kinematic synthesis are currently available, either commercially or semi-commercially. A list, with some features, follows:

- SAM: Package intended not only for kinematics, but also for static analysis. Runs only on Windows and is commercially available from *Artas Engineering Software*, of RJ Neuen, The Netherlands:

  www.artas.nl

- PRO/ENGINEER: Comprehensive package for mechanical design and analysis at large. Its PRO/MECHANICA module provides motion analysis, simulation, and animation of fairly complex mechanisms. Runs on Windows and Unix. Vendor: *Parametric Technology, Inc.* (USA):

  http://www.ptc.com/

- UNIGRAPHICS: High-end, comprehensive package with modules for finite-element analysis, CAD/CAM, and CAE (Computer-Aided Engineering). Vendor: *Siemens PLM Software*

  http://www.plm.automation.siemens.com

- CATIA: a software of production design supporting CAD, CAE, and CAM functions. Manufacturer: *Dassault Systèmes*

  https://www.3ds.com/products-services/catia/

- SolidWorks: a software tool for mechanical design, simulation and analysis. The tool has an add-on tool, *Mechanism Generator*, which targets the synthesis of four- and six-bar linkages.

  http://www.solidworks.com/

- ADAMS, a general tool for mechanism and multibody-system analysis, produced by *MSC Software Corporation*. No synthesis features are supported.
- AUTOCAD: Comprehensive package for mechanical design and geometric analysis. To be used as a CAD support for linkage synthesis. No special features for linkage synthesis available. Vendor: *Autodesk, Inc.*:

www.autodesk.com

- MATLAB: General-purpose numerical analysis package with excellent routines for equation-solving and optimization. To be used as a support for linkage synthesis. Some MATLAB routines are specifically geared to linkage analysis. Package is produced by *The MathWorks*:

  www.mathworks.com

- MACSYMA, MAPLE, and MATHEMATICA: commercial packages for symbolic computations. MAPLE:

  www.maplesoft.com

  and MATHEMATICA:

  www.wolfram.com/mathematica

  provide for numerical computations
- BERTINI: Developed by Bates et al. (2013), a C program for solving polynomial systems. Key features are as follows:
  - Finds isolated solutions by means of *homotopies*.
  - Implements a method, called "regeneration," which efficiently finds isolated solutions by introducing the equations one-by-one.
  - Finds positive-dimensional solution sets and breaks them into irreducible components.
  - Has adaptive multiprecision arithmetic for maintaining accuracy in larger problems.
  - Endgames for fast, accurate treatment of singular roots.
  - Simple input file format.
  - Provides parameter homotopy, useful for efficiently solving multiple examples in a parameterized family of systems.
  - Allows user-defined homotopies.
  - Supports parallel computing.
  - A detailed user's manual is available from SIAM Books. See Charles W. Wampler's publications in

    http://www3.nd.edu/~cwample1/Software.htm

- HOMLAB: Developed by Charles Wampler, a suite of MATLAB routines for learning about polynomial continuation. Although created for use with the book by Sommese and Wampler (2005), HomLab is a general-purpose solver, fast enough for moderately sized systems. If you are concerned about speed, numerical accuracy, and user-friendliness, try Bertini. If you want to learn the techniques of polynomial continuation from the inside, HomLab is your entry point.
- PHC: A code written in Ada, by Jan Verschelde. Key features:
  - Treatment of isolated solutions, including polyhedral homotopy (also known as the BKK approach, mixed volume, or polytope method).

  – Treatment of positive-dimensional solutions includes irreducible decomposition and diagonal homotopy.
  – The PHC pages also include a large collection of interesting examples.

- HOM4PS: Developed by T.-Y. Li, T.-L. Lee, T. Chen and N. Ovenhouse, a piece of code for polyhedral homotopy. Very fast polyhedral homotopy method.
- Ch: C/C++ code for numerical computations, of special interest to solving complex kinematics problems. It handles dual numbers.

References

Angeles, J., & López-Cajún, C. S. (1991). *Optimization of cam mechanisms*. Kluwer Academic.

Bates, D., Hauenstein, J., Sommese, A., & Wampler, C. (2013). *Bertini: A software package for numerical algebraic geometry*. Accessible at `https://bertini.nd.edu/`.

Beck, T. (1859). *Beiträge zur Geschichte des Maschinenbaues*. J. Springer.

Björck, Å., & Dahlquist, G. (1974). *Numerical methods*. Prentice-Hall Inc.

Bogolyubov, A. N. (1976). *Teoriya mekhanismov v istoricheskom razvitii (Theory of Mechanisms and its Historical Development)*. Nauka.

Borgnis, G. A. (1818). *Traité Complet de Mécanique Appliquée aux Arts*. Traité des Compositions des Machines.

Chen, C., & Angeles, J. (2008). A novel family of linkages for advanced motion synthesis. *Mechanism and Machine Theory,43*(7), 882–890.

Dudiţă, F. D., & Diaconescu, D. (1987). *Optimizarea Structurală a Mecanismelor (Optimization of Mechanisms)*. Tehnicá Publishers.

Euler, L. (1753). De machinis in genere. *Novii Comentarii Academiæ Scientiarum Petropolitanæ, III*, 254–285.

Fowler, H. W., Fowler, F. G., & Thompson, D. (1995). *The concise oxford dictionary of current english*. Clarendon Press.

French, M. E. (1992). *Form*. Structure and Mechanism.

Frolov, K. V. (1987). *Teoriya Mechanismov i Mashin*. Vyschaya Shkola.

Gleick, J. (1988). *Chaos: Making a new science* Penguin Books.

Golub, G. H., & Van Loan, C. F. (1983). *Matrix computations*. The Johns Hopkins University Press.

Hachette, J. N. (1811). *Traité Elémentaire des Machines*. J. Klostermann père et fils.

Hartenberg, R., & Denavit, J. (1964). *Kinematic synthesis of linkages*. McGraw-Hill Book Company.

IFToMM PC SoT (2003). IFToMM PC for standardization of terminology. *Mechanism and Machine Theory, 38*, 7–10.

Kempe, A. B. (1875). On a general method of describing Plane Curves of the nth degree by Linkwork. *Proceedings of the London Mathematical Society,s1-7*(1), 213–216.

Koenigs, G. (1897). *Leçons de Cinématique*. Librairie Scientifique A.

Koenigs, G. (1901). Etude critique sur la théorie générale des mécanismes. *Comptes Rendus de l'Académie des Sciences, 133,* 485.

Leupold, J. (1725). *Theatrum Machinarium, Oder: Schau-Platz der Heb-Zeuge*. gedruckt bei Christoph Zunkel.

Mandelbrot, B. B. (1983). *The fractal geometry of nature* (3rd edn.). W.H. Freeman and Company.

Merriam-Webster (2003). *Merriam-Webster's collegiate dictionary* (11th edn.). (online). https://www.merriam-webster.com/.

Norton, R. (2004). *Design of machinery: An introduction to the synthesis and analysis of mechanisms and machines*. McGraw-Hill Higher Education.

Plecnik, M. M., & McCarthy, J. M. (2013). Synthesis of a Stephenson II function generator for eight precision positions. In *Proceedings of ASME 2013 International Design Engineering Technical Conferences and Computers and Information in Engineering Conference (IDETC/CIE 2013)*. Paper DETC2013-12763.

Poncelet, J. V. (1824). *Traité de Mécanique Appliquée aux Machines*. Liège.

Reuleaux, F. (1875). *Theoretische kinematik* (Vol. 1). Vieweg.

Reuleaux, F. (1900). *Lehrbuch der Kinematik* (Vol. 2). F. Vieweg.

Robert, P. (1994). *Le Petit Robert 1*. Le Robert, Dictionnaire Alphabétique et Analogique de la Langue Française.

Salmon, G. (1964). *Higher algebra*. Chelsea Publishing Co.

Sommese, A., & Wampler, C. (2005). *The numerical solution of systems of polynomials arising in engineering and science*. World Scientific.

Stein, J. (1979). *Random house college dictionary*. Random House.

Strang, G. (1988). *Linear Algebra* (3rd edn.). Harcourt Brace Jovanovich College Publishers, Fort Worth.

Vitruvius, P. M. (28 B.C.E.). *De Architectura* (vol. X).

Wittenburg, J. (2016). *Kinematics*. Springer.

Chapter 2
The Qualitative Synthesis of Kinematic Chains

li·ai·son 1: a binding or thickening agent used in cooking
2 (a) a close bond or connection: INTERRELATIONSHIP
(b): an illicit sexual relationship: AFFAIR

Merriam Webster's Collegiate Dictionary, Tenth Edition (C)1997, 1996 Zane Publishing, Inc.

Qui pourrait ne pas frémir en songeant aux malheurs que peut causer une seule liaison dangereuse!

Lettre CLXXV. Madame DE VOLANGES à Madame DE ROSEMONDE (de Laclos, 1782).

The fundamental concepts of *motion representation* and *groups of displacements*, as pertaining to rigid bodies, are recalled in this chapter. These concepts are then applied to (a) the classification of kinematic chains according to their mobility; (b) the determination of the degree of freedom of kinematic chains; and (c) the *qualitative synthesis* of multiloop chains occurring in various types of machines, including parallel robots.

2.1 Notation

In following the notation introduced in Chap. 1, we denote with lower-case boldfaces all vectors; with upper-case boldfaces all matrices. Additionally, sets will be denoted with calligraphic fonts, e.g., $\mathcal{A}$, $\mathcal{B}$, etc., while lower kinematic pairs (LKP), to be formally introduced in Sect. 2.3, are denoted with *sans-serif* upper cases: R, P, H, C, F, and S.

J. Angeles and S. Bai, *Kinematics of Mechanical Systems*, Mathematical Engineering,
https://doi.org/10.1007/978-3-031-09544-3_2

2.2 Background

The notion of *rigid body* is fundamental in the study of kinematic chains. A rigid body is a geometric concept that stems from the more general concept of *continuum*: *A rigid body $\mathcal{B}$ is an unbounded continuum of points such that, under any possible motion, two arbitrary points of $\mathcal{B}$ remain equidistant.*

A rigid body $\mathcal{B}$ is thus a set of points that fills continuously the three-dimensional Euclidean space $\mathcal{E}$. That is, between any two distinct points of $\mathcal{B}$ we can always find an infinite, *nondenumerable set* of points of $\mathcal{B}$. A rigid body, as any set of points, is capable of undergoing transformations. In the case at hand, these transformations preserve the distance between any two points of $\mathcal{B}$; as a consequence, the same transformation preserves the angle between any two lines of the body. Any such transformation is called an *isometry*—from Greek *isos* for "equal" and *metron* for "measure." We must, however, distinguish between two kinds of isometries, as described below: Choose a *set* of any four points O, A, B, and C of the body, not lying in a plane. If, when the last three points, as viewed from O, lie in the ccw order A, B, C, the trihedron defined by segments OA, OB, and OC is said to be *right-handed*; otherwise, it is *left-handed*. If a "hand" can be attributed to a set, we refer to this feature as the *chirality* of the set—Greek: *chéir* = hand. It is apparent that under any *physically possible* motion of $\mathcal{B}$, a right-handed (left-handed) trihedron remains right-handed (left-handed). Isometries that do not preserve the hand of the trihedron are *reflections*, examples of which are the two shoes, or the two gloves, etc., of the same individual. One is a reflection, or a mirror-image, of the other. A hand-preserving isometry of $\mathcal{B}$ is implicit in a *displacement* of $\mathcal{B}$ from a reference *pose*—both position and orientation— to its current pose. To simplify matters, we will denote body and pose with the same calligraphic letter, while distinguishing among various poses of the same body by subscripts, whenever needed. Thus, $\mathcal{B}_0$ denotes a reference pose of $\mathcal{B}$, while its current pose can be represented by $\mathcal{B}$, as long as no confusion arises. Chirality-preserving isometries are involved in *rigid-body motions*.

Rigid-body pose and displacement are thus two abstract concepts. To quantify the pose, we resort to *coordinate frames*. A coordinate frame is attached to a rigid body $\mathcal{B}$. The *orientation* of $\mathcal{B}$ with respect to a reference frame is thus given by the orientation of the body-frame with respect to that of the reference frame. The position of $\mathcal{B}$, in turn, is given by that of the origin of the body-frame in the reference frame. The body pose thus comprises both body-position and body-orientation. Body-position is thus defined by the *position vector* $\mathbf{o}_B$ of the origin O_B of the body-frame, while body-orientation by the *rotation matrix* $\mathbf{Q}$ carrying the reference frame into an attitude coincident with that of the body-frame.

According to Euler's Theorem (Angeles, 2014), the displacement of a rigid body about a fixed point O, called a *pure rotation*, is fully characterized by an axis passing through O and parallel to the unit vector $\mathbf{e}$, besides an angle ϕ, as depicted in Fig. 2.1. Prior to introducing the rotation matrix, some preliminaries are invoked in the paragraphs below.

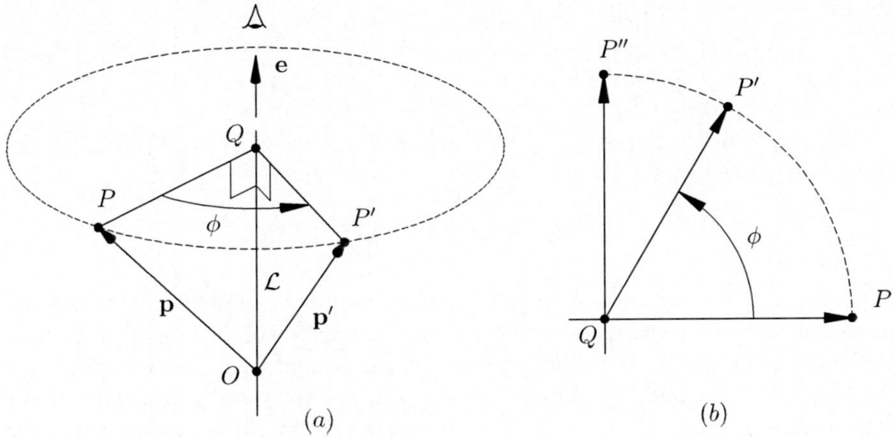

Fig. 2.1 Rotation of a rigid body about a line

Given two vectors $\mathbf{u}, \mathbf{v} \in \mathbb{R}^n$, the product $\mathbf{u}\mathbf{v}^T$ is termed a *rank-one matrix* because any $\mathbf{w} \in \mathbb{R}^n$ normal to $\mathbf{v}$ is mapped by $\mathbf{u}\mathbf{v}^T$ into $\mathbf{0}_n \in \mathbb{R}^n$. As there are $n-1$ linearly independent vectors normal to $\mathbf{v}$, the *nullity* of $\mathbf{u}\mathbf{v}^T$, i.e., the dimension of its null space, is $n-1$. According to the rank-nullity theorem, which states that the dimension of a matrix is the sum of its rank and its nullity, we thus have *rank* $(\mathbf{u}\mathbf{v}^T) = 1$.

The foregoing motion is represented algebraically by a rotation matrix, i.e., a 3×3 *proper orthogonal* matrix $\mathbf{Q}$—an orthogonal matrix is said to be proper orthogonal if its determinant is $+1$; if the said determinant is -1, then the matrix is said to be *improper orthogonal*—that adopts any of the equivalent forms given below:

$$\mathbf{Q} = e^{\phi \mathbf{E}} \tag{2.1a}$$
$$\mathbf{Q} = \mathbf{1} + \sin\phi\mathbf{E} + (1 - \cos\phi)\mathbf{E}^2 \tag{2.1b}$$
$$\mathbf{Q} = \mathbf{e}\mathbf{e}^T + \cos\phi(\mathbf{1} - \mathbf{e}\mathbf{e}^T) + \sin\phi\mathbf{E} \tag{2.1c}$$

In the above expressions, we have resorted to the matrix exponential in the first representation of $\mathbf{Q}$. Moreover, we introduced matrices $\mathbf{1}$, $\mathbf{e}\mathbf{e}^T$, and $\mathbf{E}$, that will be described presently. Matrix $\mathbf{1}$ denotes the 3×3 *identity matrix*, while $\mathbf{e}\mathbf{e}^T$ is a *symmetric, rank-one matrix*; finally, $\mathbf{E}$ denotes the *cross-product matrix* (CPM) of the unit vector $\mathbf{e}$, the CPM being defined as follows: Given *any* three-dimensional vector $\mathbf{a}$, the cross-product matrix $\mathbf{A}$ of $\mathbf{a}$ is given by

$$\mathbf{A} \equiv \frac{\partial(\mathbf{a} \times \mathbf{v})}{\partial \mathbf{v}} = \mathrm{CPM}(\mathbf{a}) \tag{2.1d}$$

for any three-dimensional vector $\mathbf{v}$. More concretely, if $\mathbf{a}$ has components $[a_1 \; a_2 \; a_3]^T$ in a given coordinate frame, then, in the same frame,

$$\mathbf{aa}^T = \begin{bmatrix} a_1^2 & a_1a_2 & a_1a_3 \\ a_1a_2 & a_2^2 & a_2a_3 \\ a_1a_3 & a_2a_3 & a_3^2 \end{bmatrix}, \quad \mathbf{A} = \begin{bmatrix} 0 & -a_3 & a_2 \\ a_3 & 0 & -a_1 \\ -a_2 & a_1 & 0 \end{bmatrix} \tag{2.2a}$$

An identity that the reader is invited to verify follows: if $\mathbf{A} = \mathrm{CPM}(\mathbf{a})$, for any three-dimensional vector $\mathbf{a}$, then

$$\mathbf{A}^2 = -\|\mathbf{a}\|^2\mathbf{1} + \mathbf{aa}^T \tag{2.3}$$

In particular, if $\|\mathbf{a}\| = 1$, then $\mathbf{A}^2 = -\mathbf{1} + \mathbf{aa}^T$. Now it should be apparent that expression (2.1c) follows from expression (2.1b) upon application of the foregoing identity.

Furthermore, given a rigid body in two poses, $\mathcal{B}_1$ and $\mathcal{B}_2$, characterized by the position vectors $\mathbf{o}_1$ and $\mathbf{o}_2$, and the rotation matrices $\mathbf{Q}_1$ and $\mathbf{Q}_2$, respectively, the displacement of the body from $\mathcal{B}_1$ to $\mathcal{B}_2$ is *represented* by (a) the vector difference $\mathbf{u} = \mathbf{o}_2 - \mathbf{o}_1$, termed the *translation*, and (b) the matrix product $\mathbf{Q} = \mathbf{Q}_1^{\mathrm{T}}\mathbf{Q}_2$, termed the *rotation*. Special cases of displacements are the *pure rotation*, as introduced above, for which $\mathbf{u} = \mathbf{0}$, and the *pure translation*, for which $\mathbf{Q} = \mathbf{1}$.

The concept inverse[1] to the CPM of a three-dimensional vector is the axial vector of a 3×3 matrix. In this regard, we recall the *Cartesian decomposition* of a $n \times n$ matrix $\mathbf{A}$, given by

$$\mathbf{A} = \mathbf{A}_S + \mathbf{A}_{SS} \tag{2.4}$$

where $\mathbf{A}_S$ is symmetric and $\mathbf{A}_{SS}$ is skew-symmetric.

The reader can readily prove that

$$\mathbf{A}_S \equiv \frac{1}{2}(\mathbf{A} + \mathbf{A}^T), \quad \mathbf{A}_{SS} \equiv \frac{1}{2}(\mathbf{A} - \mathbf{A}^T) \tag{2.5}$$

and hence, the Cartesian decomposition is *unique*.

Definition 2.2.1 For $n = 3$, let $\mathbf{v}$ be any three-dimensional vector. Then, there exists a vector $\mathbf{a}$ such that

$$\mathbf{A}_{SS}\mathbf{v} \equiv \mathbf{a} \times \mathbf{v} \tag{2.6}$$

Vector $\mathbf{a}$ is called the axial vector of $\mathbf{A}_{SS}$, which is represented as

$$\mathbf{a} = \mathrm{vect}(\mathbf{A}_{SS}) = \mathrm{vect}(\mathbf{A}) \tag{2.7}$$

Theorem 2.2.1

$$\mathbf{A}_{SS} = \mathrm{CPM}(\mathbf{a}) \tag{2.8}$$

Proof Straightforward.

[1] Not the inverse matrix of the CPM, as the CPM is necessarily singular, its nullspace generated by the original vector, $\mathbf{a}$ in the case at hand.

Theorem 2.2.2 *Operations* CPM(·) *and* vect(·) *are linear, i.e., if* **A** *and* **B** *are* 3×3 *matrices and* **a** *and* **b** *three-dimensional vectors, with* α *and* β *being real numbers, then*

$$\text{CPM}(\alpha\mathbf{a} + \beta\mathbf{b}) = \alpha\text{CPM}(\mathbf{a}) + \beta\text{CPM}(\mathbf{b}) \tag{2.9a}$$

while

$$\text{vect}(\alpha\mathbf{A} + \beta\mathbf{B}) = \alpha\text{vect}(\mathbf{A}) + \beta\text{vect}(\mathbf{B}) \tag{2.9b}$$

Proof follows directly from the definitions of CPM(·) and vect(·) plus that of the cross product.

If a_{ij} is the (i, j) entry of **A**, below we find the entries of **a**:

$$\mathbf{A}_S = \begin{bmatrix} a_{11} & (a_{12} + a_{21})/2 & (a_{13} + a_{31})/2 \\ & a_{22} & (a_{23} + a_{32})/2 \\ \text{sym} & & a_{33} \end{bmatrix} \tag{2.10a}$$

$$\mathbf{A}_{SS} = \begin{bmatrix} 0 & (a_{12} - a_{21})/2 & (a_{13} - a_{31})/2 \\ & 0 & (a_{23} - a_{32})/2 \\ \text{skew-sym} & & 0 \end{bmatrix} \tag{2.10b}$$

Further, if a_i denotes the ith component of **a** with a similar definition for v_i,

$$\mathbf{a} \times \mathbf{v} = \det\left(\begin{bmatrix} \mathbf{i} & \mathbf{j} & \mathbf{k} \\ a_1 & a_2 & a_3 \\ v_1 & v_2 & v_3 \end{bmatrix}\right) = \begin{bmatrix} a_2v_3 - a_3v_2 \\ a_3v_1 - a_1v_3 \\ a_1v_2 - a_2v_1 \end{bmatrix} \tag{2.11}$$

As well,

$$\mathbf{A}_{SS}\mathbf{v} = \frac{1}{2}\begin{bmatrix} (a_{12} - a_{21})v_2 + (a_{13} - a_{31})v_3 \\ -(a_{12} - a_{21})v_1 + (a_{23} - a_{32})v_3 \\ -(a_{13} - a_{31})v_1 - (a_{23} - a_{32})v_2 \end{bmatrix} \tag{2.12}$$

As per Eq. (2.6), equate Eqs. (2.11) and (2.12) to obtain

$$a_2 = \frac{1}{2}(a_{13} - a_{31}),\ a_3 = -\frac{1}{2}(a_{12} - a_{21}),\ a_1 = -\frac{1}{2}(a_{23} - a_{32}) \tag{2.13}$$

or

$$\mathbf{a} = \frac{1}{2}\begin{bmatrix} a_{32} - a_{23} \\ a_{13} - a_{31} \\ a_{21} - a_{12} \end{bmatrix} \tag{2.14}$$

Facts: For any $n \times n$ matrix **M** and any n-dimensional vectors **a** and **b**,

Fact 2.2.1

$$\text{tr}(\mathbf{M}) = m_{11} + m_{22} + \cdots + m_{nn}$$

Fact 2.2.2

$$\mathrm{tr}(\mathbf{a}\mathbf{b}^T) = a_1 b_1 + a_2 b_2 + \cdots + a_n b_n = \mathbf{a}^T \mathbf{b}$$

Fact 2.2.3 If $\mathbf{M} = -\mathbf{M}^T$, then all its diagonal entries vanish, and hence, $\mathrm{tr}(\mathbf{M}) = 0$. That is, the trace of a skew-symmetric matrix vanishes, a result that holds for any $n \times n$ matrix of this kind.

Theorem 2.2.3 *If* $\mathbf{A} = \mathbf{A}^T$, *then* $\mathrm{vect}(\mathbf{A}) = \mathbf{0}$

Proof follows from Eq. (2.14).

Theorem 2.2.4 *The* trace$(\cdot)$ *operation is linear, i.e.,*

$$\mathrm{tr}(\alpha \mathbf{A} + \beta \mathbf{B}) = \alpha \mathrm{tr}(\mathbf{A}) + \beta \mathrm{tr}(\mathbf{B})$$

Proof follows from its definition.

The foregoing concepts are now applied to find the *natural invariants* of $\mathbf{Q}$, i.e., the unit vector $\mathbf{e}$, its cross-product matrix $\mathbf{E}$, and angle ϕ, as per Eqs. (2.1a)–(2.1c).

$$\begin{aligned}\mathrm{tr}(\mathbf{Q}) &= \underbrace{\mathrm{tr}(\mathbf{e}\mathbf{e}^T) + \mathrm{tr}[\cos\phi(\mathbf{1} - \mathbf{e}\mathbf{e}^T)] + \mathrm{tr}(\sin\phi\mathbf{E})}_{\text{Theorem 2.2.4}} \\ &= \underbrace{\mathbf{e}^T\mathbf{e}}_{\text{Fact 2.2.2}} + \underbrace{\cos\phi[\underbrace{\mathrm{tr}(\mathbf{1})}_{3} - \underbrace{\mathbf{e}^T\mathbf{e}}_{1}]}_{\text{Theorem 2.2.4}} + \sin\phi \underbrace{\underbrace{\mathrm{tr}(\mathbf{E})}_{0}}_{\text{Fact 2.2.3}} \\ &= 1 + 2\cos\phi\end{aligned} \tag{2.15}$$

$$\begin{aligned}\mathrm{vect}(\mathbf{Q}) &= \underbrace{\mathrm{vect}(\mathbf{e}\mathbf{e}^T) + \mathrm{vect}[\cos\phi(\mathbf{1} - \mathbf{e}\mathbf{e}^T)] + \mathrm{vect}(\sin\phi\mathbf{E})}_{\text{Theorem 2.2.2}} \\ &= \underbrace{\mathbf{0}}_{\text{Theorem 2.2.3}} + \underbrace{\mathbf{0}}_{\text{Theorem 2.2.3}} + \underbrace{\sin\phi\mathrm{vect}(\mathbf{E})}_{\text{Theorem 2.2.2}} \\ &= \underbrace{\sin\phi\mathbf{e}}_{\text{Eq. (2.7)}}\end{aligned} \tag{2.16}$$

which leads to an uncertainty, as ϕ can be associated to one of two possible quadrants. Moreover, a reversal in the sign of ϕ, which does not change the direction of the axis of rotation, can be compensated with a reversal of the sign of the components of $\mathbf{e}$, to leave $\mathbf{Q}$ immutable. For concreteness, and without loss of generality, we shall assume henceforth that $\sin\phi \geq 0$, and hence, that ϕ lies either in the first or the second quadrant. The quadrant ambiguity is resolved by the sign of $\cos\phi$:

$$\text{if } \cos\phi = \frac{\mathrm{tr}(\mathbf{Q}) - 1}{2} > 0, \text{ then } \phi \text{ in 1st quadrant}$$

$$\text{if } \cos\phi = \frac{\mathrm{tr}(\mathbf{Q}) - 1}{2} < 0, \text{ then } \phi \text{ in 2nd quadrant}$$

Example 2.2.1 For $\mathbf{Q}_1$ of Eq. (2.21a), find $\mathbf{e}$ and ϕ.

$$\text{vect}(\mathbf{Q}_1) = \frac{1}{2}\begin{bmatrix} 3/2 \\ \sqrt{3}/2 \\ \sqrt{3}/2 \end{bmatrix} = \frac{1}{4}\begin{bmatrix} 3 \\ \sqrt{3} \\ \sqrt{3} \end{bmatrix} \Rightarrow \sin\phi = ||\text{vect}(\mathbf{Q}_1)|| = \sqrt{15}/4$$

$$\Rightarrow \phi = 75.52^\circ \text{ or } 104.48^\circ$$

$$\mathbf{e} = \text{vect}(\mathbf{Q}_1)/||\text{vect}(\mathbf{Q}_1)|| = \frac{\sqrt{15}}{15}\begin{bmatrix} 3 \\ \sqrt{3} \\ \sqrt{3} \end{bmatrix}$$

$$\text{tr}(\mathbf{Q}_1) = \frac{1}{2} \Rightarrow \cos\phi = -\frac{1}{4} \Rightarrow \phi = 104.48^\circ$$

thereby resolving the ambiguity.

2.3 Kinematic Pairs

The concepts and, to a great extent, the notation and nomenclature that follow are taken from the work conducted by J.M. Hervé, a professor at *Ecole Central de Paris* between 1978 and 1999.

We start by going back to basics: the kinematics of machines is studied via their underlying *kinematic chains*. A kinematic chain is the result of the coupling of rigid bodies, called *links*. Upon coupling *two* links, a *kinematic pair* is obtained. When the coupling takes place in such a way that the two links share a common surface, a *lower kinematic pair* results; when the coupling takes place along a common line or a common point of the two links, a *higher kinematic pair* is obtained. In the book, only lower kinematic pairs are of interest.[2]

If every link of a chain is coupled to at most two other links, then the chain is said to be *simple*. If all the links of a simple kinematic chain are coupled to two other links, then a *closed kinematic chain* is obtained. Moreover, this chain constitutes a *single loop*. If a simple chain has a link coupled to only one other link, then it has necessarily a second link coupled to only one other link, an *open chain* thus resulting. A multiloop chain can have open *subchains*. Single-loop kinematic chains are present in single-degree-of-freedom mechanisms, but a single-loop chain may have a degree of freedom (dof) greater than or smaller than unity. The latter are termed *overconstrained chains*, some example being included in Sects. 2.6 and 2.7. Simple kinematic chains of the open type are present in robotic manipulators of the *serial* type. Multiloop kinematic chains occur in both single-degree-of-freedom machines, e.g., in automobile suspensions, and in multi-dof robotic manipulators of

[2] An introduction to mechanisms with higher kinematic pairs is available (Angeles and López-Cajún, 1991), with pertinent references to the literature on the subject.

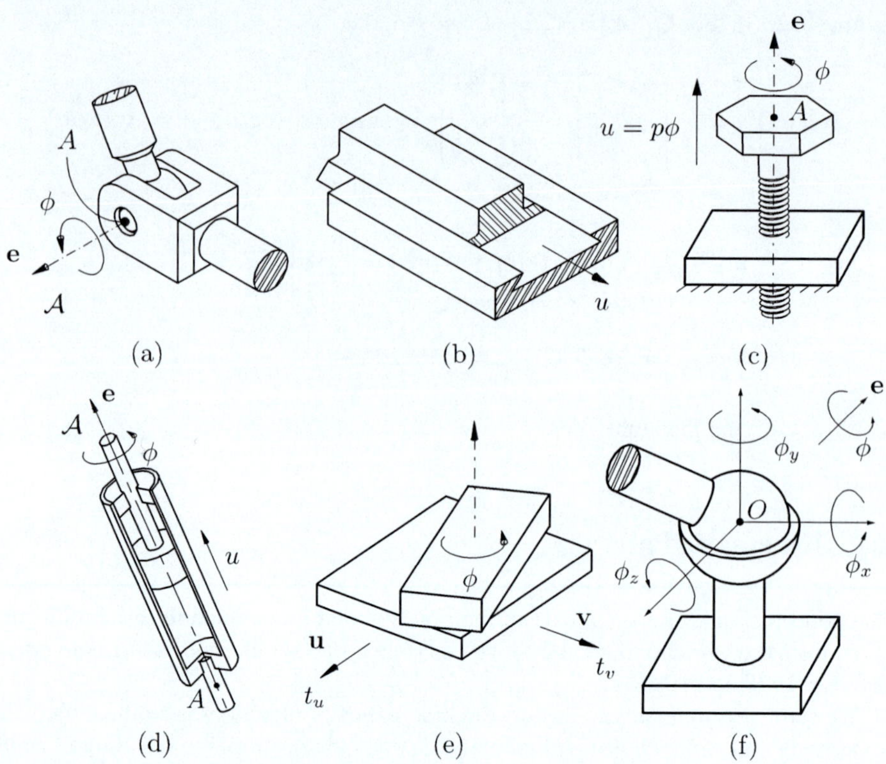

Fig. 2.2 The six lower kinematic pairs: **a** revolute (R); **b** prismatic (P); **c** screw (H); **d** cylindrical (C); **e** planar (F); and **f** spherical (S)

the *parallel type*, paradigms of which are flight simulators. We shall elaborate on these concepts in this book.

Lower kinematic pairs deserve special attention for various reasons: one is that they model fairly well the mechanical couplings in a variety of machines; one more is that they are known to occur in exactly six types, to be described presently. Higher kinematic pairs occur in cam-follower mechanisms and in gears, in which contact occurs along common lines or common points of the coupled bodies, not on common surfaces, in the case of lower kinematic pairs. The book focuses on lower kinematic pairs.

The six lower kinematic pairs, displayed in Fig. 2.2, are described as follows:

(*a*) The *revolute pair* R allows a relative rotation through an angle ϕ about one axis $\mathcal{A}$ passing through a point A of position vector **a** and parallel to a unit vector **e**;

(*b*) The *prismatic pair* P allows a relative translation u in the direction of a unit vector **e**;

(*c*) The *screw pair* H allows both a relative rotation through an angle ϕ about an axis $\mathcal{A}$ passing through a point A of position vector **a** and parallel to the unit vector **e**, plus a relative translation u in the direction of **e**. However, the rotation and the translation are not independent, for they are related by the *pitch* p of the pair: $u = p\phi$;

(*d*) The *cylindrical pair* C allows both a relative rotation through an angle ϕ about an axis $\mathcal{A}$ passing through a point A of position vector **a** and parallel to the unit vector **e**, plus a relative translation u in the direction of **e**, rotation and translation being independent;

(*e*) The *planar pair* F allows two independent translations t_u and t_v in the directions of the *distinct* unit vectors **u** and **v**, respectively, plus a rotation ϕ about an axis normal to the plane of these two vectors; and

(*f*) The *spherical pair* S allowing one independent rotation about each of three non-coplanar axes concurrent at a point O. The relative motion allowed by S is, thus, (*i*) characterized by point O and (*ii*) associated with an axis parallel to the unit vector **e** and with the angle of rotation ϕ about this axis, as per Euler's Theorem.

Remark 2.3.1 While the R, H, and C pairs are characterized by an axis, the P pair is characterized by a *direction* alone; not by an axis.[3]

2.3.1 The Π Kinematic Pair

Besides the six LKPs, the Π pair will be introduced in this chapter. This joint is a *parallelogram four-bar linkage*, as depicted in Fig. 2.3, which couples two links, 1, that is considered *fixed*, and 3. The latter moves under *pure translation*, more exactly, curvilinear translation, all its points describing circles of variable location and radius identical to the common length of links 2 and 4.

In the standard terminology, link 3 is termed the *coupler*. The linkage thus provides a kinematic pair of the coupler link with respect to the fixed link. Hervé and Sparacino (1992) termed this coupling a Π kinematic pair, a.k.a. a Π-joint. At about the same time, Wohlhart (1991, 1992) and Dietmaier (1992) reported work on the use of the same type of joints in mechanisms.

Note that the Π joint does not belong to the class of lower kinematic pairs. The Π pair can be shown to be equivalent to a P pair of continuously variable direction. Indeed, the *instant center*[4] (Uicker, Jr. et al., 2011) $\mathcal{I}$ of the coupler link 3 with respect to link 1 of the parallelogram linkage lies at infinity, in the direction of the axes of the parallel links 2 and 4 of Fig. 2.3.[5] This point thus changes continuously its location, depending on the angle of inclination of the axes of links 2 and 3 with respect to line R_1R_4. The same point is thus common to both links 1 and 3, but it is *the only common*

[3] More often than not, one encounters in the literature a *line* associated with the relative motion allowed by a prismatic pair, which can be misleading.

[4] This is the instantaneously common position of two points of corresponding rigid bodies moving under *planar motion*.

[5] Remember the Aronhold–Kennedy Theorem of elementary mechanisms courses!

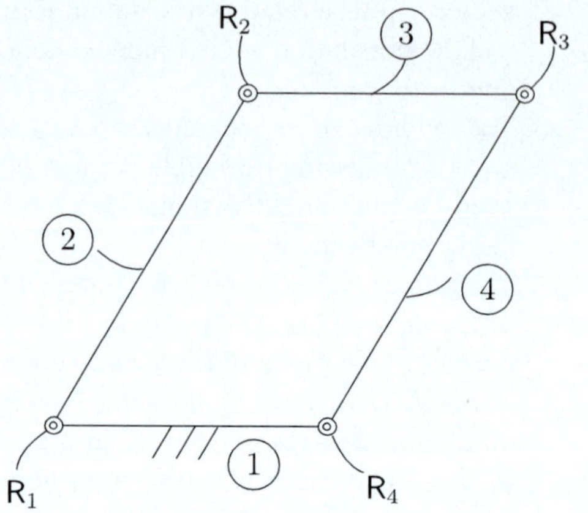

Fig. 2.3 The Π joint, a four-bar parallelogram coupling links 1 and 3

point of the two bodies. If the linkage is regarded as a three-dimensional object, then the instant center is, in fact, a line, the instant screw axis—see Theorem A.4.5—and hence, contact between the two coupled links takes place not along a surface, but along a line, which disqualifies the Π pair from being a LKP. The interest of this pair lies in its ability to generate pure translations when combined with other Π-joints or with lower kinematic pairs, as discussed below.

2.4 Graph Representation of Kinematic Chains

Figure 2.3 depicts a planar four-bar linkage in the most common way: its R joints are represented as small circles—with smaller circles inside that most often are replaced by dots or even left empty—their center denoting the projection of the R axis onto the plane of the figure. The joints are coupled by means of straight lines representing the mechanism links. These lines convey geometric information, as their lengths are the distances between neighboring-joint axes. An alternative representation of linkages, particularly useful in the preliminary stages of mechanism design, in which dimensions are yet to be found, is based on *graphs*. A graph is a two-dimensional sketch composed of *nodes* and *edges* (Harary, 1972). The nodes are objects—such as pieces of land in the famous Königsberg/Kaliningrad—related by the edges—e.g., the seven Königsberg/Kaliningrad bridges that so much attracted Leonhard Euler's attention.[6] In kinematic synthesis, the objects are the mechanism links and the edges the joints that couple them. The graph representation of the four-bar linkage of Fig. 2.3, or of any four-bar linkage for that matter, as dimensions become immaterial in a graph, is depicted in Fig. 2.4.

[6] https://www.mathworld.wolfram.com/KoenigsbergBridgeProblem.html.

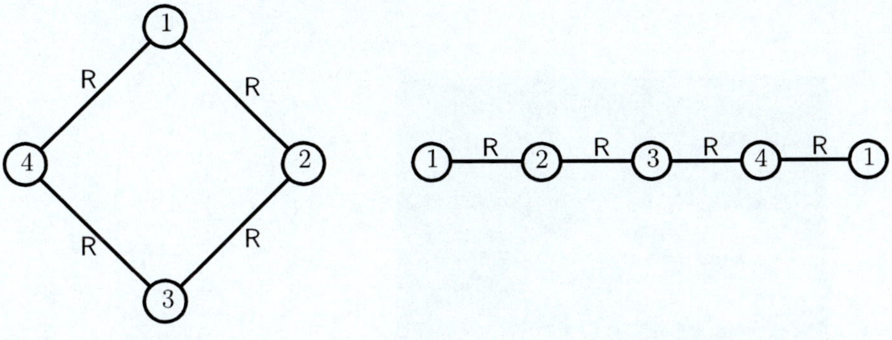

Fig. 2.4 Two alternative graphs of a four-bar linkage

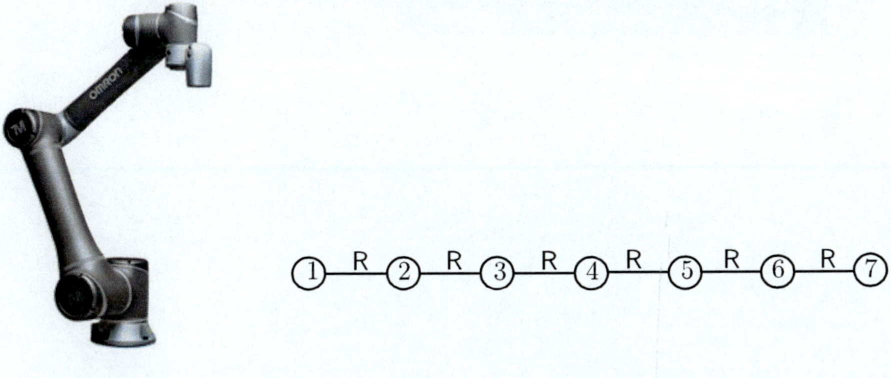

(a) (b)

Fig. 2.5 A six-axis industrial robot of the serial type: **a** its photograph (*Source* OMRON) and **b** its graph

Apparently, a mechanism graph encapsulates non-dimensional, *relational information* on the *architecture* of a linkage. This information includes connectivity—which link is coupled to which—and layout—whether the chain is closed, as in the above case, or open, as in the case of an industrial robot of the serial type, like the one shown in Fig. 2.5. In the graph representation of Fig. 2.5b, link 1 is the base and link 7 the end-effector.

The architecture of some machines is more complex in that it may contain several loops. For example, the flight simulator of Fig. 2.6a comprises one base platform ($\mathcal{B}$), the ground, and one moving platform ($\mathcal{M}$), the one carrying the aircraft cockpit, both coupled by means of six extensible limbs. Its graph, in two versions, is shown in Fig. 2.6b, c. As a matter of fact, each limb is itself a serial kinematic chain of the serial type, its architecture being depicted in Fig. 2.7, which is, again, a six-joint open chain, similar to the one of Fig. 2.5a, with one difference: its third joint from

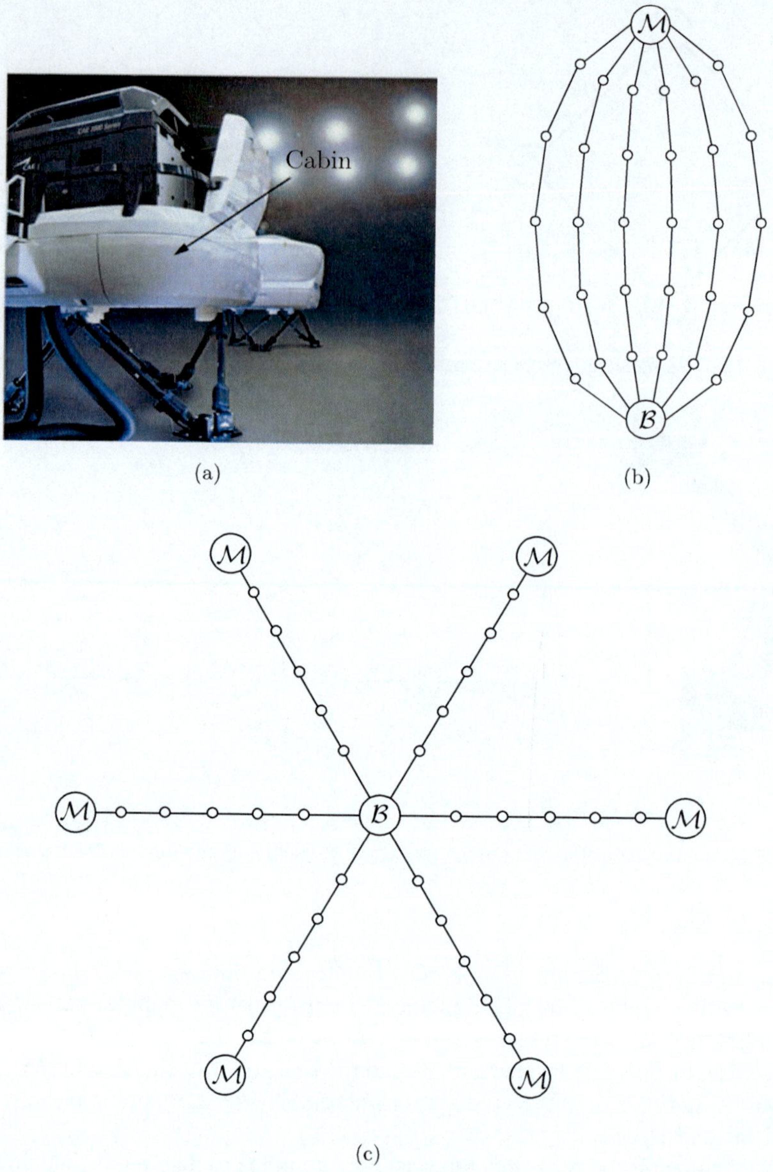

Fig. 2.6 A flight simulator: **a** its photograph and **b** and **c** two alternatives of its graph representation

the base (1) is of the P type, instead of the R type of Fig. 2.5b. One more difference, not affecting its topology but worth pointing out, is that, of all six joints of the chain shown in Fig. 2.7, only the P joint is actuated, all R joints being *passive*.

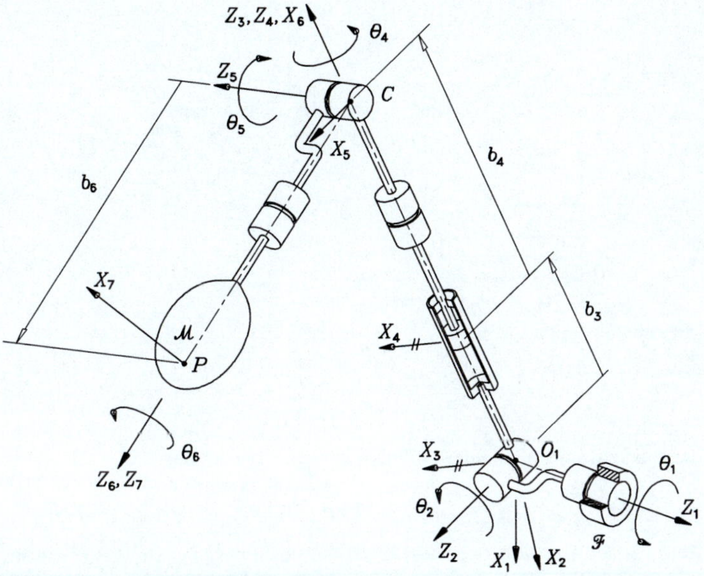

Fig. 2.7 The kinematic chain of a flight-simulator limb

2.5 Groups of Displacements

In the sequel, we shall resort to the algebraic concept of *group*. A group is a set $\mathcal{G}$ of elements related by a *binary operation* $\star$ with four properties:

P1 if a and $b \in \mathcal{G}$, then $a \star b \in \mathcal{G}$;
P2 if a, b, and $c \in \mathcal{G}$, then $a \star (b \star c) = (a \star b) \star c$;
P3 $\mathcal{G}$ contains an element ι[7] called the *identity* of $\mathcal{G}$ under $\star$, such that $a \star \iota = \iota \star a = a$; and
P4 for every $a \in \mathcal{G}$, there exists an element a^{-1}, called the *inverse of a under* $\star$ such that $a \star a^{-1} = a^{-1} \star a = \iota$.

Two types of groups are found, discrete and continuous; the former have a discrete set of elements and the latter a continuum. An example of a discrete group is the *symmetry group* of a regular polygon, defined as the set of rotations about the center of the polygon that leaves the figure unchanged. Continuous groups, or Lie groups, named after the Norwegian mathematician Sophus Lie (1842–1899), are of interest to this chapter.

If the elements of a set $\mathcal{D}$ are the displacements undergone by a rigid body, then (i) $\mathcal{D}$ is a Lie group, and (ii) we can define a binary operation $\odot$—read "o-dot"[8]—of

[7] ι is the Greek letter *iota*.

[8] we use $\odot$, henceforth, to refer specifically to displacement subgroups, while $\star$ introduced in the beginning denotes a generic group operation.

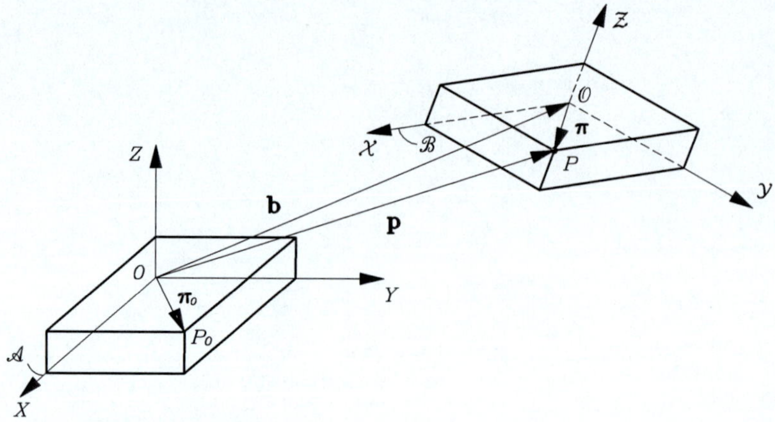

Fig. 2.8 Coordinate frames with different origins

displacements to be the *composition* of displacements: As the body undergoes first a displacement d_a, and then a displacement d_b, taking the body, successively, from pose $\mathcal{B}_0$ to pose $\mathcal{B}_a$, and then to pose $\mathcal{B}_b$, it is intuitively apparent that the composition of the two displacements, $d_a \odot d_b$, is in turn a rigid-body displacement.

More concretely, a rigid-body displacement, illustrated in Fig. 2.8, is defined by a translation $\mathbf{b}$ of a landmark point, say O in the figure, and a rotation $\mathbf{Q}$ about the same point. Point P in the displaced posture is the displaced counterpart of a point P_0—shown in the figure as well—of position vector $\boldsymbol{\pi}_0$ in the reference posture. Under these conditions, the position vector $\mathbf{p}$ of P in the reference frame $\mathcal{A}$ can be expressed as

$$\mathbf{p} = \mathbf{b} + \underbrace{\boldsymbol{\pi}}_{\mathbf{Q}\boldsymbol{\pi}_0} = \mathbf{b} + \mathbf{Q}\boldsymbol{\pi}_0 \tag{2.17}$$

which gives the displaced position of P as a sum of vectors, the second being, in turn, the product of a matrix by a vector. A terser representation can be obtained, involving only one product of a matrix by a vector, if *homogeneous coordinates* are introduced. The homogeneous coordinates of a *finite point*, i.e., a point whose Cartesian coordinates are finite real numbers, are grouped in a four-dimensional array $\{\mathbf{p}\}$, whose first three components are the Cartesian coordinates of P, the fourth being the real unity, i.e.,

$$\{\mathbf{p}\} \equiv \begin{bmatrix} \mathbf{p} \\ 1 \end{bmatrix}, \quad \{\mathbf{p}_0\} \equiv \begin{bmatrix} \boldsymbol{\pi}_0 \\ 1 \end{bmatrix} \tag{2.18}$$

$\{\mathbf{p}_0\}$ comprising the homogeneous coordinates of P_0. Now, with the aid of the above definitions and relation (2.17), the same relation can be expressed as a product of a matrix by a vector, in the form

$$\{\mathbf{p}\} = \{\mathbf{T}\}\{\mathbf{p}_0\}, \quad \{\mathbf{T}\} \equiv \begin{bmatrix} \mathbf{Q} & \mathbf{b} \\ \mathbf{0}^T & 1 \end{bmatrix} \tag{2.19}$$

where $\{\mathbf{T}\}$ is termed a *homogeneous-transformation matrix*. Notice that homogeneous transformations, as defined above, are represented by 4×4 matrices that involve four blocks: the upper left block is a proper orthogonal matrix, accounting for the rigid-body rotation; the upper-right three-dimensional vector block $\mathbf{b}$, accounting for the translation of the landmark point O, the designated origin of frame $\mathcal{A}$; the lower-left three-dimensional row block $\mathbf{0}^T$, with $\mathbf{0}$ denoting the three-dimensional zero vector; and the lower-right 1×1 "block" that is the real unity.

It is left as an exercise to the reader to prove that homogeneous transformations form a continuous group under the operation of matrix multiplication.

Hence, with the notation introduced above, for the rigid-body displacements d_a and d_b:

(a) $d_a \odot d_b \in \mathcal{D}$;
(b) if a third displacement d_c is introduced, carrying $\mathcal{B}$ from pose $\mathcal{B}_b$ to pose $\mathcal{B}_c$, then $d_a \odot (d_b \odot d_c) = (d_a \odot d_b) \odot d_c$;
(c) under no motion, any pose $\mathcal{B}$ of a rigid body is preserved, the motion undergone by the body then being represented by a displacement ι—read "iota"—that can be defined as the *identity element* of $\mathcal{D}$, such that, for any displacement d, $d \odot \iota = \iota \odot d = d$; and
(d) for any displacement d carrying the body from pose $\mathcal{B}_0$ to pose $\mathcal{B}$, the *inverse displacement* d^{-1} is defined as that bringing back the body from $\mathcal{B}$ to $\mathcal{B}_0$, i.e., $d \odot d^{-1} = d^{-1} \odot d = \iota$.

From the foregoing discussion, it should be apparent that the set of rigid-body displacements $\mathcal{D}$ has the algebraic structure of a group. Henceforth, we refer to the set of displacements of a rigid body as *group* $\mathcal{D}$. The interest in studying rigid-body displacements as algebraic groups lies in that, on the one hand, $\mathcal{D}$ includes interesting and practical subgroups that find relevant applications in the design of production-automation and prosthetic devices. On the other hand, the same subgroups can be *combined* to produce novel mechanical layouts that would be insurmountably difficult to produce by sheer intuition. The combination of subgroups, in general, can take place via the standard set operations of *union* and *intersection*. As we shall see, however, the set defined as that comprising the elements of two displacement subgroups is not necessarily a subgroup, and hence, one cannot speak of the union of displacement subgroups. On the contrary, the intersection of two displacement subgroups is always a subgroup itself, and hence, the *intersection of displacement subgroups* is a valid group operation.

Rather than the union of groups, what we have is the *product* of groups. Let $\mathcal{G}_1$ and $\mathcal{G}_2$ be two groups defined over the same binary operation $\star$; if $g_1 \in \mathcal{G}_1$ and $g_2 \in \mathcal{G}_2$, then the product of these two groups, represented by $\mathcal{G}_1 \bullet \mathcal{G}_2$, is the *set* of elements of the form $g_1 \star g_2$, where the order is important, for **commutativity is not to be taken for granted in group theory**.

The intersection of the two foregoing groups, represented by the usual set-theoretic symbol $\cap$, i.e., $\mathcal{G}_1 \cap \mathcal{G}_2$, is the group of elements g belonging to both $\mathcal{G}_1$ and $\mathcal{G}_2$, and hence, the order is immaterial.

2.5.1 *Displacement Subgroups*

A *subgroup* $\mathcal{G}_s$ of a given group $\mathcal{G}$ is a set of objects such that (a) they all belong to $\mathcal{G}$, although some objects, or elements, of $\mathcal{G}$ may not belong to $\mathcal{G}_s$, and (b) $\mathcal{G}_s$ has the algebraic structure of a group. Therefore, a subgroup $\mathcal{D}_s$ of the group of rigid-body displacements $\mathcal{D}$ is itself a group of displacements, but may lack some rigid-body displacements. If $\mathcal{D}$ includes elements not included in $\mathcal{D}_s$, then the latter is said to be a *proper subset* of the former.

The six lower kinematic pairs can be regarded as *generators* of displacement subgroups. We thus have the following:

(i) The revolute pair R of axis $\mathcal{A}$ generates the subgroup $\mathcal{R}(\mathcal{A})$ of rotations about $\mathcal{A}$. Each element of this subgroup is characterized by both the axis $\mathcal{A}$ and the angle ϕ of the corresponding rotation;

(ii) the prismatic pair in the direction $\mathbf{e}$ generates the subgroup $\mathcal{P}(\mathbf{e})$ of translations[9] along $\mathbf{e}$. Each element of $\mathcal{P}(\mathbf{e})$ is characterized by the translation u along $\mathbf{e}$;

(iii) the screw pair of axis $\mathcal{A}$ and pitch p generates the subgroup $\mathcal{H}(\mathcal{A}, p)$ of rotations ϕ about $\mathcal{A}$ and translations u along the direction of the same axis, translation and rotation being related by the pitch p in the form $u = p\phi$, as described when the screw pair was introduced. Each element of $\mathcal{H}(\mathcal{A}, p)$ can thus be characterized either by u or by ϕ;

(iv) the cylindrical pair of axis $\mathcal{A}$ generates the subgroup $\mathcal{C}(\mathcal{A})$ of independent rotations about and translations parallel to $\mathcal{A}$. Each element of $\mathcal{C}(\mathcal{A})$ is thus characterized by both the displacement u and the rotation ϕ;

(v) the planar pair generates the subgroup $\mathcal{F}(\mathbf{u}, \mathbf{v})$ of two independent translations in the directions of the *distinct* unit vectors $\mathbf{u}$ and $\mathbf{v}$ and one rotation about an axis normal to both $\mathbf{u}$ and $\mathbf{v}$. Each element of $\mathcal{F}(\mathbf{u}, \mathbf{v})$ is thus characterized by the two translations t_u, t_v and the rotation ϕ;

(vi) the spherical pair generates the subgroup $\mathcal{S}(O)$ of rotations about point O. Each element of $\mathcal{S}(O)$, a rotation about O, is characterized by the axis of rotation passing through O in the direction of a unit vector $\mathbf{e}$ and through an angle ϕ. Alternatively, each rotation about O can be characterized by the independent rotations about three designated non-coplanar axes, e.g., the well-known *Euler angles*.

Besides the six foregoing subgroups, we can define six more, namely,

---

[9] Sometimes, a P joint is represented as associated with a line, but this is confusing. Reason: a rigid-body translation has a direction, but no axis!

(vii) The *identity subgroup* $\mathcal{I}$, whose single element is the identity displacement ι introduced above;

$(viii)$ the *planar-translation subgroup* $\mathcal{T}_2(\mathbf{u}, \mathbf{v})$ of translations in the directions of the two distinct unit vectors $\mathbf{u}$ and $\mathbf{v}$. Each element of this group is thus characterized by two translations, t_u and t_v;

(ix) the *translation subgroup* $\mathcal{T}_3$ of translations in $\mathcal{E}$. Each element of the subgroup is characterized by three independent translations t_u, t_v, and t_w, *along three non-coplanar directions*;

(x) the subgroup $\mathcal{Y}(\mathbf{e}, p) = \mathcal{T}_2(\mathbf{u}, \mathbf{v}) \bullet \mathcal{H}(\mathcal{A}, p)$, with $\mathbf{u} \times \mathbf{v} = \mathbf{e} \parallel \mathcal{A}$ of motions allowed by a screw of pitch p and axis parallel to $\mathbf{e}$ undergoing arbitrary translations in a direction normal to $\mathbf{e}$. Each element of this subgroup is thus characterized by the two independent translations t_u, t_v of the axis and either the rotation ϕ about this axis or the translation t_w along the axis. *Faute-de-mieux*, we shall call this subgroup the *translating-screw group*[10];

(xi) the subgroup $\mathcal{X}(\mathbf{e}) = \mathcal{F}(\mathbf{e}) \bullet \mathcal{P}(\mathbf{e})$, resulting of the product of the planar subgroup of plane normal to $\mathbf{e}$ and the prismatic subgroup of direction $\mathbf{e}$. Each element of this subgroup is thus characterized by the two translations t_u, t_v and the angle ϕ of the planar subgroup, plus the translation t_w in the direction of $\mathbf{e}$. Moreover, note that $\mathcal{F}(\mathbf{e}) \bullet \mathcal{P}(\mathbf{e}) = \mathcal{P}(\mathbf{e}) \bullet \mathcal{F}(\mathbf{e})$. This subgroup, named after the German mathematician and mineralogist Arthur Moritz Schönflies (1853–1928), is known as the *Schönflies subgroup* and is generated most commonly by what is known as SCARA systems, for *Selective-Compliance Assembly Robot Arm*;

(xii) the group $\mathcal{D}$ itself. Each element of this subgroup is characterized by three independent translations and three independent rotations.

It is thus apparent that each subgroup includes a set of displacements with a specific *degree of freedom*. We shall need below an extension of the concept of dof, for which reason we term the dof of each subgroup its *dimension*, and denote the dimension of any subgroup $\mathcal{G}_s$ by $\dim[\mathcal{G}_s]$. Thus,

$$\dim[\mathcal{I}] = 0 \tag{2.20a}$$
$$\dim[\mathcal{R}(\mathcal{A})] = \dim[\mathcal{P}(\mathbf{e})] = \dim[\mathcal{H}(\mathcal{A}, p)] = 1 \tag{2.20b}$$
$$\dim[\mathcal{T}_2(\mathbf{u}, \mathbf{v})] = \dim[\mathcal{C}(\mathcal{A})] = 2 \tag{2.20c}$$
$$\dim[\mathcal{T}_3] = \dim[\mathcal{F}(\mathbf{e})] = \dim[\mathcal{S}(O)] = \dim[\mathcal{Y}(\mathbf{e}, p)] = 3 \tag{2.20d}$$
$$\dim[\mathcal{X}(\mathbf{e})] = 4 \tag{2.20e}$$
$$\dim[\mathcal{D}] = 6 \tag{2.20f}$$

The foregoing list of *twelve* displacement subgroups is *exhaustive*, none *being of dimension five*. The reader may wonder whether displacement products are missing from the list that might be subgroups. However, any displacement product not appearing in the above list *is not a subgroup*. As a matter of fact, any set of displace-

[10] Think of the tray of a waiter in a restaurant, carrying the wine bottle, the glasses, and a cork-screw.

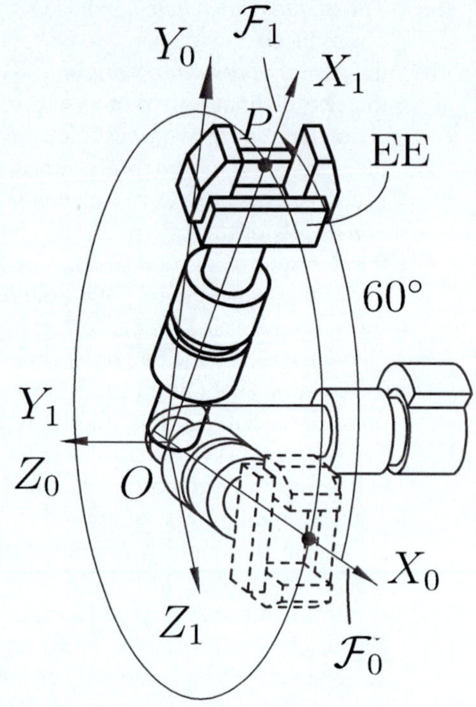

Fig. 2.9 A pitch-roll wrist producing a first rotation

ments including rotations about *two non-parallel axes*, and no more than two, fails to have a group structure. Consider, for example, the set of rotations $\mathcal{W}$ produced by a (two-dof) *pitch-roll wrist* (PRW), as depicted in Fig. 2.9. With reference to this figure, $\mathcal{F}_0$, the *reference frame*, is defined with X_0 along the roll axis, passing through the *operation point* P, while Z_0 is defined along the pitch axis.

Now we introduce a first rotation $\mathbf{Q}_1$ (not a coordinate transformation!): Define $\mathcal{F}_1$ with X_1 along the displaced roll axis, passing through P in the displaced pose of the end-effector (EE), and making an angle of 60° with X_0, Z_1 lying in the X_0–Y_0 plane, and making an angle of $-30°$ with X_0, when measured in the direction of Z_0. Matrix $\mathbf{Q}_1$, rotating $\mathcal{F}_0$ into $\mathcal{F}_1$ as illustrated in Fig. 2.9, is obtained, in $\mathcal{F}_0$-coordinates, upon assigning to its first column the $\mathcal{F}_0$-components of the unit vector $\mathbf{i}_1$, parallel to X_1, its second column being given by the $\mathcal{F}_0$-components of the unit vector $\mathbf{j}_1$, parallel to Y_1; its third column follows the same pattern.[11] The rotation $\mathbf{Q}_1$ carrying $\mathcal{F}_0$ into $\mathcal{F}_1$ as depicted in Fig. 2.9 is, hence,

$$\mathbf{Q}_1 = \begin{bmatrix} 1/2 & 0 & \sqrt{3}/2 \\ \sqrt{3}/2 & 0 & -1/2 \\ 0 & 1 & 0 \end{bmatrix} \tag{2.21a}$$

[11] The matrix transforming $\mathcal{F}_0$-coordinates into $\mathcal{F}_1$-coordinates is $\mathbf{Q}_1^T$.

Fig. 2.10 A pitch-roll wrist producing a second rotation

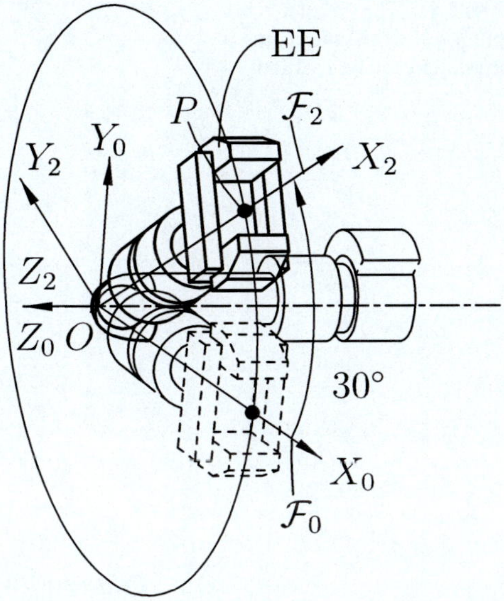

Next, we introduce a second rotation $\mathbf{Q}_2$: Define a new frame $\mathcal{F}_2$ with X_2 as shown in Fig. 2.10, along the roll axis, passing through P in the displaced pose of the EE, and making an angle of 30° with X_0, when measured in the direction of Z_0, Z_2 coinciding with Z_0. Hence, in $\mathcal{F}_0$-coordinates as well,

$$\mathbf{Q}_2 = \begin{bmatrix} \sqrt{3}/2 & -1/2 & 0 \\ 1/2 & \sqrt{3}/2 & 0 \\ 0 & 0 & 1 \end{bmatrix} \tag{2.21b}$$

Moreover, let $\mathbf{Q}_3 = \mathbf{Q}_1\mathbf{Q}_2$, a third rotation obtained as the product of the first two, namely,

$$\mathbf{Q}_3 = \begin{bmatrix} \sqrt{3}/4 & -1/4 & \sqrt{3}/2 \\ 3/4 & -\sqrt{3}/4 & -1/2 \\ 1/2 & \sqrt{3}/2 & 0 \end{bmatrix} \tag{2.21c}$$

which yields a third attitude of the EE, as depicted in Fig. 2.11.

We note that, in an arbitrary configuration, the roll axis remains normal to Z_0. Hence, any rotation produced by the PRW takes the EE to a pose in which the roll axis is normal to Z_0, i.e., the set of possible displaced configurations of the roll axis is a pencil of lines passing through the origin and normal to Z_0. The roll axis in the displaced pose of the EE thus lies in the X_0–Y_0 plane. Any EE pose whereby the roll axis lies outside of the X_0–Y_0 plane is attained by a rotation *outside* of $\mathcal{W}$.

As it turns out, the roll axis is carried by $\mathbf{Q}_3$ into a configuration parallel to $\mathbf{i}_3$, the image of $\mathbf{i}_0$ under $\mathbf{Q}_3$, as depicted in Fig. 2.11, i.e.,

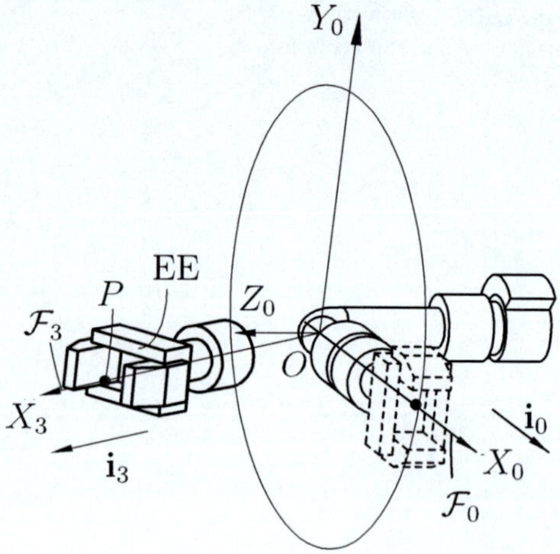

Fig. 2.11 The EE of a pitch-roll wrist undergoing a third, unfeasible, rotation

$$\mathbf{i}_3 = \mathbf{Q}_3\mathbf{i}_0 = \left[\sqrt{3}/4 \;\; 3/4 \;\; 1/2\right]^T \tag{2.22}$$

which, apparently, is not normal to Z_0—its Z_0-component is $1/2 \neq 0$—and, hence, $\mathbf{Q}_3$ lies outside of the set $\mathcal{W}$ of feasible rotations produced by the PRW. Therefore, *the set of rotations produced by a PRW does not have the algebraic structure of a group*.

2.6 Kinematic Bonds

Displacement subgroups can be combined to produce new sets of displacements that may or may not be displacement subgroups themselves. To combine subgroups, we resort to the group operations of product ($\bullet$) and intersection ($\cap$).

Now we introduce the concept of *kinematic bond*, which is a generalization of kinematic pair, as first proposed by Hervé (1978), who called this concept *liaison cinématique* in French. This concept has been termed *kinematic liaison* (Angeles, 1982) or *mechanical connection* (Hervé, 1999) in English. Since "liaison" in English is usually applied to human relations, the term "bond" seems more appropriate and is thus adopted here.

We illustrate the concept with an example: Let us assume three links, numbered from 1 to 3, and coupled by two kinematic pairs generating the two subgroups $\mathcal{G}_1$ and $\mathcal{G}_2$, where these two subgroups are instanced by specific displacement subgroups below. We then have

$$\mathcal{G}_1 \bullet \mathcal{G}_2 = \mathcal{R}(\mathcal{A}) \bullet \mathcal{P}(\mathbf{e}) = \mathcal{C}(\mathcal{A}), \quad \text{for } \mathbf{e} \parallel \mathcal{A} \tag{2.23a}$$

$$\mathcal{G}_1 \bullet \mathcal{G}_2 = \mathcal{R}(\mathcal{A}) \bullet \mathcal{T}_2(\mathbf{u}, \mathbf{v}) = \mathcal{F}(\mathbf{e}), \quad \text{for } \mathbf{e},\ \mathcal{A} \perp \mathbf{u},\ \mathbf{v} \tag{2.23b}$$

$$\mathcal{G}_1 \bullet \mathcal{G}_2 = \mathcal{R}(\mathcal{A}) \bullet \mathcal{R}(\mathcal{B}) = \mathcal{L}(1, 3) \tag{2.23c}$$

$$\mathcal{G}_1 \cap \mathcal{G}_2 = \mathcal{R}(\mathcal{A}) \cap \mathcal{C}(\mathcal{A}) = \mathcal{R}(\mathcal{A}) \tag{2.23d}$$

$$\mathcal{G}_1 \cap \mathcal{G}_2 = \mathcal{R}(\mathcal{A}) \cap \mathcal{S}(O) = \mathcal{R}(\mathcal{A}), \quad \text{for } O \in \mathcal{A} \tag{2.23e}$$

$$\mathcal{G}_1 \cap \mathcal{G}_2 = \mathcal{R}(\mathcal{A}) \cap \mathcal{P}(\mathbf{e}) = \mathcal{I}, \quad \text{for any } \mathcal{A},\ \mathbf{e} \tag{2.23f}$$

All of the above examples, except for the third one, amount to a displacement subgroup. This is why no subgroup symbol is attached to that set. Instead, we have used the symbol $\mathcal{L}(1, 3)$ to denote the kinematic bond between the first and third links of the chain. In general, a kinematic bond between links i and n of a kinematic chain, when no ambiguity is possible, is denoted by $\mathcal{L}(i, n)$. When the chain connecting these two links is not unique, such as in a closed chain, where these two links can be regarded as connected by two, or more, possible *paths*, a subscript will be used, e.g., $\mathcal{L}_I(i, j), \mathcal{L}_{II}(i, j)$, etc. A kinematic bond is, thus, a set of displacements, as stemming from a binary operation of displacement subgroups, although *the bond itself need not be a subgroup*. Obviously, the 12 subgroups described above are themselves special cases of kinematic bonds.

The kinematic bond between links i and n can be conceptualized as the product of the various subgroups associated with the kinematic pairs between the ith and the nth links. To keep the discussion general enough, we shall denote the subgroup associated with the kinematic-pair coupling links i and $i + 1$ as $\mathcal{L}(i, i + 1)$, with a similar notation for all other kinematic-pair subgroups. Thus,

$$\mathcal{L}(i, n) = \mathcal{L}(i, i + 1) \bullet \mathcal{L}(i + 1, i + 2) \bullet \cdots \bullet \mathcal{L}(n - 1, n) \tag{2.24}$$

For example, in a six-axis serial manipulator, like the one shown in Fig. 2.5, we can set $i = 1, n = 7$, all six kinematic pairs in-between being revolutes of independent axes $\mathcal{R}(\mathcal{A}_1), \mathcal{R}(\mathcal{A}_2), \ldots, \mathcal{R}(\mathcal{A}_6)$. Then,

$$\mathcal{L}(1, 7) = \mathcal{D}$$

That is, the manipulator is a generator of the general six-dimensional group of rigid-body displacements $\mathcal{D}$.

As an example of group-intersection, let us consider the *Sarrus mechanism*, depicted in Fig. 2.12.

A less common realization of the Sarrus mechanism is depicted in Fig. 2.13. This is a $\Pi\Pi\Pi\Pi$ closed kinematic chain, modeled as a *compliant mechanism*, which bears a *monolithic* structure, made of a polymer. The R joints of the mechanism are realized by removing material at the joint locations, so as to render these areas much more compliant than the other areas. The mechanism is designed so as to serve as a uniaxial accelerometer.

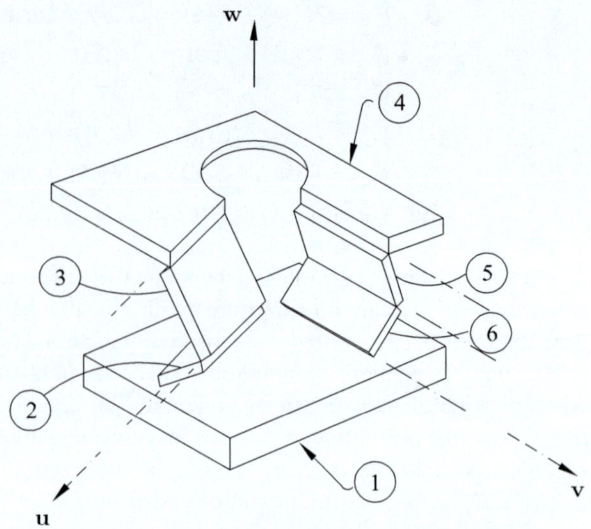

Fig. 2.12 The Sarrus mechanism

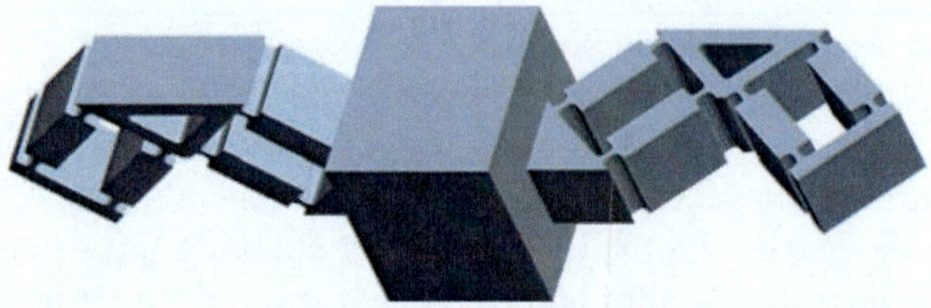

Fig. 2.13 An alternative realization of the Sarrus mechanism

In the Sarrus mechanism of Fig. 2.12, we have six links, coupled by six revolute pairs. Moreover, the revolute pairs occur in two triplets, each on one leg of the mechanism. The axes of the three revolute pairs of each leg are parallel to each other. The bond $\mathcal{L}(1,4)$, apparently, is not unique, for it can be defined by traversing any of the two legs. Let the leg of links 1, 2, 3, and 4, coupled by revolutes of axes parallel to the unit vector $\mathbf{u}$, be labeled I; the other leg, of links 4, 5, 6, and 1, coupled by revolutes of axes parallel to the unit vector $\mathbf{v}$, is labeled II. It is apparent that, upon traversing leg I, we obtain

$$\mathcal{L}_I(1,4) = \mathcal{F}(\mathbf{u})$$

Moreover, upon traversing leg II,

$$\mathcal{L}_{II}(1,4) = \mathcal{F}(\mathbf{v})$$

That is, leg I is a generator of the planar subgroup $\mathcal{F}$ of plane normal to vector $\mathbf{u}$, while leg II is that of the subgroup $\mathcal{F}$ of plane normal to vector $\mathbf{v}$. Therefore, the intersection $\mathcal{L}_I(1,4) \cap \mathcal{L}_{II}(1,4)$ is the set of displacements common to the two $\mathcal{F}$-subgroups, namely, the prismatic subgroup of translations in the direction $\mathbf{w} = \mathbf{v} \times \mathbf{u}$, i.e.,

$$\mathcal{L}_I(1,4) \cap \mathcal{L}_{II}(1,4) = \mathcal{P}(\mathbf{w})$$

The Sarrus mechanism is thus a revolute realization of the prismatic joint.

2.7 The Chebyshev–Grübler–Kutzbach–Hervé Formula

Finding the degree of freedom (dof) f of a given kinematic chain has been an elusive task for over a century. Here we adopt the methodology proposed by Hervé (1978), based on the concept of *groups of displacements*.

Essentially, Hervé considers whether the *topology* of a kinematic chain suffices to predict its dof or not. The topology of a kinematic chain pertains to the number of links and joints as well as their layouts, regardless of the values of the geometric parameters of the chain, such as distances and angles between pair axes and the like. Accordingly, kinematic chains can be classified into three categories, based on their mobility:

(a) *Trivial*, when all the possible kinematic bonds between any pair of links is a subset of one particular subgroup of $\mathcal{D}$, including $\mathcal{D}$ itself, but excluding $\mathcal{I}$. Moreover, if the common subgroup of interest is $\mathcal{D}$ itself, then the chain is trivial *if the product of the subgroups of all the foregoing kinematic pairs yields* $\mathcal{D}$. The dof of a trivial kinematic chain can be determined with the aid of the formula derived below, which takes into account only the topology of the chain. **Caveat**: An algorithm to decide, with knowledge only of the topology of a linkage, whether the linkage is trivial, is still to be devised.

(b) If the chain (i) is not trivial; (ii) all its possible kinematic bonds between any pair of its links form a subset $\mathcal{D}_s$ of $\mathcal{D}$ that does not bear the group property; and (iii) the bond between two of its links can be expressed as the *intersection* $\mathcal{D}_g$ of a number of kinematic bonds that is a subgroup of $\mathcal{D}$, then the chain is said to be *exceptional*. In this case, the dof of the chain is the dimension of $\mathcal{D}_g$.

(c) *Paradoxical*, when the topology of the kinematic chain alone does not suffice to determine the chain dof. In this case, special quantitative relations among the various geometric parameters of the chain yield a mobility that would be absent under general values of those parameters, for the same topology.

2.7.1 Trivial Chains

Regarding trivial chains, let $\mathcal{G}_m$ be the subgroup of the *least dimension* d_m, containing all possible bonds between any pair of links of the chain. $\mathcal{G}_m$ can thus be thought of as a kind of *least common multiple* of all possible bonds of the chain. Moreover, let d_i be the dimension of the subgroup associated with the ith kinematic pair, and $r_i \equiv d_m - d_i$ be its degree of constraint, termed its *restriction* for brevity. In determining the dof of a chain, we are interested in the *relative* motion capability of the chain, and hence, we consider arbitrarily one link *fixed*. It is immaterial which specific link is the designated fixed one. If the chain is composed of l links and p kinematic pairs, then its dof f is given by the difference between its total dof before coupling and the sum of its restrictions, i.e.,

$$f = d_m(l - 1) - \sum_{i=1}^{p} r_i \tag{2.25}$$

The above relation can be termed a *generalized Chebyshev–Grübler–Kutzbach* (CGK) formula in that it generalizes the concept involved in the parameter d_m above. Conventional CGK formulas usually consider that d_m can attain one of two possible values, 3 for planar and spherical chains and 6 for spatial chains. In the generalized formula, d_m can attain any of the values 2, 3, 4, or 6. Moreover, rather than considering only three subgroups of displacements, we consider all 12 described above, none of which is of dimension five.

As an example of the application of the above formula, we consider the *vise mechanism*, displayed in Fig. 2.14. In that figure, we distinguish three links and three LKPs. The links are the frame 1, the crank 2, and the slider 3, which define three bonds, namely

$$\mathcal{L}(1,2) = \mathcal{R}(\mathcal{A}), \quad \mathcal{L}(2,3) = \mathcal{H}(\mathcal{A}), \quad \mathcal{L}(3,1) = \mathcal{P}(\mathbf{a})$$

in which $\mathcal{A}$ is the common axis of the R and the H pairs, while **a** is the unit vector parallel to $\mathcal{A}$. In this case, it is apparent that all three bonds lie in the $\mathcal{C}(\mathcal{A})$ subgroup, and hence, $d_m = 2$. Moreover, if we number the three joints in the order R, H, P, and notice that the dimension d_i associated with each of the three joints is unity, then $r_i = 1$, for $i = 1, 2, 3$. Application of the generalized CGK formula (2.25) yields

$$f = 2(3 - 1) - 3 \times 1 = 4 - 3 = 1$$

which is indeed the correct value of the vise dof.

While the generalized CGK formula is more broadly applicable and less error-prone than its conventional counterpart, it is not error-free. Indeed, let us consider the HHHRRH closed chain of Fig. 2.15, first proposed by Hervé (1978). The four H pairs of this figure have distinct pitches.

It is apparent that all links move in parallel planes and that these planes also translate along their common normal direction. The displacement subgroup containing all possible kinematic bonds of the mechanism under study, of minimum dimension, is thus the Schönflies subgroup $\mathcal{X}(\mathbf{u})$, and hence, $d_m = 4$. Since we have six links and six joints, each of *restriction* $r_i = d_m - f_i$, for $f_i = 1$ and $i = 1, \ldots, 6$, the dof of the mechanism is obtained from the CGK formula as

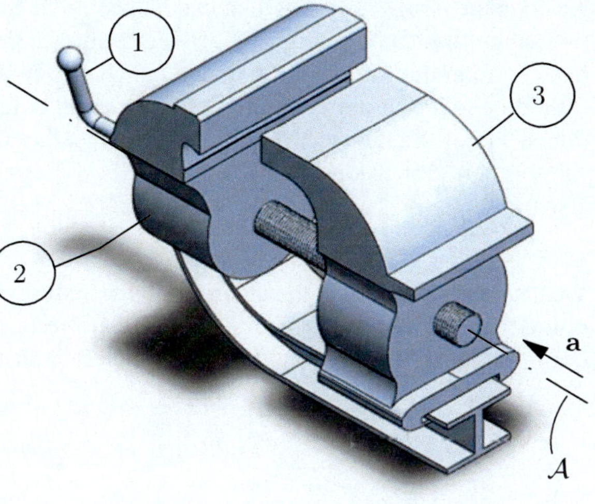

Fig. 2.14 The well-known vise mechanism

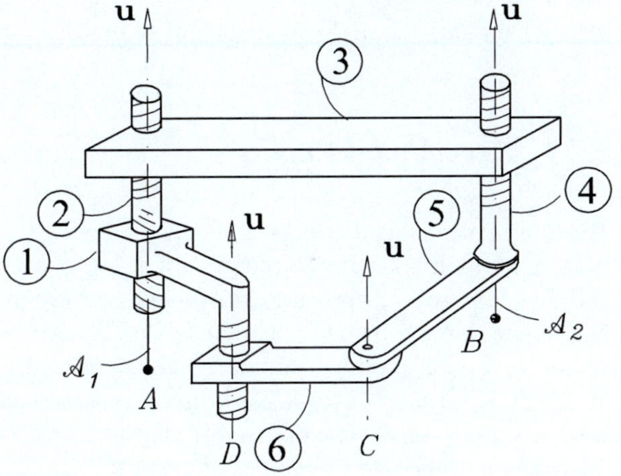

Fig. 2.15 The HHHRRH mechanism

$$f = 4(6 - 1) - 6 \times 3 = 2$$

However, *the above result is wrong*, for it predicts a too large dof. Indeed, the mechanism has one idle dof, as can be readily shown by means of a bond analysis: Let us compute $\dim[\mathcal{L}(1, 5)]$:

$$\mathcal{L}(1, 5) = \underbrace{\mathcal{L}(1, 2) \bullet \mathcal{L}(2, 3)}_{\mathcal{C}(\mathcal{A}_1)} \bullet \underbrace{\mathcal{L}(3, 4) \bullet \mathcal{L}(4, 5)}_{\mathcal{C}(\mathcal{A}_2)}$$

where $\mathcal{A}_1$ and $\mathcal{A}_2$ are axes parallel to vector **u** and pass through points A and B, respectively, of Fig. 2.15. Now we find the above-mentioned idle dof. To this end, we compute $\dim[\mathcal{L}(1, 5)]$, which may appear to be the sum of the dimensions of the two subgroups, $\mathcal{C}(\mathcal{A}_1)$ and $\mathcal{C}(\mathcal{A}_2)$. However, notice that these two subgroups include one common translation along $\mathbf{u}$, and hence, in computing the said dimension, care should be taken in not counting this translation twice. What this means is that the dimension of the intersection of the above two factors must be subtracted from the sum of their dimensions, i.e.,

$$\dim[\mathcal{L}(1, 5)] = \dim[\mathcal{C}(\mathcal{A}_1)] + \dim[\mathcal{C}(\mathcal{A}_2)] - \dim[\mathcal{C}(\mathcal{A}_1) \cap \dim[\mathcal{C}(\mathcal{A}_2)] = 2 + 2 - 1 = 3$$

We have thus shown that the chain entails one idle dof. In order to obtain the correct dof of the chain from the generalized CGK formula, then the total number m of idle dof must be subtracted from the dof predicted by that formula, i.e.,

$$f = d_m(n-1) - \sum_{i=1}^{p} r_i - m \tag{2.26}$$

which can be fairly called the *Chebyshev–Grübler–Kutzbach–Hervé* formula. In the case at hand, $m = 1$, and hence, the dof of the chain of Fig. 2.15 is unity.

2.7.2 Exceptional Chains

The Sarrus mechanism of Figs. 2.12 and 2.13 is an example of an *exceptional chain*. Indeed, all its links undergo motions of either one of two planar subgroups, $\mathcal{F}(\mathbf{u})$ and $\mathcal{F}(\mathbf{v})$. Moreover, the product of these two subgroups does not yield the group $\mathcal{D}$—notice that the linkage has two sets of R pairs, each parallel to a distinct unit vector, $\mathbf{u}$ or $\mathbf{v}$. The dof of this mechanism can still be found, but not with the aid of the CGK formula of Eq. (2.25), for all its kinematic bonds do not belong to the same subgroup of $\mathcal{D}$. The dof in question is found, rather, as the dimension of the intersection of the two foregoing subgroups, i.e.,

$$f = \dim[\mathcal{F}(\mathbf{u}) \cap \mathcal{F}(\mathbf{v})] = \dim[\mathcal{P}(\mathbf{u} \times \mathbf{v})] = 1$$

Another example of an exceptional chain is the familiar slider-crank mechanism of either internal combustion engines or compressors, as shown in Fig. 2.16. It is customary to represent this mechanism as a planar RRRP mechanism. However, a close look at the coupling of the piston with its chamber reveals that this coupling is not via a prismatic, but rather via a cylindrical pair. It is thus apparent that the displacements of all the links lie not in one single subgroup of $\mathcal{D}$, but rather in a subset that can be decomposed into two kinematic bonds, which happen to be subgroups of $\mathcal{D}$, the $\mathcal{F}(\mathbf{e})$ subgroup of motions generated by the RRR subchain and the $\mathcal{C}(\mathcal{A})$ subgroup of the piston–chamber coupling C. Here, $\mathcal{A}$ is the axis of the cylindrical chamber and $\mathbf{e}$ the unit vector parallel to the axes of the three R pairs.

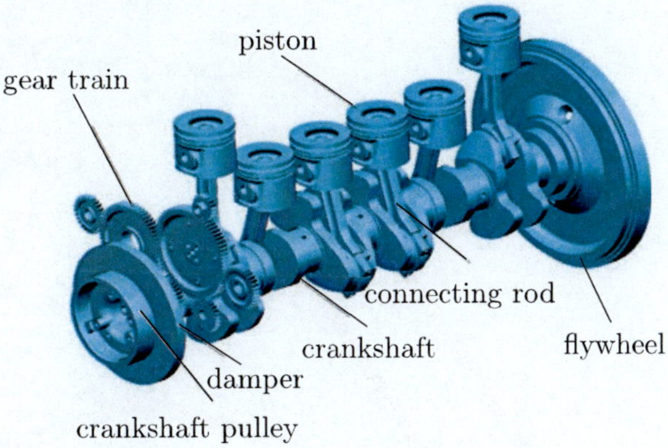

Fig. 2.16 The slider-crank mechanism as a key component of an internal combustion engine: a power-generation system with six cylinders in line (courtesy of MWM International Motores, Campinas, Brazil)

Apparently, the product of these two subgroups does not generate all of $\mathcal{D}$, for it is short of rotations about an axis normal to both **e** and $\mathcal{A}$. Nevertheless, the dof of this chain can be determined as the dimension of the intersection of the two subgroups, i.e.,

$$f = \dim[\mathcal{F}(\mathbf{e}) \cap \mathcal{C}(\mathcal{A})] = \dim[\mathcal{P}(\mathbf{u})] = 1, \quad \mathbf{u} \parallel \mathcal{A}$$

Now, why would such a simple planar mechanism—the slider-crank—as portrayed in elementary books on mechanisms, be built with a spatial structure? The answer to this question lies in the *assemblability* of the mechanism: a planar RRRP mechanism requires a highly accurate machining of the crankshaft, connecting rod, piston, and chamber, in order to guarantee that the axes of the three R pairs are indeed parallel and that the axis of the cylindrical chamber is normal to the three R axes. This is by no means a simple task!

One more example of an exceptional chain is the parallel robot of Fig. 2.17, consisting of four identical limbs that join a base $A_I A_{II} A_{III} A_{IV}$ with a moving plate $D_I D_{II} D_{III} D_{IV}$. Each limb, moreover, is a PRΠRRchain (Altuzarra et al., 2009).

The kinematic chain thus contains five joints per limb and 18 links: the base plate, the mobile plate, and four intermediate links per limb. If the CGK formula is applied to the general kinematic chain, with $d_m = 6$, $l = 18$, $r_i = 5$, for $i = 1, \ldots, 18$, the dof f thus resulting turns out to be

$$f = 6(18 - 1) - 20 \times 5 = 102 - 100 = 2$$

which is not what the authors claim, namely four. In order to elucidate the apparent contradiction, we conduct below a group-theoretic analysis of the chain mobility: first, let $\mathcal{R}(P, \mathbf{e})$ denote the subgroup generated by a R joint of axis passing through point P and parallel to the unit vector **e**; then, let $\mathcal{L}_J$ denote the kinematic bond of the Jth limb, which is the product of five simple bonds, each with a dimension equal to unity, namely

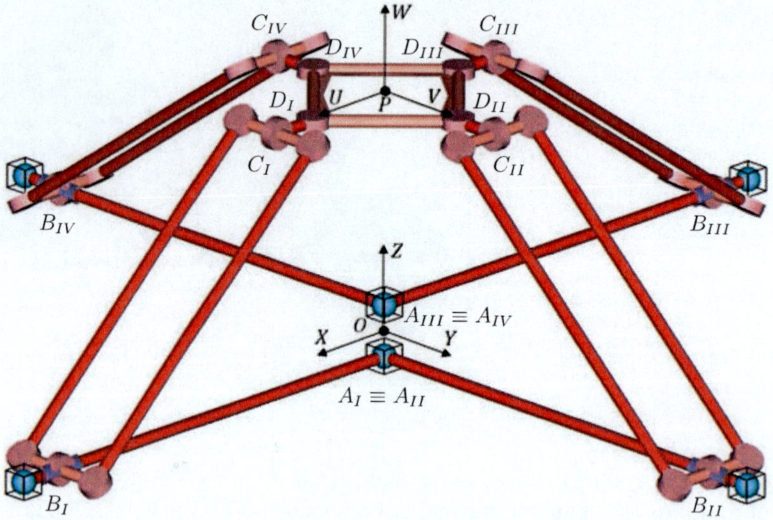

Fig. 2.17 The Schönflies-motion generator developed at the University of the Basque Country, Bilbao, Spain

1. either the prismatic subgroup $\mathcal{P}(\mathbf{i})$ of displacements parallel to vector **i**, for $J = I, III$, or its counterpart $\mathcal{P}(\mathbf{j})$ of displacements parallel to **j**, for $J = II, IV$;
2. the rotation subgroup $\mathcal{R}(B_J, \mathbf{j})$, of axis of rotation passing through point B_J and parallel either to **j**, for $J = I, III$, or its counterpart $\mathcal{R}(B_J, \mathbf{i})$, for $J = II, IV$;
3. the subset of displacements $\mathcal{D}_{\Pi}(\mathbf{n}_J)$ associated with the Π-joint, characterized by translations along circles of radius $\overline{B_J C_J}$ lying in the plane of the Jth parallelogram, of normal $\mathbf{n}_J$;
4. the rotation subgroup $\mathcal{R}(C_J, \mathbf{j})$, of axis of rotation passing through point C_J and parallel either to **j**, for $J = I, III$ or to **i** for $J = II, IV$;
5. the rotation subgroup $\mathcal{R}(D_J, \mathbf{k})$ of axis of rotation passing through D_J and parallel to **k**.

Therefore,

$$\mathcal{L}_J = \underbrace{\mathcal{P}(\mathbf{i}) \bullet \mathcal{R}(B_J, \mathbf{j}) \bullet \mathcal{D}_{\Pi}(\mathbf{n}_J) \bullet \mathcal{R}(C_J, \mathbf{j})}_{\mathcal{X}(\mathbf{j})} \bullet \mathcal{R}(D_J, \mathbf{k}) = \mathcal{X}(\mathbf{j}) \bullet \mathcal{R}(D_J, \mathbf{k}), \quad J = I, III$$

Likewise,

$$\mathcal{L}_J = \mathcal{X}(\mathbf{i}) \bullet \mathcal{R}(D_J, \mathbf{k}), \quad J = II, IV$$

Notice that none of the four bonds derived above is a subgroup of $\mathcal{D}$, which disqualifies the multiloop kinematic chain from being trivial. However, notice also that

$$\mathcal{X}(\mathbf{j}) \bullet \mathcal{R}(D_J, \mathbf{k}) = \mathcal{X}(\mathbf{k}) \bullet \mathcal{R}(C_J, \mathbf{j}), \quad J = I, III$$

and

$$\mathcal{X}(\mathbf{i}) \bullet \mathcal{R}(D_J, \mathbf{k}) = \mathcal{X}(\mathbf{k}) \bullet \mathcal{R}(C_J, \mathbf{i}), \quad J = II, IV$$

Therefore,

$$\mathcal{L}_J \cap \mathcal{L}_K = \mathcal{X}(\mathbf{k}), \quad J,\ K = I, \ldots IV, \quad J \neq K$$

thereby proving that, indeed, the intersection of all limb bonds is a subgroup of $\mathcal{D}$, namely the Schönflies subgroup $\mathcal{X}(\mathbf{k})$. The dof f of the robot at hand is, thus,

$$f = \dim[\mathcal{X}(\mathbf{k})] = 4$$

and, according to Hervé's classification, the multiloop chain is *exceptional*.

2.7.3 *Paradoxical Chains*

Examples of paradoxical chains are documented in the literature (Bricard, 1927; Angeles, 1982). These include the *Bennett mechanism* and the *Bricard mechanism*, among others.

2.8 Applications to Robotics

The foregoing concepts are now applied to the *qualitative* synthesis of parallel robotic architectures. By qualitative we mean here the determination of the topology of the kinematic chain, not including the corresponding dimensions. These dimensions are found at a later stage, by means of methods of *quantitative synthesis*, which Hartenberg and Denavit (1964a) term dimensional synthesis, the subject of Chaps. 3–6. The full synthesis of the kinematic chain, including dimensions, yields what is known as the *architecture* of the robotic system at hand.

2.8.1 *The Synthesis of Robotic Architectures and Their Drives*

The first robotic architecture with Π-joints was proposed by Clavel in what he called the *Delta Robot* (Clavel, 1988). The kinematic chain of this robot is displayed in Fig. 2.18. Delta is a generator of the $\mathcal{T}_3$ displacement subgroup; it is capable of three-dof translations.

The kinematic chain of the Delta robot is composed of two triangular plates, the top ($\mathcal{A}$) and the bottom ($\mathcal{B}$) plates in Figs. 2.18. The top plate supports the three (direct-drive) motors, the bottom plate the gripper, and hence, constitutes the moving platform (MP) of the robot. The MP is capable of translating in three-dimensional space with respect to the upper plate, which remains fixed. The two plates are coupled by means of three legs, each with a RRΠR chain.

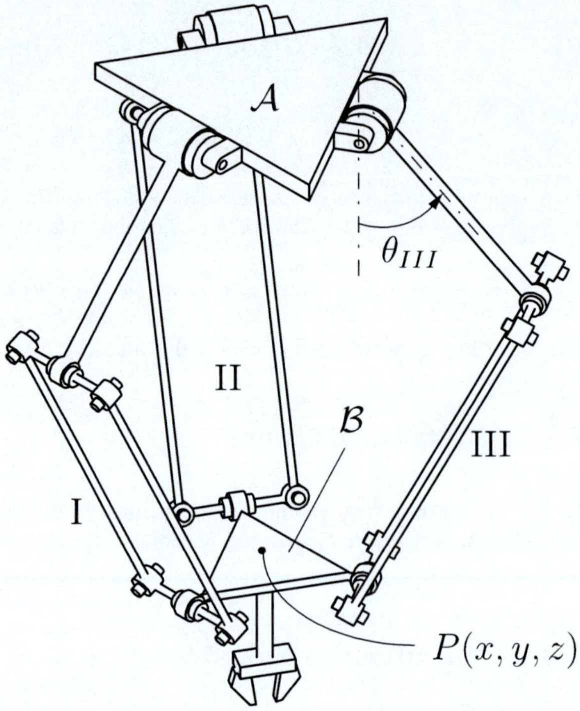

Fig. 2.18 Kinematic chain of the Clavel delta robot

To be true, the Π-joints of the actual Delta are not composed of R joints, but rather of *orientable pin joints*, kinematically equivalent to S joints. The reason is that providing parallelism between any pair of R axes is quite a challenge. To allow for *assemblability*, then a *margin of manoeuvre* must be provided. This is achieved by means of the S joints.

While Clavel did not claim any group-theoretical reasoning behind his ingenious design, an analysis in this framework will readily explain the principle of operation of the robot. This analysis is conducted on the *ideal* kinematic chain displayed in Fig. 2.18.

The ith leg is a generator of the Schönflies $\mathcal{X}(\mathbf{e}_i)$ subgroup, with $\mathbf{e}_i$ denoting the unit vector parallel to the axis of the ith motor. That is, the ith leg generates a Schönflies subgroup of displacements comprising translations in three-dimensional space and one rotation about an axis parallel to $\mathbf{e}_i$. The subset of EE displacements generated by the Delta robot is thus the intersection of the three subgroups $\mathcal{X}(\mathbf{e}_i)$, for $i = 1, 2, 3$, i.e., the subgroup $\mathcal{T}_3$. Therefore, the EE is capable of pure translations in three-dimensional space. This kinematic chain is, thus, of the *exceptional* type.

One second example of application is the microfinger of Japan's Mechanical Engineering Laboratory (MEL) at Tsukuba (Arai et al., 1996), as displayed in Fig. 2.19. In the MEL design, the authors use a structure consisting of two plates that translate with respect to each other by

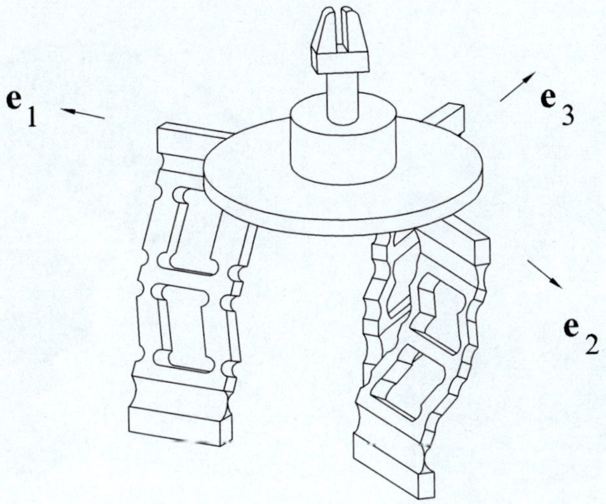

Fig. 2.19 The MEL microfinger

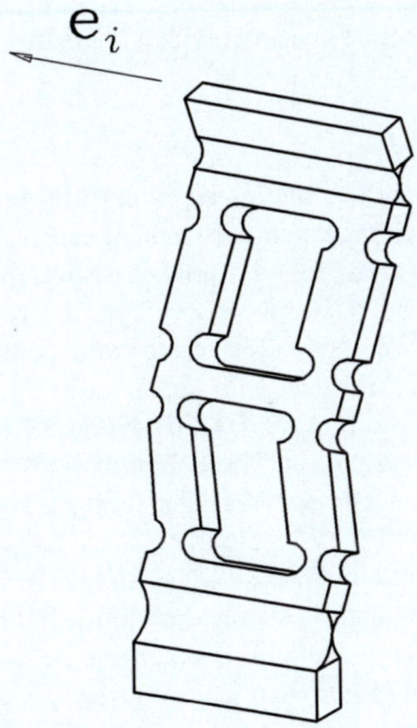

Fig. 2.20 The *i*th leg of the MEL microfinger

means of three legs coupling the plates. The *i*th leg entails a RΠΠR chain, shown in Fig. 2.20, that generates the Schönflies subgroup in the direction of a unit vector $\mathbf{e}_i$, for $i = 1, 2, 3$. The three unit vectors, moreover, are coplanar and make angles of 120° pairwise. The motion of

(a)

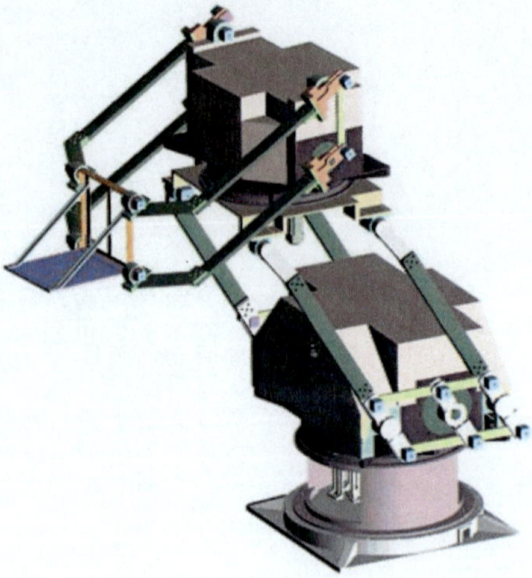

(b)

Fig. 2.21 A serial Schönflies-motion generator with a RΠRΠ architecture: **a** Prototype; **b** CAD rendering

the moving plate is thus the result of the intersection of these three subgroups, which is, in turn, the $\mathcal{T}_3$ subgroup. Moreover, the kinematic chain of each leg is made of an elastic material in one single piece, in order to allow for micrometric displacements.

Yet another example is the Y-Tristar robot, developed at *Ecole Centrale de Paris* by Hervé and Sparacino (1992). One more application of the same concepts is the four-dof SCARA-motion generator proposed by Morozov and Angeles (2007) and displayed in Fig. 2.21.[12] This robot entails a kinematic chain of the RΠRΠ type with two vertical revolutes and two Π-pairs lying in distinct, vertical planes. The Schönflies subgroup generated by this device carries one vertical axis. While Delta and Y-Tristar robots are made up of Schönflies-motion generators, the intersection of all these is the translation subgroup $\mathcal{T}_3$. A Schönflies-motion generator with parallel architecture is possible, as shown in Fig. 2.22. This architecture is the result of coupling two identical Schönflies-motion generators of the type displayed in Fig. 2.21, each generating the same Schönflies subgroup. As a result, the two-legged parallel robot generates the intersection of two identical subgroups, which is *the same subgroup*. Yet another application of the Π pair is found in the four-degree-of-freedom parallel robot Quattro patented by Pierrot et al. (2001) and now owned by Omron under the trademark Quattro iX4-650H/HS. A photograph of this robot is displayed in Fig. 2.23.

[12] In subfigure (a) of this figure, the robot designer, Dr. A. Morozov, is included, to indicate the dimensions of the prototype.

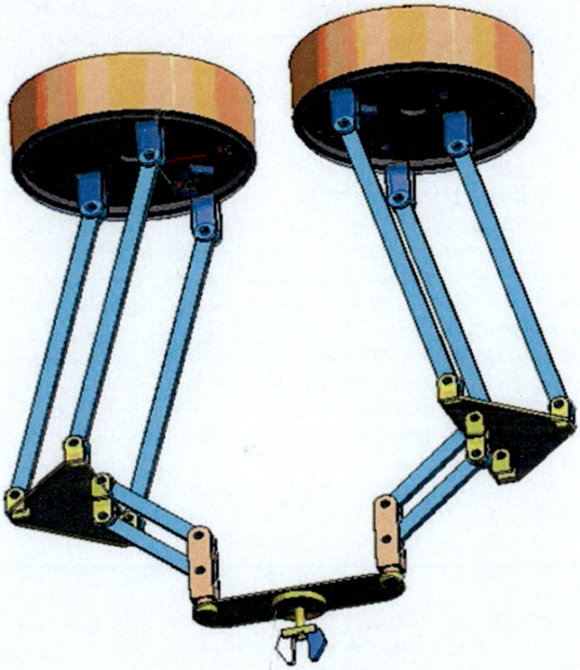

Fig. 2.22 A parallel Schönflies-motion generator composed of two RΠRΠ legs

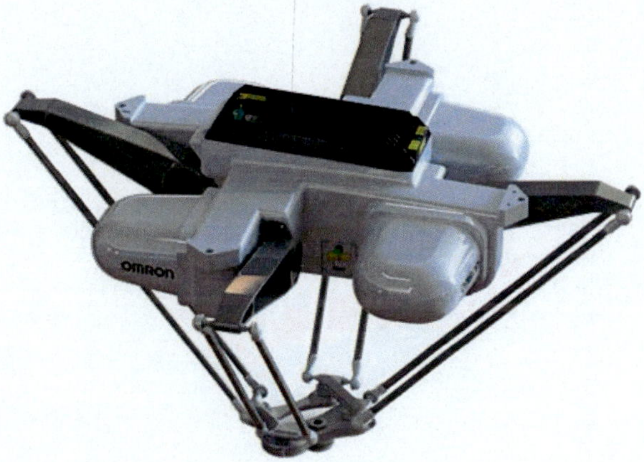

Fig. 2.23 Omron's Adept Quattro robot, a parallel Schönflies-motion generator (*Source* Omron)

References

Altuzarra, O., Salgado, O., Hernández, A., & Angeles, J. (2009). Multiobjective optimum design of a symmetric parallel schönflies-motion generator. *ASME Journal of Mechanical Design,131*(3), 031002–1–031002–11.

Angeles, J. (1982). *Spatial kinematic chains. Analysis, synthesis, optimization*. Springer.

Angeles, J. (2014). *Fundamentals of robotic mechanical systems: Theory, methods, and algorithms* (4th ed.). Springer.

Angeles, J., & López-Cajún, C. S. (1991). *Optimization of cam mechanisms*. Kluwer Academic.

Arai, T., Hervé, J. M., % Tanikawa, T. (1996). Development of 3 DOF micro finger. In *Proceedings of IROS' 96* (pp. 981–987). Osaka.

Bricard, R. (1927). *Leçons de Cinématique*. Vols I & II, Gauthier-Villars et Cie. Publishers, Paris, France.

Clavel, R. (1988). Delta, a fast robot with parallel geometry. In *Proceedings of 18th International Sympoisum Industrial Robots* (pp. 91–100). Lausanne.

Dietmaier, P. (1992). Inverse kinematics of manipulators with 3 revolute and 3 parallelogram joints. In *Proceedings of ASME 22nd Biennial Mechanisms Conference* (vol. 45, pp. 35–40). Scottsdale.

Harary, F. (1972). *Graph theory*. Addison-Wesley Publishing Company.

Hartenberg, R., & Denavit, J. (1964). *Kinematic synthesis of linkages*. McGraw-Hill Book Company.

Hervé, J. (1978). Analyse structurelle des mécanismes par groupes des déplacements. *Mechanism and Machine Theory, 13*(4), 437–450.

Hervé, J. M. (1999). The Lie group of rigid body displacements, a fundamental tool for mechanism design. *Mechanism and Machine theory, 34*(5), 719–730.

Hervé, J., & Sparacino, F. (1992). Star, a new concept in robotics. In *Proceedings of 3rd International Workshop on Advances in Robot Kinematics* (pp. 176–183). Ferrara.

Morozov, A., & Angeles, J. (2007). The mechanical design of a novel Schönflies-motion generator. *Robotics and Computer-Integrated Manufacturing, 23*(1), 82–93.

Pierrot, F., Company, O., Shibukawa, T., & Morita, K. (2001). Four-degree-of-freedom parallel robot. European Patent EP1084802.

Wohlhart, K. (1991). Der homogene paralleltrieb-mechanismus. *Mathematica Pannonica, 2*(2), 59–76.

Wohlhart, K. (1992). Displacement analysis of the general spatial parallelogram manipulator. In *Proceedings of 3rd International Workshop on Advances in Robot Kinematics* (pp. 104–111). Ferrara, Italy.

Chapter 3
Linkage Synthesis for Function Generation

3.1 Introduction

In the realm of linkage synthesis the simplest problem is the coordination of the single-degree-of-freedom (single-dof) motion of one link, termed the *output link*, with the, likewise, single-dof motion of another link, termed the *input link*. Moreover, the two foregoing links are not coupled directly, but via an intermediate link, termed the *coupler link*. The input link, in the most common application, is driven by one single motor, either *rotational* or *translational*, the motor being supported by one fourth link, which is the *fixed link*, resting on the linkage frame. In this chapter we study the synthesis of a four-bar linkage for the production of a single-dof motion of the output link, as *remotely driven*, i.e., by means of the *input link*, ususally moving at a *constant* angular or translational velocity, as the case may be. This synthesis task is known as *function generation*.

In the best known applications, motion is produced by a motor, usually running at a constant rpm, and coupled by means of a speed reducer—gear train, harmonic drive, or similar—to the *input link*. Under these conditions, the input link moves at a constant speed as well.

In some applications, a linkage is human powered. Examples of this class are numerous, and sometimes taken for granted, e.g., when cutting a sheet of paper with scissors, when riding a bicycle, etc. In the case of scissors, the two blades of this tool form a *two-link open chain* coupled by a R pair. When held by a user, this chain is coupled to a second, similar chain formed by the proximal phalanx of the thumb and the intermediate phalanx of the index finger, thereby forming a *four-bar linkage*. Likewise, in the case of a bicycle, the frame and one of the two pedals form a two-link open chain, which couples with a second, similar chain, formed by the calf and the thigh of a human user, coupled by the R pair of the knee, thereby forming, again, a four-bar linkage.

One more application of the concepts studied in this chapter involves *parameter identification*, whereby a linkage exists but is not accessible for measurements, and

J. Angeles and S. Bai, *Kinematics of Mechanical Systems*, Mathematical Engineering,
https://doi.org/10.1007/978-3-031-09544-3_3

we want to know its dimensions. Take the case of the human subtalar and ankle-joint complex, which is known to entail a closed kinematic chain, i.e., a linkage, but its joints are not readily accessible for measurement. We can cite here a case in which a series of experiments was conducted, measuring input and output angles, from which linkage dimensions were estimated by fitting the measurements to a linkage kinematic model (Wright et al., 1964). The application, in this case, was the design of a prothesis to replace a limb. This introduction is intended to make the reader aware of the *omnipresence* of linkages and their applications.

3.2 Input-Output (IO) Functions

The subject of this chapter is the *simplest* problem in the synthesis of four-bar linkages, which is, likewise, the *simplest* linkage in the realm of mechanical systems.

The synthesis of planar, spherical and spatial four-bar linkages for *function-generation* is discussed in this chapter. As shown in the sequel, the *Freudenstein parameters* of the linkage under synthesis are first found from a system of three synthesis equations for planar linkages, four in the case of spherical linkages. The synthesis of spatial linkages proceeds in two steps: first, the linkage parameters representing angles are obtained exactly as in the case of spherical linkages, using exactly the same equations; in the second step, with the angular parameters of the linkage in question known, the parameters representing lengths are obtained from the remaining four equations, whereby the unknowns are all lengths.

Moreover, under the assumption that the number of unknown linkage parameters equals the number of synthesis equations, and realizing that the synthesis procedure relies on the solution of *linear equations*, the solution obtained, *if feasible*, is unique.

3.2.1 *Planar Four-Bar Linkages*

The classical problem of function generation was first formulated algebraically by Freudenstein in a seminal paper that has been recognized as the origin of *modern* kinematics (Freudenstein, 1955).[1] In the paper, Freudenstein set to finding the link lengths $\{ a_i \}_1^4$ of the planar four-bar linkage displayed in Fig. 3.1 so as to obtain a prescribed relation between the angles ψ and ϕ.

In Fig. 3.1, ψ denotes the *input angle* of the linkage, ϕ the *output angle*, θ the *coupler angle*, and μ the *transmission angle*, which will be studied in Sect. 3.6. We state below the *function-generation problem* associated with the linkage of the above figure:

[1] In a nutshell, the emphasized qualifier is commonly used to denote the breakaway point in which kinematics turned to computers solve the numerical problems arising in the realm of analysis and synthesis.

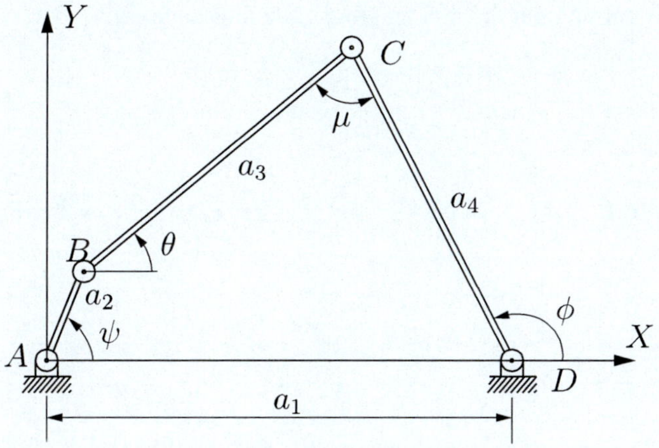

Fig. 3.1 A four-bar linkage for function generation

Find $\{a_k\}_1^4$ *that will allow the linkage to produce the set of m* input-output pairs $\{\psi_k,\ \phi_k\}_1^m$.

In the foregoing statement an algebraic relation between the two angles, ψ and ϕ, known as the *input-output (IO) equation*, is assumed to be available in the form of an *implicit function*, namely,

$$F(\psi,\ \phi) = 0 \tag{3.1}$$

In formulating the input-output equation, we introduce four two-dimensional vectors, all represented in the coordinate frame of Fig. 3.1:

$$\mathbf{r}_1 \equiv \overrightarrow{AB} = a_2 \begin{bmatrix} \cos\psi \\ \sin\psi \end{bmatrix}, \quad \mathbf{r}_2 \equiv \overrightarrow{BC} = a_3 \begin{bmatrix} \cos\theta \\ \sin\theta \end{bmatrix}, \tag{3.2a}$$

$$\mathbf{r}_3 \equiv \overrightarrow{AD} = a_1 \begin{bmatrix} 1 \\ 0 \end{bmatrix}, \quad \mathbf{r}_4 \equiv \overrightarrow{DC} = a_4 \begin{bmatrix} \cos\phi \\ \sin\phi \end{bmatrix} \tag{3.2b}$$

From Fig. 3.1 follows that

$$\mathbf{r}_1 + \mathbf{r}_2 = \mathbf{r}_3 + \mathbf{r}_4 \tag{3.3}$$

Apparently, we need a *scalar* relation between the *input and output angles*, but we have above a vector equation. Notice, however, that the angles of interest appear in $\mathbf{r}_1$ and $\mathbf{r}_4$; $\mathbf{r}_3$ remains constant throughout the linkage motion; and $\mathbf{r}_2$ contains an *unwanted unknown*, θ. This is eliminated below:

From Eq. (3.3),

$$\mathbf{r}_2 = \mathbf{r}_3 + \mathbf{r}_4 - \mathbf{r}_1 \tag{3.4}$$

Now, the right-hand side of the above equation is independent of θ. If we take the *Euclidean norm*, a.k.a. the *magnitude*, of both sides of Eq. (3.4), then angle θ is eliminated, for the magnitude of $\mathbf{r}_2$ is independent of this angle; in fact, this magnitude is nothing but the link length a_3. We thus have

$$\|\mathbf{r}_2\|^2 = \|\mathbf{r}_3 + \mathbf{r}_4 - \mathbf{r}_1\|^2 \tag{3.5}$$

Upon expansion,

$$\|\mathbf{r}_2\|^2 = \|\mathbf{r}_3\|^2 + \|\mathbf{r}_4\|^2 + \|\mathbf{r}_1\|^2 + 2\mathbf{r}_3^T\mathbf{r}_4 - 2\mathbf{r}_3^T\mathbf{r}_1 - 2\mathbf{r}_4^T\mathbf{r}_1 \tag{3.6}$$

where

$$\|\mathbf{r}_1\|^2 = a_2^2, \quad \|\mathbf{r}_2\|^2 = a_3^2, \quad \|\mathbf{r}_3\|^2 = a_1^2, \quad \|\mathbf{r}_4\|^2 = a_4^2$$
$$\mathbf{r}_3^T\mathbf{r}_4 = a_1a_4\cos\phi, \quad \mathbf{r}_3^T\mathbf{r}_1 = a_1a_2\cos\psi, \quad \mathbf{r}_4^T\mathbf{r}_1 = a_2a_4\cos(\phi-\psi)$$

Plugging the foregoing expressions into Eq. (3.6) yields a form of the input-output (IO) equation:

$$a_3^2 = a_1^2 + a_4^2 + a_2^2 + 2a_1a_4\cos\phi - 2a_1a_2\cos\psi - 2a_2a_4\cos(\phi-\psi) \tag{3.7}$$

which is already a scalar relation between the input and the output angles, with the link lengths as parameters, free of any other linkage variable, like θ and μ. However, this relation is not yet in the most suitable form for our purposes. Indeed, it is apparent that a scaling of the link lengths by the same factor does not change the input-output relation, and hence, the above equation cannot yield all four link lengths. This means that we can only obtain the relative values of the link lengths for a set of prescribed input-output angles. One more remark is in order: the link lengths appear as unknowns when a pair of input-output angles is given; moreover, these unknowns appear quadratically in that equation. Thus, simply dividing the two sides of the equation by any link length will still yield a quadratic equation in the link-length ratios. Freudenstein (1955) cleverly realized that, by means of a suitable *nonlinear mapping* from link lengths into nondimensional parameters, a *linear equation* in these parameters can be produced. To this end, both sides of Eq. (3.7) are divided by $2a_2a_4$. Once this is done, the definitions below are introduced:

$$k_1 \equiv \frac{a_1^2 + a_2^2 - a_3^2 + a_4^2}{2a_2a_4}, \quad k_2 \equiv \frac{a_1}{a_2}, \quad k_3 \equiv \frac{a_1}{a_4} \tag{3.8}$$

which are the *Freudenstein parameters* of the linkage at hand. The inverse relations are readily derived, if in terms of one of the link lengths that cannot possibly vanish, which leads to the obvious candidate, a_1, and hence,

$$a_2 = \frac{1}{k_2}a_1, \quad a_4 = \frac{1}{k_3}a_1, \quad a_3 = \sqrt{a_1^2 + a_2^2 + a_4^2 - 2k_1a_2a_4} \tag{3.9}$$

The IO equation (3.7) then becomes

$$k_1 + k_2 \cos \phi - k_3 \cos \psi = \cos(\phi - \psi) \tag{3.10}$$

thereby obtaining the *Freudenstein equation*. Notice that, upon writing this equation in homogeneous form, we obtain $F(\psi, \phi)$ of Eq. (3.1), namely,

$$F(\psi, \phi) \equiv k_1 + k_2 \cos \phi - k_3 \cos \psi - \cos(\phi - \psi) = 0 \tag{3.11}$$

If we now write Eq. (3.10) for the m input-output pairs $\{\psi_k, \phi_k\}_1^m$, we obtain m linear equations in the three Freudenstein parameters, arrayed in vector **k**, namely,

$$\mathbf{Sk} = \mathbf{b} \tag{3.12}$$

where **S** is the $m \times 3$ *synthesis matrix*; **k** is the 3-dimensional vector of unknown Freudenstein parameters; and **b** is a m-dimensional vector of known components, i.e.,

$$\mathbf{S} \equiv \begin{bmatrix} 1 & \cos\phi_1 & -\cos\psi_1 \\ 1 & \cos\phi_2 & -\cos\psi_2 \\ \vdots & \vdots & \vdots \\ 1 & \cos\phi_m & -\cos\psi_m \end{bmatrix}, \quad \mathbf{k} \equiv \begin{bmatrix} k_1 \\ k_2 \\ k_3 \end{bmatrix}, \quad \mathbf{b} \equiv \begin{bmatrix} \cos(\phi_1 - \psi_1) \\ \cos(\phi_2 - \psi_2) \\ \vdots \\ \cos(\phi_m - \psi_m) \end{bmatrix} \tag{3.13}$$

Three cases arise:

$m < 3$: Case $m = 1$ reduces to the synthesis of a quadrilateron with two given angles, which admits infinitely many solutions. Case $m = 2$ seldom occurs in practice without additional conditions, that render the problem more complex, e.g., in the synthesis of *quick-return mechanisms*;

$m = 3$: The number of equations coincides with the number of unknowns, and hence, the problem admits one *unique* solution—unless the synthesis matrix is **singular**. We are here in the case of *exact synthesis*;

$m > 3$: The number of equations exceeds the number of unknowns, which leads to an *overdetermined* system of equations. Hence, no solution is possible, in general, but an *optimum solution* can be found that best approximates the synthesis equations in the least-square sense, for example. Problem falls in the category of *approximate synthesis*.

Examples of exact and approximate synthesis will follow in Sects. 3.3 and 3.5.

3.2.2 The Denavit-Hartenberg Notation

Prior to deriving the IO equations of spherical and spatial linkages, we introduce the *Denavit-Hartenberg (DH) notation*, which is extremely useful in the analysis of kinematic chains in three dimensions.

In order to uniquely describe the *architecture* of a kinematic chain, i.e., the relative location and orientation of its neighbouring-pair axes, the Denavit-Hartenberg notation (Hartenberg and Denavit, 1964a) is introduced. To this end, we assume a *simple kinematic chain*, open or closed, with links numbered $1, \ldots, n$, the ith pair being defined as that coupling the $(i-1)$st link with the ith link. Next, a coordinate frame $\mathcal{F}_i$ is defined with origin O_i and axes X_i, Y_i, Z_i. This frame is attached to the $(i-1)$st link—**not** to the ith link!—for $i = 1, \ldots, n$. This is done by following the rules given below:

1. Z_i is the axis of the ith pair. Notice that there are two possible definitions of the positive direction of this axis, since each pair axis is only a line, not a directed segment. Moreover, the Z_i axis of a prismatic pair can be located arbitrarily, since the translation allowed by this pair is defined by one *direction*, not by a line, as in the case of the revolute pair.
2. X_i is defined as the common perpendicular to Z_{i-1} and Z_i, directed from the former to the latter, as shown in Fig. 3.2a. Notice that, if these two axes intersect, the positive direction of X_i is undefined and hence, can be *freely assigned*. Henceforth, we will follow the *right-hand* rule in this case. This means that if unit vectors $\mathbf{i}_i$, $\mathbf{k}_{i-1}$, and $\mathbf{k}_i$ are attached to axes X_i, Z_{i-1}, and Z_i, respectively—as indicated in Fig. 3.2b—$\mathbf{i}_i$ is defined as the unit vector in the direction of $\mathbf{k}_{i-1} \times \mathbf{k}_i$. Moreover, if Z_{i-1} and Z_i are parallel, the location of X_i is undefined. In order to define it uniquely, we will specify X_i as passing through the origin of the $(i-1)$st frame, as shown in Fig. 3.2c.
3. The *distance* between Z_i and Z_{i+1} is denoted a_i, which is thus *nonnegative*.
4. The Z_i-coordinate of the intersection O_i' of Z_i with X_{i+1} is denoted d_i. Since this quantity is a coordinate, it can be either positive or negative. Its absolute value is the distance between X_i and X_{i+1}, also called the *offset* between successive common perpendiculars.
5. The angle between Z_i and Z_{i+1} is defined as α_i and is measured about the positive direction of X_{i+1}. This item is known as the *twist angle* between two successive axes.
6. The angle between X_i and X_{i+1} is defined as θ_i, measured about the positive direction of Z_i.

3.2.3 Spherical Four-Bar-Linkages

A spherical four-bar linkage occurs when all its four *revolute* axes intersect at one common point, O in Fig. 3.3, as intended for *function generation*. In this case we are interested, as in the planar case, in deriving a relation between the input angle ψ and the output angle ϕ, that should include the *linkage dimensions* $\{ \alpha_i \}_1^4$ as parameters. To this end, we introduce the unit vectors $\{ \mathbf{e}_i \}_1^4$, directed along the concurrent axes of the four revolutes, as depicted in Fig. 3.3. Notice that, in order to bring the notation adopted for planar four-bar linkages, as proposed by Freudenstein and displayed in

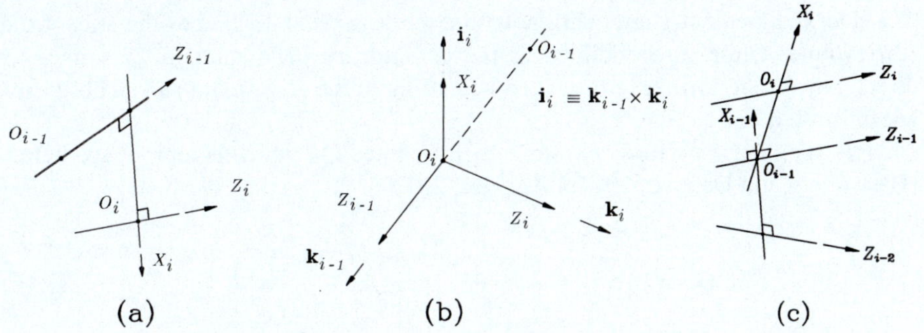

Fig. 3.2 Definition of X_i when Z_{i-1} and Z_i: **a** are skew; **b** intersect; and **c** are parallel

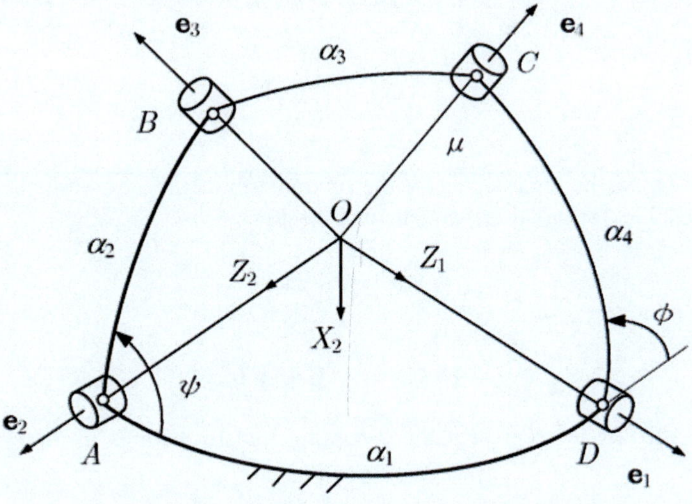

Fig. 3.3 A spherical four-bar linkage for function generation with DH notations. Only X_2, Z_2, and Z_1 are displayed for clarity

Fig. 3.1, in line with the Denavit-Hartenberg notation, we have placed Z_1 along the axis of the output joint, Z_2 along that of the input joint.

Deriving the desired relation is now a simple matter, for we have one geometric relation at our disposal, namely,

$$\mathbf{e}_3 \cdot \mathbf{e}_4 = \cos \alpha_3 \tag{3.14}$$

Moreover, in order to compute the foregoing dot product, we need its two factors (i) *in the same coordinate frame*, and (ii) in terms of the input and output angles. Apparently, if we choose $\mathcal{F}_2$, the coordinate frame fixed to the mechanism frame, to represent the two unit vectors in question, we will have the desired expression.

Under the Denavit-Hartenberg notation, the Z_i-axis is defined as the axis of the ith revolute, while X_i is defined as the common perpendicular to Z_{i-1} and Z_i, directed from the former to the latter, according to the *right-hand rule*. These axes are illustrated in Fig. 3.3.

Now, the matrix rotating $\mathcal{F}_i$ into $\mathcal{F}_{i+1}$ is denoted $\mathbf{Q}_i$, which is reproduced below (Hartenberg and Denavit, 1964a; Angeles, 2014):

$$\mathbf{Q}_i \equiv [\,\mathbf{Q}_i\,]_i \equiv \begin{bmatrix} \cos\theta_i & -\lambda_i \sin\theta_i & \mu_i \sin\theta_i \\ \sin\theta_i & \lambda_i \cos\theta_i & -\mu_i \cos\theta_i \\ 0 & \mu_i & \lambda_i \end{bmatrix} \tag{3.15}$$

where $\lambda_i \equiv \cos\alpha_i$ and $\mu_i \equiv \sin\alpha_i$, while θ_i was already defined in Sect. 3.2.2. Apparently, vector $\mathbf{e}_i$ in $\mathcal{F}_i$, denoted $[\,\mathbf{e}_i\,]_i$, is given by

$$[\,\mathbf{e}_i\,]_i = \begin{bmatrix} 0 \\ 0 \\ 1 \end{bmatrix} \tag{3.16}$$

Moreover, $\mathbf{Q}_i$ can be regarded as the matrix transforming $\mathcal{F}_{i+1}$-coordinates into $\mathcal{F}_i$-coordinates, i.e., for *any* three-dimensional vector $\mathbf{v}$,

$$[\,\mathbf{v}\,]_i = \mathbf{Q}_i[\,\mathbf{v}\,]_{i+1} \tag{3.17}$$

and hence,

$$[\,\mathbf{v}\,]_{i+1} = [\,\mathbf{Q}_i^T\,]_i[\,\mathbf{v}\,]_i \tag{3.18}$$

More specifically, we represent the foregoing transformations in the abbreviated form:

$$\mathbf{Q}_1\colon \mathcal{F}_1 \to \mathcal{F}_2, \quad \mathbf{Q}_2\colon \mathcal{F}_2 \to \mathcal{F}_3, \quad \mathbf{Q}_3\colon \mathcal{F}_3 \to \mathcal{F}_4, \quad \mathbf{Q}_4\colon \mathcal{F}_4 \to \mathcal{F}_1 \tag{3.19a}$$

$$\mathbf{Q}_1\colon [\,\cdot\,]_2 \to [\,\cdot\,]_1, \quad \mathbf{Q}_2\colon [\,\cdot\,]_3 \to [\,\cdot\,]_2, \quad \mathbf{Q}_3\colon [\,\cdot\,]_4 \to [\,\cdot\,]_3, \quad \mathbf{Q}_4\colon [\,\cdot\,]_1 \to [\,\cdot\,]_4 \tag{3.19b}$$

$$\mathbf{Q}_1^T : \mathcal{F}_2 \to \mathcal{F}_1, \quad \mathbf{Q}_2^T : \mathcal{F}_3 \to \mathcal{F}_2, \quad \mathbf{Q}_3^T : \mathcal{F}_4 \to \mathcal{F}_3, \quad \mathbf{Q}_3^T : \mathcal{F}_1 \to \mathcal{F}_4 \tag{3.19c}$$

$$\mathbf{Q}_1^T : [\,\cdot\,]_1 \to [\,\cdot\,]_2, \quad \mathbf{Q}_2^T : [\,\cdot\,]_2 \to [\,\cdot\,]_3, \quad \mathbf{Q}_3^T : [\,\cdot\,]_3 \to [\,\cdot\,]_4, \quad \mathbf{Q}_4^T : [\,\cdot\,]_4 \to [\,\cdot\,]_1 \tag{3.19d}$$

In particular, given expression (3.16) for $[\,\mathbf{e}_i\,]_i$, it is apparent that the third column of $\mathbf{Q}_i^T$ or, equivalently, the third row of $\mathbf{Q}_i$, is $[\,\mathbf{e}_i\,]_{i+1}$. By the same token, the third column of $\mathbf{Q}_i$ is $[\,\mathbf{e}_{i+1}\,]_i$, i.e.,

$$[\,\mathbf{e}_i\,]_{i+1} = \begin{bmatrix} 0 \\ \mu_i \\ \lambda_i \end{bmatrix}, \quad [\,\mathbf{e}_{i+1}\,]_i = \begin{bmatrix} \mu_i \sin\theta_i \\ -\mu_i \cos\theta_i \\ \lambda_i \end{bmatrix} \tag{3.20}$$

The vector representations required in this sequel are derived below. We do this by recalling that $[\,\mathbf{e}_3\,]_2$ is the third column of $\mathbf{Q}_2$, $[\,\mathbf{e}_4\,]_1$ the third row of $\mathbf{Q}_4$. Hence,

$$[\,\mathbf{e}_3\,]_2 = \begin{bmatrix} \mu_2 \sin\theta_2 \\ -\mu_2 \cos\theta_2 \\ \lambda_2 \end{bmatrix} \tag{3.21}$$

$$[\,\mathbf{e}_4\,]_2 = \mathbf{Q}_1^T [\,\mathbf{e}_4\,]_1 \equiv \begin{bmatrix} \cos\theta_1 & \sin\theta_1 & 0 \\ -\lambda_1 \sin\theta_1 & \lambda_1 \cos\theta_1 & \mu_1 \\ \mu_1 \sin\theta_1 & -\mu_1 \cos\theta_1 & \lambda_1 \end{bmatrix} \begin{bmatrix} 0 \\ \mu_4 \\ \lambda_4 \end{bmatrix} \tag{3.22}$$

$$= \begin{bmatrix} \mu_4 \sin\theta_1 \\ \mu_4\lambda_1 \cos\theta_1 + \lambda_4\mu_1 \\ -\mu_4\mu_1 \cos\theta_1 + \lambda_4\lambda_1 \end{bmatrix} \tag{3.23}$$

which are the expressions sought. Therefore,

$$\begin{aligned} [\,\mathbf{e}_3^T\,]_2[\,\mathbf{e}_4\,]_2 = {} & \mu_2\mu_4 \sin\theta_1 \sin\theta_2 - \mu_2 \cos\theta_2(\mu_4\lambda_1 \cos\theta_1 + \lambda_4\mu_1) \\ & + \lambda_2(-\mu_4\mu_1 \cos\theta_1 + \lambda_4\lambda_1) \end{aligned} \tag{3.24}$$

Upon substituting the above expression into Eq. (3.14), we obtain

$$\begin{aligned} \lambda_1\lambda_2\lambda_4 - \lambda_3 - \lambda_4\mu_1\mu_2 \cos\theta_2 - \lambda_1\mu_2\mu_4 \cos\theta_1 \cos\theta_2 - \lambda_2\mu_1\mu_4 \cos\theta_1 \\ + \mu_2\mu_4 \sin\theta_1 \sin\theta_2 = 0 \end{aligned} \tag{3.25}$$

which is a form of the input-output equation sought. This equation can be simplified upon realizing that the last coefficient of its left-hand side cannot vanish, least one of the input and output links, or even both, shrinks to one point on the sphere—a consequence of at least one of α_2 and α_4 vanishing or equating π. In this light, we can safely divide both sides of the above equation by $\mu_2\mu_4$. Moreover, in order to render the same equation terser, we introduce below the *Freudenstein parameters for the spherical linkage*:

$$k_1 \equiv \frac{\lambda_1\lambda_2\lambda_4 - \lambda_3}{\mu_2\mu_4}, \quad k_2 = \frac{\lambda_4\mu_1}{\mu_4}, \quad k_3 = \lambda_1, \quad k_4 = \frac{\lambda_2\mu_1}{\mu_2} \tag{3.26}$$

Now the two sides of Eq. (3.25) are divided by $\mu_2\mu_4$, definitions (3.26) then introduced in the equation thus resulting, which leads to

$$k_1 - k_2 \cos\theta_2 - k_3 \cos\theta_1 \cos\theta_2 - k_4 \cos\theta_1 + \sin\theta_1 \sin\theta_2 = 0 \tag{3.27}$$

This equation, however, involves a relation between angles θ_1 and θ_2 of the DH notation, which are different from, although related to, the input and the output angles ψ and ϕ. To better understand the relation between the two pairs of angles, we sketch these in Fig. 3.4. From this figure, it is apparent that

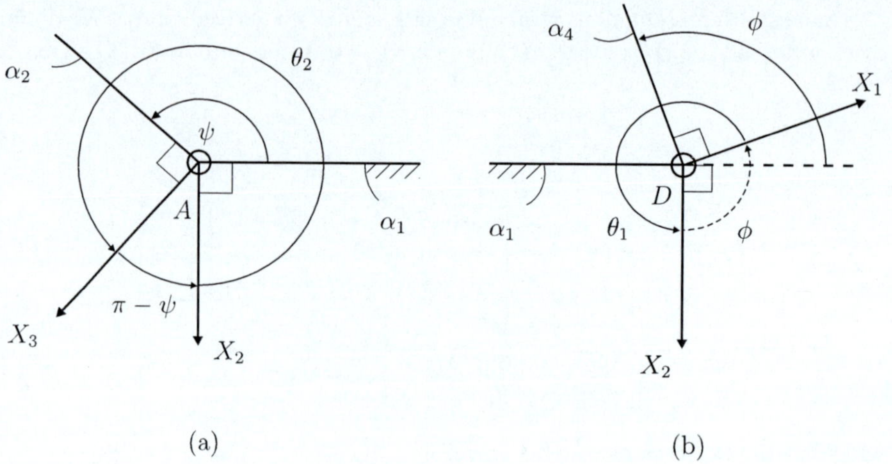

Fig. 3.4 Relation between input and output angles with their counterparts in the DH notation: **a** ψ and θ_2; and **b** ϕ and θ_1

$$\theta_2 = \psi + \pi, \quad \theta_1 = 2\pi - \phi \quad \text{or} \quad \theta_1 = -\phi \tag{3.28}$$

Hence,

$$\cos\theta_2 = -\cos\psi, \quad \sin\theta_2 = -\sin\psi, \quad \cos\theta_1 = \cos\phi, \quad \sin\theta_1 = -\sin\phi \tag{3.29}$$

Substitution of relations (3.29) into Eq. (3.27) leads to

$$F(\psi, \phi) \equiv k_1 + k_2\cos\psi + k_3\cos\psi\cos\phi - k_4\cos\phi + \sin\psi\sin\phi = 0 \tag{3.30}$$

which is the *input-output equation for spherical linkages*, written in terms of the input and output angles ψ and ϕ, respectively. Moreover, Eq. (3.30) along with definitions (3.26), can be fairly termed the *Freudenstein equation* for spherical linkages. Either Eq. (3.27) or Eq. (3.30) can be used to find the output angle ϕ for a given linkage and a given value of the input angle ψ, which constitutes the *analysis problem*. The same equation is to be used for *synthesis*, as described below.

In a synthesis problem, we aim to calculate the set of unknown linkage angles $\{\alpha_i\}_1^4$, for a *given* set of pairs $\{(\psi_i, \phi_i)\}_1^m$ of IO angle values that the linkage is to meet. In order to obtain the synthesis equations allowing us to compute the set of Freudenstein parameters leading to the desired linkage, we proceed as in the planar case and write Eq. (3.30) for the given set of pairs of angle values, thus obtaining the *synthesis equations* in the form of Eq. (3.12), as derived for planar four-bar linkages. Obviously, the pertinent synthesis matrix **S** and vectors **b** and **k** are

$$
\mathbf{S} \equiv \begin{bmatrix} 1 & \cos\psi_1 & \cos\psi_1\cos\phi_1 & -\cos\phi_1 \\ 1 & \cos\psi_2 & \cos\psi_2\cos\phi_2 & -\cos\phi_2 \\ \vdots & \vdots & \vdots & \\ 1 & \cos\psi_m & \cos\psi_m\cos\phi_m & -\cos\phi_m \end{bmatrix}, \quad \mathbf{k} \equiv \begin{bmatrix} k_1 \\ k_2 \\ k_3 \\ k_4 \end{bmatrix}, \quad \mathbf{b} \equiv \begin{bmatrix} -\sin\psi_1\sin\phi_1 \\ -\sin\psi_2\sin\phi_2 \\ \vdots \\ -\sin\psi_m\sin\phi_m \end{bmatrix} \tag{3.31}
$$

Similar to the planar case, we have exact synthesis when the number m of given pairs of input-output angular values equals the number of Freudenstein parameters at hand, which in this case happens when $m = 4$. When $m > 4$, we have a problem of *approximate synthesis*.

However, Eq. (3.30) only provides values for the linkage parameters $\{k_i\}_1^4$. Hence, we need a means to convert the latter into the linkage dimensions.

Notice that the relations between the two sets, $\{\alpha_i\}_1^4$ and $\{k_i\}_1^4$, are *nonlinear*, and hence, solving Eqs. (3.26) for the former needs careful planning. For starters, the foregoing equations are rewritten explicitly in terms of the linkage dimensions $\{\alpha_i\}_1^4$, and reordered conveniently:

$$c\alpha_1 - k_3 = 0 \tag{3.32a}$$

$$c\alpha_4 s\alpha_1 - k_2 s\alpha_4 = 0 \tag{3.32b}$$

$$c\alpha_2 s\alpha_1 - k_4 s\alpha_2 = 0 \tag{3.32c}$$

$$c\alpha_1 c\alpha_2 c\alpha_4 - c\alpha_3 - k_1 s\alpha_2 s\alpha_4 = 0 \tag{3.32d}$$

In applications, α_1 is given—not up to the designer to find it!—and hence, the problem of finding the remaining DH parameters simplifies tremendously. This can be done sequentially, from Eqs. (3.32b) to (3.32d), in this order.

In a semigraphical method, based on *contour-intersection* and favored in this book, all but two of the unknowns are first eliminated from the set of nonlinear equations, thereby ending up with a reduced number of equations in the two remaining unknowns. Each of these equations is then plotted in the plane of the two unknowns, which yields one contour per bivariate equation, in that plane. All real solutions are then found *by inspection*, from the intersections of all the contours. Notice that, if the reduced system comprises more than two bivariate equations, then the system entails *algebraic redundancy*, which is good news, as this adds *robustness* to the system. What the emphasized term means is *robustness to roundoff error*. This means, in turn, low sensitivity to unavoidable error brought about by numerical computations.

In the particular case at hand, the transformation sought can be most readily found by noticing the structure of Eqs. (3.32a)–(3.32d): The first equation involves one single unknown, α_1; the second only *one new* unknown, α_4; the third only *one new* unknown, α_2; and the fourth only one new unknown as well, α_3. Hence, the *computational algorithm* follows:

(*i*) From Eq. (3.32a), compute $\alpha_1 = \cos^{-1}(k_3) \Rightarrow$ two possible values of α_1;
(*ii*) From Eq. (3.32b), compute $\alpha_4 = \tan^{-1}(\sin\alpha_1/k_2) \Rightarrow$ two possible values of α_4 for each value of α_1;

(*iii*) From Eq. (3.32c), compute $\alpha_2 = \tan^{-1}(\sin\alpha_1/k_4) \Rightarrow$ two possible values of α_2 for each value of α_1 *and its corresponding* α_4, thereby leading to four possible values of α_2;
(*iv*) From Eq. (3.32d), compute $\alpha_3 = \cos^{-1}(\cos\alpha_1\cos\alpha_2\cos\alpha_4 - k_1\sin\alpha_2\sin\alpha_4)$ $\Rightarrow$ two possible values of α_3 for each pair of values of α_1, α_2 and α_4.

In summary, we have: two possible values of α_1; four possible values of α_4; four possible values of α_2, which can be combined in $2 \times 4 \times 4 = 32$ forms. Hence, we end up with up to $2 \times 32 = 64$ values of α_3. Therefore, we can expect up to 64 sets of $\{\alpha_i\}_1^4$ values for one single set $\{k_i\}_1^4$. However, some of the 64 sets of linkage dimensions found above may be complex, and hence, uninteresting. The problem may also not admit any single real solution, for example, if $|k_3| > 1$. As well, α_3 becomes complex when $|\cos\alpha_1\cos\alpha_2\cos\alpha_4 - k_1\sin\alpha_2\sin\alpha_4| > 1$. That is, for a feasible linkage, two conditions must be imposed on the Freudenstein parameters $\{k_i\}_1^4$:

$$|k_3| \le 1, \quad |\cos\alpha_1\cos\alpha_2\cos\alpha_4 - k_1\sin\alpha_2\sin\alpha_4| \le 1 \tag{3.33}$$

Notice that the semigraphical method *filters* all complex solutions and is guaranteed to yield all real solutions if *i*) the problem admits any real solutions and (*ii*) the two contours in question are plotted inside a square of side π centered at the origin of the α_2-α_3 plane. In order to implement the semigraphical method, first, four pairs of values (α_1, α_4) are computed from Eqs. (3.32a) and (3.32d). Each of these pairs is then substituted into Eqs. (3.32c) and (3.32d), thereby obtaining four pairs of contours in the α_2-α_3 plane. The intersections of each pair of contours, which can be estimated by inspection, yield one subset of real solutions. Each of these estimates of α_3 and α_4 values can then be used as an *initial guess* for a Newton-Raphson solution of the two equations. Due to the proximity of each estimate from the pair of real roots, the Newton-Raphson method should converge in a pair of iterations for a reasonable tolerance. Once all four pairs of contours have been exhausted, *all real roots* of the problem have been computed. Alternatively, all real solutions can be obtained via contour-intersection upon *zooming-in* at every visible intersection.

Finally, notice that any *spherical triangle* and, in fact, any *spherical polygon* defined on the surface of the unit sphere has an *antipodal* counterpart. In this light, then, even if we end up with a full set of feasible linkage dimensions, only 32 four-bar linkages defined by this set are distinct.

3.2.4 Spatial Four-Bar-Linkages

The analysis of spatial four-bar linkages relies heavily on the algebra of *dual numbers*, which is extensively discussed in Appendix A. What we should recall now is (a) the usual representation of dual quantities, by means of a "hat" (^) on top of the variable in question and (b) the definition of the *dual unit*, ϵ, via its two properties

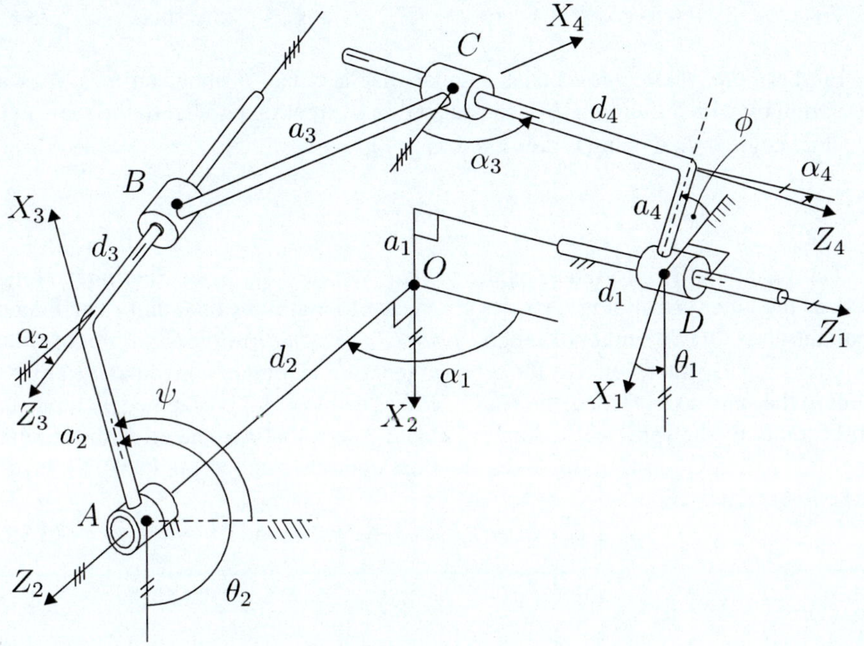

Fig. 3.5 A RCCC linkage for function generation

$$\epsilon \neq 0, \quad \epsilon^2 = 0 \tag{3.34}$$

A general layout of a spatial four-bar linkage is included in Fig. 3.5, in which we use the Denavit-Hartenberg notation, introduced in Sect. 3.2.2. Similar to that subsection, we have laid the ouput axis along Z_1, in order to comply with both the DH notation and that adopted in Figs. 3.1 and 3.3. In this case, ψ and ϕ denote the input and the output angles, as in Sect. 3.2.3, their relations with angles θ_1 and θ_2 of the DH notation being exactly as in the spherical case, namely,

$$\psi = \theta_2 - \pi, \quad \phi = 2\pi - \theta_1 \tag{3.35}$$

The IO equation of the RCCC linkage is most readily derived by resorting to the *Principle of Transference* (Dimentberg, 1965; Rico Martínez and Duffy, 1993), which is cited below:

> The kinematics and statics relations of spatial linkages and cam mechanisms can be derived upon replacing the real variables occurring in the corresponding relations for spherical linkages or cam mechanisms, as the case may be, by dual numbers.

Put quite simply, the IO equation of the RCCC linkage can be derived from that of the spherical RRRR linkage upon "putting hats" on both the variables and the (Freudenstein) parameters occurring in Eq. (3.30), thereby obtaining

$$\hat{F}(\hat{\psi}, \hat{\phi}) \equiv \hat{k}_1 + \hat{k}_2 \cos\hat{\psi} + \hat{k}_3 \cos\hat{\psi}\cos\hat{\phi} - \hat{k}_4 \cos\hat{\phi} + \sin\hat{\psi}\sin\hat{\phi} = 0 \quad (3.36)$$

Furthermore, *dualization* of an angular displacement θ about an axis $\mathcal{A}$ is an operation by which a sliding d_o is introduced in a direction parallel to the same axis, the dual angle $\hat{\theta}$ then being represented as

$$\hat{\theta} = \theta + \epsilon d_o \quad (3.37)$$

The variable θ is referred to as the *primal part* of $\hat{\theta}$, d_o as its *dual part*. Notice that, as the latter is a sliding, its units are those of length; the dual unit ϵ can then be thought of as "having units of length-inverse." The *dualization operation* can thus be kinematically interpreted as the replacement of a R joint of a given axis $\mathcal{A}$ by a C joint of the same axis. Hence, the RCCC linkage of Fig. 3.5 is obtained by replacing all R joints of the spherical linkage of Fig. 3.3, but that associated with the input angle ψ, with a C joint. In this light, the dual quantities appearing in Eq. (3.36) are to be interpreted as

$$\hat{F}(\hat{\psi}, \hat{\phi}) = F(\hat{\psi}, \hat{\phi}) + \epsilon F_o(\hat{\psi}, \hat{\phi}) \quad (3.38a)$$

with

$$\hat{\psi} = \psi + \epsilon d_2, \quad \hat{\phi} = \phi + \epsilon d_1, \quad \hat{k}_i \equiv k_i + \epsilon k_{oi}, \; i = 1, \ldots, 4 \quad (3.38b)$$

Consistently, the k_i parameters are dimensionless, while their dual counterparts k_{oi}, for $i = 1, \ldots, 4$, have all units of length. The former are displayed in Eq. (3.26) for the spherical four-bar linkage; they are the same for its spatial counterpart. The latter are obtained below upon extracting them from the dual version of the expressions displayed in the equation cited above. The computations are lengthy and tedious, for which reason the dual expressions were obtained using computer algebra.[2] In deriving the desired expressions, the relation that gives the dual version $\hat{f}(\hat{x})$ of a real function $f(x)$ of a real variable x, whereby the dual argument is defined as $\hat{x} \equiv x + \epsilon x_o$, is given in Eq. (A.3), and reproduced below for quick reference.

$$\hat{f}(\hat{x}) = \hat{f}(x + \epsilon x_o) \equiv f(x) + \epsilon x_o \frac{df(x)}{dx}$$

Hence,

$$\cos\hat{\psi} \equiv \cos\psi - \epsilon d_2 \sin\psi, \quad \sin\hat{\psi} \equiv \sin\psi + \epsilon d_2 \cos\psi \quad (3.39a)$$

$$\cos\hat{\phi} \equiv \cos\phi - \epsilon d_1 \sin\phi, \quad \sin\hat{\phi} \equiv \sin\phi + \epsilon d_1 \cos\phi \quad (3.39b)$$

in which d_2 is constant, while d_1 is variable, as per Fig. 3.5.

[2] See the Maple 15 worksheet `DualRCCC-IO131224.mw`.

Likewise, for the dual angles $\hat{\alpha}_i \equiv \alpha_i + \epsilon a_i$,

$$\cos\hat{\alpha}_i \equiv \cos\alpha_i - \epsilon a_i \sin\alpha_i, \quad \sin\hat{\alpha}_i \equiv \sin\alpha_i + \epsilon a_i \cos\alpha_i \tag{3.40}$$

Further, the primal parts $\{ k_i \}_1^4$ of the dual Freudenstein parameters $\{ \hat{k}_i \}_1^4$ are identical to those of the spherical linkage, displayed in Eq. (3.26). For quick reference, the corresponding parameters are reproduced below:

$$k_1 \equiv \frac{\lambda_4\lambda_1\lambda_2 - \lambda_3}{\mu_2\mu_4}, \quad k_2 = \frac{\lambda_4\mu_1}{\mu_4}, \quad k_3 = \lambda_1, \quad k_4 = \frac{\mu_1\lambda_2}{\mu_2} \tag{3.41}$$

their dual counterparts being

$$\begin{aligned} k_{o1} &= -\frac{a_1\lambda_2\lambda_4\mu_1\mu_2\mu_4 + a_2(\lambda_1\lambda_4 - \lambda_2\lambda_3)\mu_4 - a_3\mu_2\mu_3\mu_4 + a_4(\lambda_1\lambda_2 - \lambda_3\lambda_4)\mu_2}{\mu_2^2\mu_4^2} \\ k_{o2} &= \frac{a_1\lambda_1\lambda_4\mu_4 - a_4\mu_1}{\mu_4^2}, \quad k_{o3} = -a_1\mu_1, \quad k_{o4} = \frac{a_1\lambda_1\lambda_2\mu_2 - a_2\mu_1}{\mu_2^2} \end{aligned} \tag{3.42}$$

Moreover, if the angle of rotation θ_i, associated with the ith R joint of Fig. 3.3 under the DH notation, for $i = 1, 3, 4$, is dualized, the *dual angles* thus resulting become, in the notation of Fig. 3.5,

$$\hat{\theta}_1 = \theta_1 + \epsilon d_1 = -\phi + \epsilon d_1, \; \hat{\theta}_2 = \theta_2 = \psi + \pi, \; \hat{\theta}_3 = \theta_3 + \epsilon d_3, \; \hat{\theta}_4 = \theta_4 + \epsilon d_4 \tag{3.43}$$

where

$$-\infty < d_i < \infty, \quad i = 1, 3, 4$$

which is an unbounded real number, with units of length. Notice that d_i is not a "length," properly speaking, because a length is positive, while d_i can be negative, exactly the same as a joint angle. Moreover, $F_o(\psi, \phi, d_1)$ is given below:

$$\begin{aligned} F_o(\psi, \phi, d_1) = k_{o1} &- k_2 d_2 \sin(\psi) + k_{o2}\cos(\psi) - k_3 d_1 \cos(\psi)\sin(\phi) \\ &- k_3 d_2 \sin(\psi)\cos(\phi) + k_{o3}\cos(\psi)\cos(\phi) + k_4 d_1 \sin(\phi) \\ &- k_{o4}\cos(\phi) + d_1 \sin(\psi)\cos(\phi) + d_2\cos(\psi)\sin(\phi) = 0 \end{aligned} \tag{3.44}$$

Further, the synthesis equations for the spatial four-bar linkage can be readily set up by dualizing those derived for the spherical case, with the synthesis matrix $\mathbf{S}$ and the right-hand side $\mathbf{b}$ of Eq. (3.31) substituted by their "hatted" counterparts, namely,

$$\hat{\mathbf{S}}\hat{\mathbf{k}} = \hat{\mathbf{b}} \tag{3.45}$$

where, as usual,

$$\hat{\mathbf{S}} = \mathbf{S} + \epsilon\mathbf{S}_o, \quad \hat{\mathbf{k}} = \mathbf{k} + \epsilon\mathbf{k}_o, \quad \hat{\mathbf{b}} = \mathbf{b} + \epsilon\mathbf{b}_o \tag{3.46}$$

the primal part $\mathbf{k}$ of vector $\hat{\mathbf{k}}$, it is recalled, being the same as that of the spherical linkage, and displayed componentwise in Eq. (3.26). The dual part is displayed likewise in Eqs. (3.42). The primal and dual parts of $\mathbf{S}$ and $\mathbf{b}$ are derived below: Upon dualizing $\mathbf{S}$ of Eq. (3.31) componentwise, the relation below is obtained:

$$\hat{\mathbf{S}} = \begin{bmatrix} 1 & c\psi_1 - \epsilon d_2 s\psi_1 & c\psi_1(c\phi_1 - \epsilon u_1 s\phi_1) & -c\phi_1 + \epsilon u_1 s\phi_1 \\ 1 & c\psi_2 - \epsilon d_2 s\psi_2 & c\psi_2(c\phi_2 - \epsilon u_2 s\phi_2) & -c\phi_2 + \epsilon u_2 s\phi_2 \\ \vdots & \vdots & \vdots & \vdots \\ 1 & c\psi_m - \epsilon d_2 s\psi_m & c\psi_m(c\phi_m - \epsilon u_m s\phi_m) & -c\phi_m + \epsilon u_m s\phi_m \end{bmatrix} \tag{3.47}$$

or

$$\hat{\mathbf{S}} = \underbrace{\begin{bmatrix} 1 & c\psi_1 & c\psi_1 c\phi_1 & -c\phi_1 \\ 1 & c\psi_2 & c\psi_2 c\phi_2 & -c\phi_2 \\ \vdots & \vdots & \vdots & \vdots \\ 1 & c\psi_m & c\psi_m c\phi_m & -c\phi_m \end{bmatrix}}_{\mathbf{S}} + \epsilon \underbrace{\begin{bmatrix} 0 & -d_2 s\psi_1 & -u_1 c\psi_1 s\phi_1 & u_1 s\phi_1 \\ 0 & -d_2 s\psi_2 & -u_2 c\psi_2 s\phi_2 & u_2 s\phi_2 \\ \vdots & \vdots & \vdots & \vdots \\ 0 & -d_2 s\psi_m & -u_m c\psi_m s\phi_m & u_m s\phi_m \end{bmatrix}}_{\mathbf{S}_o} \tag{3.48}$$

where the definition $u_i \equiv (d_1)_i$, the displacement d_1 specified at ith data point, has been introduced, to avoid a double subscript. Likewise,

$$\hat{\mathbf{k}} = \underbrace{\begin{bmatrix} k_1 \\ k_2 \\ k_3 \\ k_4 \end{bmatrix}}_{\mathbf{k}} + \epsilon \underbrace{\begin{bmatrix} k_{o1} \\ k_{o2} \\ k_{o3} \\ k_{o4} \end{bmatrix}}_{\mathbf{k}_o}, \quad \hat{\mathbf{b}} = \underbrace{\begin{bmatrix} -s\psi_1 s\phi_1 \\ -s\psi_2 s\phi_2 \\ \vdots \\ -s\psi_m s\phi_m \end{bmatrix}}_{\mathbf{b}} + \epsilon \underbrace{\begin{bmatrix} -u_1 s\psi_1 c\phi_1 - d_2 c\psi_1 s\phi_1 \\ -u_2 s\psi_2 c\phi_2 - d_2 c\psi_2 s\phi_2 \\ \vdots \\ -u_m s\psi_m c\phi_m - d_2 c\psi_m s\phi_m \end{bmatrix}}_{\mathbf{b}_o} \tag{3.49}$$

Now, upon equating the primal and the dual parts of Eq. (3.45), two real vector equations are obtained, namely,

$$\mathbf{S}\mathbf{k} = \mathbf{b}, \quad \mathbf{S}\mathbf{k}_o + \mathbf{S}_o\mathbf{k} = \mathbf{b}_o \tag{3.50}$$

As the reader can readily verify, the first of the two foregoing equations is identical to that derived for spherical linkages in Eq. (3.31). That is, the problem of synthesis of a spatial function generator has been *decoupled* into two, the synthesis procedure then being straightforward:

1. Synthesize first a spherical linkage for the angular input-output data given at the outset;
2. Substitute vector $\mathbf{k}$, as obtained from step 1, along with the additional data $\{(d_1)_i\}_1^m$, with $u_i \leftarrow (d_1)_i$, for $i = 1, \ldots, m$, and solve the second vector equation of (3.50) for $\mathbf{k}_o$, thereby completing the solution of the synthesis problem.

Remark 3.2.1 Given that the output involves a sliding variable d_1, besides the angle ϕ, two sets of data-points must be prescribed: $\{\, \psi_i,\ \phi_i \,\}_1^m$ and $\{\, \psi_i,\ (d_1)_i \,\}_1^m$.

3.3 Exact Synthesis

The main assumption in this section is that the number of synthesis equations at hand equals that of linkage parameters available when defining a linkage of a given kind. We thus have three kinds of problems, one for each of the three kinds of linkages, i.e., planar, spherical and spatial four-bar linkages.

3.3.1 Planar Linkages

We have $m = 3$ in this case, and hence, the synthesis equations look like

$$\begin{bmatrix} 1 & c\phi_1 & -c\psi_1 \\ 1 & c\phi_2 & -c\psi_2 \\ 1 & c\phi_3 & -c\psi_3 \end{bmatrix} \begin{bmatrix} k_1 \\ k_2 \\ k_3 \end{bmatrix} = \begin{bmatrix} c(\phi_1 - \psi_1) \\ c(\phi_2 - \psi_2) \\ c(\phi_3 - \psi_3) \end{bmatrix} \tag{3.51a}$$

where

$$c(\cdot) \equiv \cos(\cdot) \quad \text{and} \quad s(\cdot) \equiv \sin(\cdot) \tag{3.51b}$$

Solving *numerically* for $\{k_i\}_1^3$ is straightforward, if Gaussian elimination, i.e., LU-decomposition, is applied—as, implemented, e.g., in Matlab. Given the simple structure of the system at hand, however, a solution in *closed form* is also possible: To this end, subtract the first equation from the second and third equations:

$$\begin{bmatrix} 1 & c\phi_1 & -c\psi_1 \\ 0 & c\phi_2 - c\phi_1 & -c\psi_2 + c\psi_1 \\ 0 & c\phi_3 - c\phi_1 & -c\psi_3 + c\psi_1 \end{bmatrix} \begin{bmatrix} k_1 \\ k_2 \\ k_3 \end{bmatrix} = \begin{bmatrix} c(\phi_1 - \psi_1) \\ c(\phi_2 - \psi_2) - c(\phi_1 - \psi_1) \\ c(\phi_3 - \psi_3) - c(\phi_1 - \psi_1) \end{bmatrix} \tag{3.52}$$

with the second and third equations free of k_1. Hence, one can readily solve the foregoing equations first for k_2 and k_3:

$$\begin{bmatrix} c\phi_2 - c\phi_1 & -c\psi_2 + c\psi_1 \\ c\phi_3 - c\phi_1 & -c\psi_3 + c\psi_1 \end{bmatrix} \begin{bmatrix} k_2 \\ k_3 \end{bmatrix} = \begin{bmatrix} c(\phi_2 - \psi_2) - c(\phi_1 - \psi_1) \\ c(\phi_3 - \psi_3) - c(\phi_1 - \psi_1) \end{bmatrix} \tag{3.53}$$

The above 2×2 system can thus be solved for k_2 and k_3 if we recall Fact 1.4.2:

$$\begin{bmatrix} k_2 \\ k_3 \end{bmatrix} = \frac{1}{\Delta} \begin{bmatrix} -c\psi_3 + c\psi_1 & c\psi_2 - c\psi_1 \\ -c\phi_3 + c\phi_1 & c\phi_2 - c\phi_1 \end{bmatrix} \begin{bmatrix} c(\phi_2 - \psi_2) - c(\phi_1 - \psi_1) \\ c(\phi_3 - \psi_3) - c(\phi_1 - \psi_1) \end{bmatrix} \tag{3.54a}$$

where

$$\Delta \equiv \det \begin{bmatrix} c\phi_2 - c\phi_1 & -c\psi_2 + c\psi_1 \\ c\phi_3 - c\phi_1 & -c\psi_3 + c\psi_1 \end{bmatrix}$$
$$= (c\phi_2 - c\phi_1)(-c\psi_3 + c\psi_1) + (c\psi_2 - c\psi_1)(c\phi_3 - c\phi_1) \quad (3.54b)$$

With k_2 and k_3 obtained from Eqs. (3.54a) and (3.54b), k_1 is derived from the first of Eqs. (3.52). The final result is

$$k_i = \frac{N_i}{\Delta}, \quad i = 1, 2, 3 \quad (3.55)$$

with numerators N_i calculated *sequentially* as

$$N_2 = (-c\psi_3 + c\psi_1)[c(\phi_2 - \psi_2) - c(\phi_1 - \psi_1]$$
$$+ (c\psi_2 - c\psi_1)[c(\phi_3 - \psi_3) - c(\phi_1 - \psi_1] \quad (3.56a)$$
$$N_3 = (-c\phi_3 + c\phi_1)[c(\phi_2 - \psi_2) - c(\phi_1 - \psi_1)]$$
$$+ (c\phi_2 - c\phi_1)[c(\phi_3 - \psi_3) - c(\phi_1 - \psi_1)] \quad (3.56b)$$
$$N_1 = c(\phi_1 - \psi_1)\Delta - c\phi_2 N_2 + c\psi_1 N_3 \quad (3.56c)$$

The foregoing problem is apparently quite simple to solve. We just showed how to solve it *in closed form*. However, the solution obtained must be correctly interpreted. Indeed, upon looking at definitions (3.8), it is apparent that, all link lengths being positive, k_2 and k_3 should be positive as well, while k_1 is capable of taking any finite positive or negative real values. However, nothing in the above formulation prevents k_2 and k_3 from turning out to be negative or zero. Negative values of these parameters are not to be discarded, for they have a geometric interpretation: Notice that, in Eq. (3.10), if ϕ is changed to $\phi + \pi$, then the sign of the second term of the left-hand side of that equation is reversed. Ditto the third term if ψ is changed to $\psi + \pi$. The conclusion then follows:

> A negative k_2 (k_3) indicates that the input (output) angle ψ (ϕ) should not be measured as indicated in Fig. 3.1, but all the way down to the extension of link O_1O_2 (O_4O_3).

If the solution to the synthesis problem leads to $k_2 = 0$, then $a_2 \to \infty$, which means that the input link is of infinite length. The interpretation now is that the first joint of the linkage is of the P type, i.e., we end up with a PRRR linkage. Likewise, if $k_3 = 0$, then $a_4 \to \infty$, and we end up with a RRRP linkage.

Finally, even in the presence of nonzero values of the Freudenstein parameters, nothing guarantees that the link lengths derived from them will yield a feasible linkage. Indeed, for a linkage to be possible, the link lengths must satisfy the *feasibility condition*:

> Any link length must be smaller than the sum of the three other link lengths.

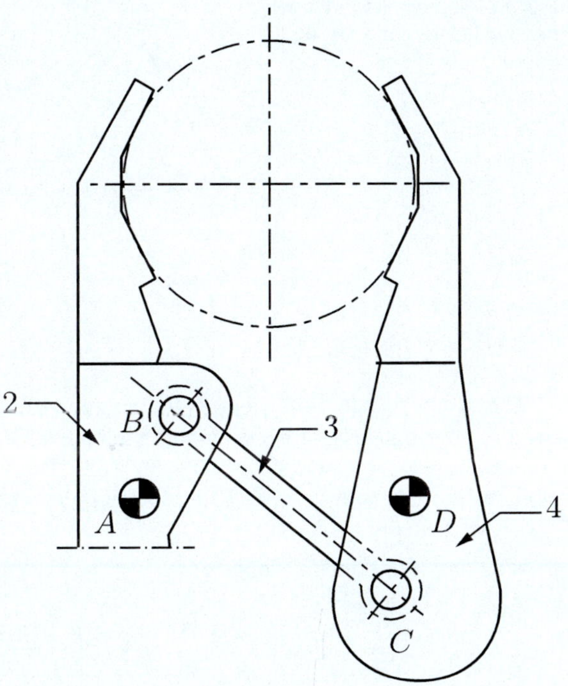

Fig. 3.6 The conceptual design of the actuation mechanism for a robotic gripper

Table 3.1 Input-output values for exact function generation with three data points

j	ψ [°]	ϕ [°]
1	30.0	240.0
2	45.0	225.0
3	60.0	210.0

Example 3.3.1 (*Synthesis of a Robotic Gripper*) Shown in Fig. 3.6 is a concept proposed by Dudiţă et al. (1989) to serve as the actuation mechanism of a robotic gripper. This is a four-bar linkage with a *symmetric architecture*, $a_2 = a_4$, as the two fingers of the gripper are intended to open and close with symmetric motions. Determine the link lengths that will produce the pairs of input-output values given in Table 3.1.

Solution: The 3×3 synthesis matrix $\mathbf{S}$ and the 3-dimensional vector $\mathbf{b}$ of the synthesis equation are first determined with the data of Table 3.1:

$$\mathbf{S} = \begin{bmatrix} 1 & -0.5000 & -0.8660 \\ 1 & -0.7071 & -0.7071 \\ 1 & -0.8660 & -0.5000 \end{bmatrix}, \quad \mathbf{b} = \begin{bmatrix} -0.8660 \\ -1.000 \\ -0.8660 \end{bmatrix}$$

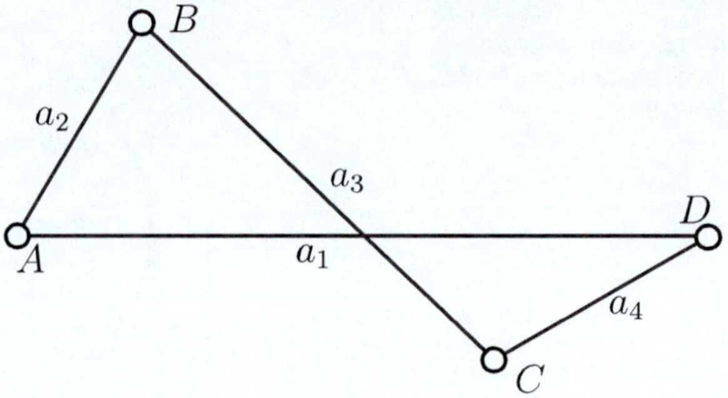

Fig. 3.7 The synthesized four-bar linkage for a robotic gripper

The synthesis equation was solved using the LU-decomposition, as implemented by computer-algebra software , which yielded, with four decimals,

$$\mathbf{k} = \left[2.9319,\ 2.7802,\ 2.7802\right]^T$$

thereby obtaining the link lengths:

$$a_1 = 1.0, \quad a_2 = 0.3597, \quad a_3 = 0.7072, \quad a_4 = 0.3597$$

which indeed produce a symmetric linkage, as $a_2 = a_4$. The linkage thus synthesized is displayed in Fig. 3.7.

Bloch Synthesis

A special kind of linkage synthesis occurs when input-output relations are not specified at three distinct values of the input and output angles, but rather at one single value of these angles, to which *velocity and acceleration conditions* are adjoined. The problem thus arising is known as *Bloch synthesis*. Besides its special nature, this problem becomes relevant because of its revelation:

> The simultaneous vanishing of velocity and acceleration of the output link, i.e., second-order rest of the output link, cannot be obtained with a planar four-bar linkage whose input link turns at a constant angular velocity.

The foregoing claim will be made apparent in the sequel. As a matter of fact, *second-order rest cannot be obtained with any linkage*, but good approximations can be obtained with six-bar linkages producing a short-duration *dwell*.

The problem at hand then can be stated as: *Synthesize a four-bar linkage that meets conditions on position, velocity and acceleration at a given position of the input link.*

In order to formulate this problem, we differentiate both sides of the Freudenstein equation, Eq. (3.10), with respect to time. After rearrangement of terms and a reversal of signs, this gives

$$\dot{\phi} s\phi k_2 - \dot{\psi} s\psi k_3 = (\dot{\phi} - \dot{\psi}) s(\phi - \psi) \tag{3.57a}$$

$$(\ddot{\phi} s\phi + \dot{\phi}^2 c\phi) k_2 - (\ddot{\psi} s\psi + \dot{\psi}^2 c\psi) k_3 = (\ddot{\phi} - \ddot{\psi}) s(\phi - \psi) + (\dot{\phi} - \dot{\psi})^2 c(\phi - \psi) \tag{3.57b}$$

Next, we write Eqs. (3.10) and (3.57a and 3.57b) at $\psi = \psi_1$, and cast them in vector form:

$$\mathbf{A}\mathbf{k} = \mathbf{b} \tag{3.58a}$$

where $\mathbf{A}$ and $\mathbf{b}$ are given below:

$$\mathbf{A} \equiv \begin{bmatrix} 1 & c\phi_1 & -c\psi_1 \\ 0 & \dot{\phi}_1 s\phi_1 & -\dot{\psi}_1 s\psi_1 \\ 0 & \ddot{\phi}_1 s\phi_1 + \dot{\phi}_1^2 c\phi_1 & -\ddot{\psi}_1 s\psi_1 - \dot{\psi}_1^2 c\psi_1 \end{bmatrix} \tag{3.58b}$$

$$\mathbf{b} \equiv \begin{bmatrix} c(\phi_1 - \psi_1) \\ (\dot{\phi}_1 - \dot{\psi}_1) s(\phi_1 - \psi_1) \\ (\ddot{\phi}_1 - \ddot{\psi}_1) s(\phi_1 - \psi_1) + (\dot{\phi}_1 - \dot{\psi}_1)^2 c(\phi_1 - \psi_1) \end{bmatrix} \tag{3.58c}$$

Notice, from Eq. (3.58b), that, if both $\dot{\phi}_1$ and $\ddot{\phi}_1$ vanish, then the first and the second columns of $\mathbf{A}$ become linearly dependent, $\mathbf{A}$ thus becoming singular. This outcome has been discarded at the outset, as both velocity and acceleration at the output joint cannot vanish simultaneously in a four-bar linkage.

Since the second and third of equations (3.58a) are free of k_1, they can be decoupled from the first equation to solve for k_2 and k_3, namely,

$$\begin{bmatrix} \dot{\phi}_1 s\phi_1 & -\dot{\psi}_1 s\psi_1 \\ \ddot{\phi}_1 s\phi_1 + \dot{\phi}_1^2 c\phi_1 & -(\ddot{\psi}_1 s\psi_1 + \dot{\psi}_1^2 c\psi_1) \end{bmatrix} \begin{bmatrix} k_2 \\ k_3 \end{bmatrix} = \begin{bmatrix} (\dot{\phi}_1 - \dot{\psi}_1) s(\phi_1 - \psi_1) \\ (\ddot{\phi}_1 - \ddot{\psi}_1) s(\phi_1 - \psi_1) \\ + (\dot{\phi}_1 - \dot{\psi}_1)^2 c(\phi_1 - \psi_1) \end{bmatrix} \tag{3.59}$$

In solving the above system, we shall need the determinant Δ of the above 2×2 matrix, which is computed below:

$$\begin{aligned} \Delta &\equiv \det \begin{bmatrix} \dot{\phi}_1 s\phi_1 & -\dot{\psi}_1 s\psi_1 \\ \ddot{\phi}_1 s\phi_1 + \dot{\phi}_1^2 c\phi_1 & -\ddot{\psi}_1 s\psi_1 - \dot{\psi}_1^2 c\psi_1 \end{bmatrix} \\ &= -\dot{\phi}_1 s\phi_1 (\ddot{\psi}_1 s\psi_1 + \dot{\psi}_1^2 c\psi_1) + \dot{\psi}_1 s\psi_1 (\ddot{\phi}_1 s\phi_1 + \dot{\phi}_1^2 c\phi_1) \end{aligned} \tag{3.60}$$

thereby making it apparent that, indeed, if $\dot{\phi}_1 = 0$ and $\ddot{\phi}_1 = 0$, then $\Delta = 0$, and the 2×2 matrix in Eq. (3.59) is singular. As a consequence, $\mathbf{A}$ itself is singular as well, thereby proving that *a four-bar linkage cannot produce zero velocity and zero acceleration concurrently at the output link when its input link turns at a constant rpm*.

Now we recall expression (1.4.2) to invert the 2×2 matrix coefficient of vector $[\, k_2, \; k_3 \,]^T$, thus obtaining

$$\begin{bmatrix} k_2 \\ k_3 \end{bmatrix} = \frac{1}{\Delta} \begin{bmatrix} -(\ddot{\psi}_1 s\psi_1 + \dot{\psi}_1^2 c\psi_1) & \dot{\psi}_1 s\psi_1 \\ -(\ddot{\phi}_1 s\phi_1 + \dot{\phi}_1^2 c\phi_1) & \dot{\phi}_1 s\phi_1 \end{bmatrix} \begin{bmatrix} (\dot{\phi}_1 - \dot{\psi}_1)s(\phi_1 - \psi_1) \\ (\ddot{\phi}_1 - \ddot{\psi}_1)s(\phi_1 - \psi_1) \\ + (\dot{\phi}_1 - \dot{\psi}_1)^2 c(\phi_1 - \psi_1) \end{bmatrix} \tag{3.61}$$

Hence,

$$k_2 = \frac{N_2}{\Delta}, \quad k_3 = \frac{N_3}{\Delta} \tag{3.62a}$$

with

$$\begin{aligned} N_2 &\equiv -(\ddot{\psi}_1 s\psi_1 + \dot{\psi}_1^2 c\psi_1)(\dot{\phi}_1 - \dot{\psi}_1)s(\phi_1 - \psi_1) + \dot{\psi}_1 s\psi_1[(\ddot{\phi}_1 - \ddot{\psi}_1)s(\phi_1 - \psi_1) \\ &\quad + (\dot{\phi}_1 - \dot{\psi}_1)^2 c(\phi_1 - \psi_1)] \qquad (3.62b) \\ N_3 &\equiv -(\ddot{\phi}_1 s\phi_1 + \dot{\phi}_1^2 c\phi_1)(\dot{\phi}_1 - \dot{\psi}_1)s(\phi_1 - \psi_1) + \dot{\phi}_1 s\phi_1[(\ddot{\phi}_1 - \ddot{\psi}_1)s(\phi_1 - \psi_1) \\ &\quad + (\dot{\phi}_1 - \dot{\psi}_1)^2 c(\phi_1 - \psi_1)] \qquad (3.62c) \end{aligned}$$

Once k_2 and k_3 are known, we can calculate k_1 from the first of Eqs. (3.58a). After simplification,

$$k_1 = \frac{c(\phi_1 - \psi_1)\Delta - N_2 c\phi_1 + N_3 c\psi_1}{\Delta} \tag{3.62d}$$

thereby completing the solution of the problem at hand.

3.3.2 *Spherical Linkages*

In this case, the synthesis matrix of Eq. (3.31) becomes of 4×4, while vector $\mathbf{b}$ of the same equation becomes four-dimensional, the synthesis equations thus taking the form

$$\begin{bmatrix} 1 & \cos\psi_1 & \cos\psi_1\cos\phi_1 & -\cos\phi_1 \\ 1 & \cos\psi_2 & \cos\psi_2\cos\phi_2 & -\cos\phi_2 \\ 1 & \cos\psi_3 & \cos\psi_3\cos\phi_3 & -\cos\phi_3 \\ 1 & \cos\psi_4 & \cos\psi_4\cos\phi_4 & -\cos\phi_4 \end{bmatrix} \begin{bmatrix} k_1 \\ k_2 \\ k_3 \\ k_4 \end{bmatrix} = \begin{bmatrix} -\sin\psi_1\sin\phi_1 \\ -\sin\psi_2\sin\phi_2 \\ -\sin\psi_3\sin\phi_3 \\ -\sin\psi_4\sin\phi_4 \end{bmatrix} \tag{3.63}$$

The structure of the synthesis matrix is strikingly similar to that of the planar case, with the entries of its first column being all unity. Hence, similar to the planar case of Sect. 3.3.1, the equations can be reduced by *elementary operations on the synthesis matrix* to a subsystem of three equations in three unknowns. This is readily done upon subtracting the first equation from the remaining three, which is equivalent to subtracting the first row of the synthesis matrix from its remaining three rows, and subtracting the first component of vector $\mathbf{b}$ from its remaining three components, namely,

$$\begin{bmatrix} 1 & c\psi_1 & c\psi_1 c\phi_1 & -c\phi_1 \\ 0 & c\psi_2 - c\psi_1 & c\psi_2 c\phi_2 - c\psi_1 c\phi_1 & -c\phi_2 + c\phi_1 \\ 0 & c\psi_3 - c\psi_1 & c\psi_3 c\phi_3 - c\psi_1 c\phi_1 & -c\phi_3 + c\phi_1 \\ 0 & c\psi_4 - c\psi_1 & c\psi_4 c\phi_4 - c\psi_1 c\phi_1 & -c\phi_4 + c\phi_1 \end{bmatrix} \begin{bmatrix} k_1 \\ k_2 \\ k_3 \\ k_4 \end{bmatrix} = \begin{bmatrix} -s\psi_1 s\phi_1 \\ -s\psi_2 s\phi_2 + s\psi_1 s\phi_1 \\ -s\psi_3 s\phi_3 + s\psi_1 s\phi_1 \\ -s\psi_4 s\phi_4 + s\psi_1 s\phi_1 \end{bmatrix} \tag{3.64}$$

The foregoing system can be now cast in a more suitable form:

$$\begin{bmatrix} 1 & \mathbf{a}^T \\ \mathbf{0}_3 & \mathbf{A}_3 \end{bmatrix} \mathbf{k} = \begin{bmatrix} b_1 \\ \mathbf{b}_3 \end{bmatrix} \tag{3.65a}$$

with blocks defined as

$$\mathbf{a} \equiv \begin{bmatrix} c\psi_1 \\ c\psi_1 c\phi_1 \\ -c\phi_1 \end{bmatrix}, \quad \mathbf{A}_3 \equiv \begin{bmatrix} c\psi_2 - c\psi_1 & c\psi_2 c\phi_2 - c\psi_1 c\phi_1 & -c\phi_2 + c\phi_1 \\ c\psi_3 - c\psi_1 & c\psi_3 c\phi_3 - c\psi_1 c\phi_1 & -c\phi_3 + c\phi_1 \\ c\psi_4 - c\psi_1 & c\psi_4 c\phi_4 - c\psi_1 c\phi_1 & -c\phi_4 + c\phi_1 \end{bmatrix}, \tag{3.65b}$$

$$b_1 \equiv -s\psi_1 s\phi_1, \quad \mathbf{b}_3 \equiv \begin{bmatrix} -s\psi_2 s\phi_2 + s\psi_1 s\phi_1 \\ -s\psi_3 s\phi_3 + s\psi_1 s\phi_1 \\ -s\psi_4 s\phi_4 + s\psi_1 s\phi_1 \end{bmatrix} \tag{3.65c}$$

and $\mathbf{0}_3$ denoting the three-dimensional zero vector. We can thus identify in the above system a reduced system of three equations in three unknowns, that has been decoupled from the original system of four equations, namely,

$$\mathbf{A}_3 \mathbf{k}_3 = \mathbf{b}_3, \quad \mathbf{k}_3 \equiv \begin{bmatrix} k_2 \\ k_3 \\ k_4 \end{bmatrix} \tag{3.66}$$

Further, $\mathbf{A}_3$ is partitioned columnwise as done in Sect. 1.4.3:

$$\mathbf{A}_3 = \begin{bmatrix} \mathbf{c}_1 & \mathbf{c}_2 & \mathbf{c}_3 \end{bmatrix} \tag{3.67}$$

Therefore, the inverse of $\mathbf{A}_3$ can be computed symbolically in this case, by means of *reciprocal bases*, as per Eq. (1.9a):

$$\mathbf{A}_3^{-1} = \frac{1}{\Delta} \begin{bmatrix} (\mathbf{c}_2 \times \mathbf{c}_3)^T \\ (\mathbf{c}_3 \times \mathbf{c}_1)^T \\ (\mathbf{c}_1 \times \mathbf{c}_2)^T \end{bmatrix} \tag{3.68a}$$

where

$$\Delta \equiv \mathbf{c}_1 \times \mathbf{c}_2 \cdot \mathbf{c}_3 \tag{3.68b}$$

and hence,

$$\mathbf{k}_3 = \frac{1}{\Delta} \begin{bmatrix} (\mathbf{c}_2 \times \mathbf{c}_3)^T \mathbf{b}_3 \\ (\mathbf{c}_3 \times \mathbf{c}_1)^T \mathbf{b}_3 \\ (\mathbf{c}_1 \times \mathbf{c}_2)^T \mathbf{b}_3 \end{bmatrix} \tag{3.68c}$$

thereby computing k_2, k_3 and k_4. The remaining unknown, k_1, is computed from the first equation of the array (3.64):

$$k_1 + k_2 c\psi_1 + k_3 c\psi_1 c\phi_1 - k_4 c\phi_1 = -s\psi_1 s\phi_1$$

Therefore,

$$k_1 = -s\psi_1 s\phi_1 - k_2 c\psi_1 - k_3 c\psi_1 c\phi_1 + k_4 c\phi_1 \tag{3.68d}$$

all unknowns having thus been found.

Remark 3.3.1 The foregoing closed-form solution of the exact synthesis problem at hand is apparently elegant and gives some insight into the relations among the variables involved. For example, the problem has no solution when the three columns of $\mathbf{A}_3$, or its three rows for that matter, represent coplanar vectors. Moreover, the numerical evaluation of the Freudenstein parameters is exactly that obtained with *Cramer's rule*, which is notorious for being inefficient and prone to roundoff-error amplification. In our case, all four components of $\mathbf{k}$ have been obtained symbolically, and hence, can be safely evaluated using computer algebra.

Remark 3.3.2 Given that $k_3 = \lambda_1 = \cos\alpha_1$, the computed k_3 cannot be greater than unity in absolute value, and hence, any solution with $|k_3| > 1$ must be rejected. By the same token, $c\alpha_3 = c\alpha_1 c\alpha_2 c\alpha_4 - k_1 s\alpha_2 s\alpha_4$, and hence, the absolute value of the foregoing difference cannot be greater than unity.

3.3.3 *Spatial Linkages*

The two synthesis matrices, $\mathbf{S}$ and $\mathbf{S}_o$, of Eq. (3.48) become now of 4×4, while vectors $\mathbf{b}$ and $\mathbf{b}_o$ of Eq. (3.49) become four-dimensional. As a matter of fact, $\mathbf{S}$, $\mathbf{k}$ and $\mathbf{b}$ are exactly the same as their counterparts in the spherical case, the first synthesis equation of Eq. (3.50) thus being identical to Eq. (3.63).

The second synthesis equation of the same Eq. (3.50) is rewritten below in the standard form in which the left-hand side includes only the terms in the unknowns, namely,

$$\mathbf{S}\mathbf{k}_o = \mathbf{b}_o - \mathbf{S}_o\mathbf{k} \tag{3.69}$$

where $\mathbf{S}$ and $\mathbf{S}_o$ are 4×4 instances of the more general matrices of Eq. (3.48), while $\mathbf{b}$ and $\mathbf{b}_o$ are 4-dimensional instances of their counterparts in Eq. (3.49). It is noteworthy that $\mathbf{k}_o$ is the only unknown in these equations, as the solutions to its primal counterpart $\mathbf{k}$ was already discussed in Sect. 3.3.2.

In the spherical case, the system (3.63) was solved in closed form upon reducing it to a system of three equations in three unknowns, which was done by subtracting the first equation from the other three. Hence, we end up with a new, reduced system of three equations in three unknowns, Eq. (3.66). The reader is invited to obtain the reduced equations that would allow for a closed-form solution of both the primal Freudenstein-parameter vector $\mathbf{k}$ and its dual counterpart $\mathbf{k}_o$.

It is noteworthy that the foregoing computations lead to the solution of a system of three equations in three unknowns, which can be solved symbolically by means of *reciprocal bases*. Remarks 3.3.1 and 3.3.2 apply in this case as well.

3.4 Analysis of the Synthesized Linkage

Once a linkage has been synthesized, its performance should be evaluated, which is done by means of *analysis*. The first step in analyzing a linkage synthesized for function generation is to produce its *Denavit-Hartenberg parameters*, for all we have is its *Freudenstein parameters*. Below we derive analysis algorithms for planar, spherical and spatial four-bar linkages.

3.4.1 Planar Linkages

We start by recalling the inverse relations of Eqs. (3.11), which we reproduce below for quick reference:

$$a_2 = \frac{1}{k_2} a_1, \quad a_4 = \frac{1}{k_3} a_1, \quad a_3 = \sqrt{a_1^2 + a_2^2 + a_4^2 - 2k_1 a_2 a_4} \tag{3.70}$$

Two remarks are in order:

(*i*) The link lengths are given in terms of a_1, which is thus the link length that determines the scale of the linkage, but any other length can be used for the same purpose; and

(*ii*) all lengths are positive. However, negative signs for k_2 and k_3 can occur, that hence can lead to negative values of a_2 or, correspondingly, a_4. As we saw in Sect. 3.3.1, negative values of any of these variables, or of both for that matter, bear a straightforward interpretation.

We now proceed to derive an algorithm for the fast and reliable computation of the output values of ϕ corresponding to a) a given linkage of feasible link lengths $\{a_i\}_1^4$ and b) a given input value ψ. We can do this in several ways. We start by recalling the IO equation of the planar four-bar linkage in homogeneous form, Eq. (3.11):

$$k_1 + k_2 \cos\phi - k_3 \cos\psi - \cos(\phi - \psi) = 0$$

Upon expansion of the fourth term in the left-hand side, the foregoing equation is rewritten as

$$A(\psi)\cos\phi + B(\psi)\sin\phi + C(\psi) = 0 \tag{3.71a}$$

with coefficients defined below:

$$A(\psi) = k_2 - \cos\psi, \quad B(\psi) = -\sin\psi, \quad C(\psi) = k_1 - k_3\cos\psi \tag{3.71b}$$

One approach to solving this equation for ϕ consists in transforming it into an algebraic equation, i.e., a polynomial equation. This is done by means of the tan-half identities, which are recalled below, as applied to angle ϕ:

$$\cos\phi \equiv \frac{1-T^2}{1+T^2}, \quad \sin\phi \equiv \frac{2T}{1+T^2}, \quad T \equiv \tan\left(\frac{\phi}{2}\right) \tag{3.72}$$

Upon substitution of the foregoing identities into Eq. (3.71a), a quadratic equation in T is obtained:

$$D(\psi)T^2 + 2E(\psi)T + F(\psi) = 0 \tag{3.73a}$$

with coefficients

$$D(\psi) \equiv k_1 - k_2 + (1-k_3)\cos\psi \tag{3.73b}$$
$$E(\psi) \equiv -\sin\psi \tag{3.73c}$$
$$F(\psi) \equiv k_1 + k_2 - (1+k_3)\cos\psi \tag{3.73d}$$

Now ϕ can be readily computed once the two roots of Eq. (3.73a) are available. Here, a caveat is in order: during the linkage motion, the three coefficients $D(\psi)$, $E(\psi)$ and $F(\psi)$ vary as ψ does. Now, given that the above-mentioned roots are computed *automatically* using scientific code, the algorithm implementing the computations must account for *pitfalls* brought about by special numerical conditions. Thus, rather than naively using *verbatim* the formula for the roots of the quadratic equation, we follow here a *robust* approach, as suggested by Forsythe (1970): in order to avoid *catastrophic cancelations* when $E^2 >> DF$, that would lead to an erroneous zero root, we first compute the root with the *larger* absolute value, namely,

$$T_1 = \frac{-E - \text{sgn}(E)\sqrt{E^2 - DF}}{D}, \quad \phi_1 = 2\tan^{-1}(T_1) \tag{3.74}$$

where sgn$(\cdot)$ is the *signum function* introduced in Sect. 1.4 when we studied Householder reflections.

Now we distinguish two cases:

1. Assume $B(\psi) \equiv E(\psi) = -\sin\psi \neq 0$. In this case, T_1 is calculated as in Eq. (3.74), while T_2 is calculated upon recalling that the independent term is proportional to the product T_1T_2, and hence,

$$T_2 = \frac{F}{DT_1}, \quad \phi_2 = 2\tan^{-1}(T_2) \tag{3.75a}$$

2. Now, we assume $B(\psi) \equiv E(\psi) = -\sin\psi = 0$. Under this condition, the term linear in T in Eq. (3.73a) vanishes, the two roots being symmetric, namely,

$$T_{1,2} = \pm\sqrt{-\frac{F(\psi)}{D(\psi)}} \tag{3.75b}$$

Thus, cancellations are avoided upon *first* computing T_1; then, T_2 is computed safely because the denominator appearing in Eq. (3.75a) has the largest possible absolute value. However, notice that the quadratic equation can degenerate into a linear equation under two cases: (a) $F(\psi) = 0$ or (b) $D(\psi) = 0$. The first case simply means that one root, T_1, is zero, T_2 being computed from the linear equation derived upon dividing the two sides of that quadratic by T. The second case is a bit more elusive, but it can be handled as the limiting case $D(\psi) \to 0$. To this end, let us divide both sides of Eq. (3.73a) by T^2:

$$D(\psi) + \frac{2E(\psi)}{T} + \frac{F(\psi)}{T^2} = 0$$

Now, upon taking the limit of both sides of the above equation when $D(\psi) \to 0$, we obtain

$$\lim_{D(\psi)\to 0} T \to \infty$$

and hence,

$$\lim_{D(\psi)\to 0} \phi = \pi \tag{3.76}$$

In any event, the two possible solutions of the quadratic equation obtained above lead to one of three possible cases:

1. The two roots $\{T_i\}_1^2$ are real and distinct: the corresponding angles $\{\phi_i\}_1^2$ provide the two *conjugate* postures of the linkage, as shown in Fig. 3.8a. As the linkage moves, the two conjugate postures generate, correspondingly, two *conjugate branches* of the linkage motion;
2. The two roots $\{T_i\}_1^2$ are real and identical: the corresponding single value of $\phi_1 = \phi_2$ indicates the merging of the two branches. This indicates in turn that the output link reached one extreme position, which is known as a *deadpoint*, as illustrated in Fig. 3.8b;
3. The two roots are complex conjugate: this indicates two possibilities:
   (a) The link lengths are unfeasible: they do not define a quadrilateron; or
   (b) The linkage is feasible, but its input link does not move through a full turn, i.e., it is a *rocker*, the given value of ψ lying outside of its range of motion.

Because of the *two conjugate branches* of the planar four-bar linkage, the linkage is said to be *bimodal*.

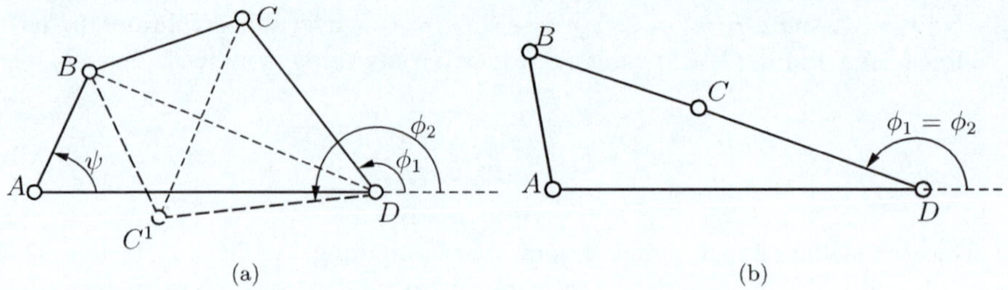

Fig. 3.8 Postures of a four-bar linkage: **a** two distinct, conjugate postures; **b** deadpoint, where the two conjugate postures merge

It is apparent that the quadratic-equation approach to the input-output analysis of the four-bar linkage must be handled with care, especially when writing code to implement it. As an alternative, we can pursue a more geometric, straightforward approach, free of the *singularity* $T \to \infty$ of the transformation (3.72), as described below: We go back to Eq. (3.71a), and rewrite it in a slightly different form

$$\mathcal{L}: \quad A(\psi)u + B(\psi)v + C(\psi) = 0 \tag{3.77a}$$

where

$$u \equiv \cos\phi, \quad v \equiv \sin\phi \tag{3.77b}$$

and hence, u and v are subject to the constraint

$$\mathcal{C}: \quad u^2 + v^2 = 1 \tag{3.77c}$$

The input-output equation thus defines a line $\mathcal{L}$ in the u-v plane, while the constraint (3.77c) defines a unit circle $\mathcal{C}$ centered at the origin of the same plane. The circle is fixed, but the location of the line in the u-v plane depends on both the linkage parameters k_1, k_2, k_3 and the input angle ψ. Therefore, depending upon the linkage at hand and the position of its input link, the line may intersect the circle or not. If it does, then, additionally, the line either intersects the circle at two distinct points or, as a special case, at one single point, in which case the line is tangent to the circle, the two intersection points thus merging into a single one. In the absence of intersections, either the linkage is unfeasible or its input link is a rocker, the given input angle lying outside of its *mobility range*. In the case of two distinct intersections, these determine the two conjugate postures of the linkage. In the case of tangency, the linkage is at a deadpoint. Figure 3.9 depicts the case of two distinct intersection points.

Let the distance of the line $\mathcal{L}$ to the origin be denoted by d. Apparently, we have the three cases below:

$d < 1$: $\mathcal{L}$ intersects $\mathcal{C}$ at two distinct points;
$d = 1$: $\mathcal{L}$ is tangent to $\mathcal{C}$;

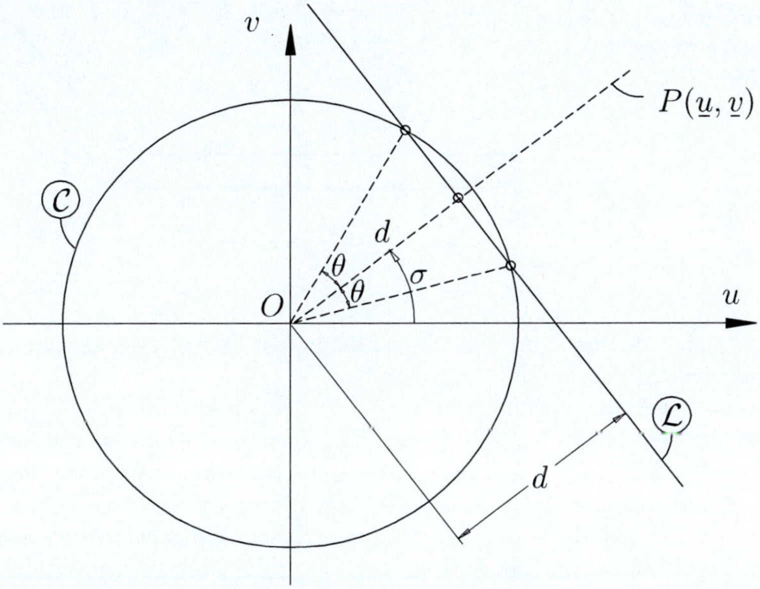

Fig. 3.9 Line and circle in the u-v plane

$d > 1$: $\mathcal{L}$ does not intersect $\mathcal{C}$.

The distance d can be readily found to be

$$d = \frac{|C(\psi)|}{S(\psi)} \tag{3.78a}$$

where $C(\psi)$ was defined in Eq. (3.71b) and

$$S(\psi) \equiv \sqrt{A(\psi)^2 + B(\psi)^2} = \sqrt{(k_2 - \cos\psi)^2 + \sin^2\psi} \tag{3.78b}$$

An interesting *singularity* occurs whereby the foregoing calculations break down: If coefficients $A(\psi)$, $B(\psi)$, and $C(\psi)$ in Eq. (3.77a) all vanish, then the line $\mathcal{L}$ disappears and *any* value of ϕ satisfies the input-output equation for the given value of ψ. The vanishing of these three coefficients is written below:

$$k_2 - \cos\psi = 0 \tag{3.79a}$$

$$\sin\psi = 0 \tag{3.79b}$$

$$k_1 - k_3\cos\psi = 0 \tag{3.79c}$$

The second equation leads to $\psi = 0$ or π. For $\psi = 0$, the first equation yields $k_2 = 1$ and the third equation $k_1 = k_3$. Now, $k_2 = 1$ means $a_2 = a_1$, which, together with $k_1 = k_3$, means $a_4 = a_3$, the result being a set of linkage postures whereby joint

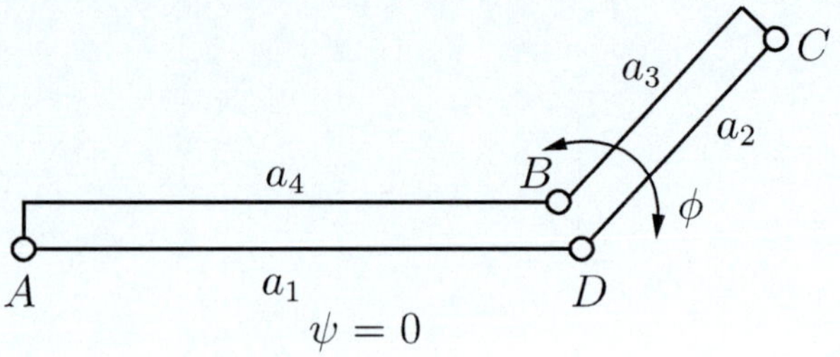

Fig. 3.10 Pathological case of a four-bar linkage with special link lengths

centers B and D coincide, the coupler and the output links thus being free to turn about joint center D as one single rigid-body.

For $\psi = \pi$, the first equation yields $k_2 = -1$, which leads to $a_2 = -a_1$, i.e., a "negative" link length. As we saw in Sect. 3.3.1, a negative a_2 means that the input angle should be measured "all the way down to the extension of the input link," and we fall into the case $\psi = 0$.

Notice that this pathological case, or *singularity*, is not apparent from the quadratic equation. It is illustrated in Fig. 3.10.

Furthermore, in order to compute the two conjugate values ϕ_1 and ϕ_2, we calculate first the intersection of $\mathcal{L}$ with its normal $\mathcal{N}$ from the origin. The intersection point P has coordinates $(\underline{u},\ \underline{v})$, given below:

$$\underline{u} = \frac{C(\psi)(k_2 - \cos\psi)}{S(\psi)^2}, \quad \underline{v} = \frac{-C(\psi)\sin\psi}{S(\psi)^2} \tag{3.80a}$$

Now, the angle σ that $\mathcal{N}$ makes with the u axis, when $d > 0$, and angle θ, half the angle subtended by the chord defined by the intersections of $\mathcal{L}$ with $\mathcal{C}$, are given by

$$\sigma = \arctan\left(\underline{v}/\underline{u}\right) \tag{3.80b}$$

$$\theta = \arccos(d) \tag{3.80c}$$

When $d = 0$, σ cannot be calculated from the above expression, but rather as $\arctan(-1/m)$, where m is the slope of $\mathcal{L}$. Nevertheless, in this case σ is not needed, for the two conjugate values of the output angle can be calculated directly, as shown below:

$$\phi_1 = \sigma + \theta, \quad \phi_2 = \sigma - \theta, \quad \text{for} \quad d > 0 \tag{3.80d}$$

$$\phi_2 = \arctan\left(\frac{k_2 - \cos\psi}{\sin\psi}\right), \quad \phi_1 = \phi_2 + \pi, \quad \text{for} \quad d = 0 \tag{3.80e}$$

We thus have devised an algorithm for computing the two conjugate values of the output angle[3]:

Algorithm `pl4bar-io(k,input)`

```
This algorithm computes the intersection of one line L and
the unit circle centered at the origin of the cos(phi)-sin(phi)
plane. The intersection points, when they exist, are returned
in array out, with out[1] and out[2] denoting the two conjugate
values of the output angle phi
  read k[1], k[2], k[3], input;
  >  pl4bar-io:=proc(k,input) #Use this
  >  procedure only if you are sure that your linkage is feasible

  >  local dpoint,feasible,pathos,D_d,N_d,d,u,v,sigma,theta; global
  >  out;

  >  dpoint:=false; #we assume that we are not in the presence of a
  >  deadpoint

  >  feasible:=true; #we assume that current psi-value is feasible

  >  pathos:=false; #we assume that we are not in the presence of
  pathological case whereby linkage becomes a one-dof open chain. if
  k[2]=1
  and psi=0 then pathos:=true; print(patholo=pathos); return;
  >  fi; #if k[2]=1, then a[1]=a[2] D_d:=k[2]*(k[2]-2*cos(input))+1:
  N_d:=-k[1]+k[3]*cos(input): d:=abs(N_d)/sqrt(D_d): #d is distance
  of line L to origin. if d>1.0 then feasible:=false; print(feas=
  unfeasible); return; fi;
  >  if d=1.0 then dpoint:=true; print(dead=dpoint); theta=0; fi;
  u:=(N_d/D_d^2)*(k[2]-cos(input)): v:=-(N_d/D_d^2)*sin(input):
  #coordinates of intersection of line L and its normal N passing
  through the origin.
  >  sigma:=arctan(v,u): #angle that normal makes with u-axis if
  dpoint=false then theta:=arccos(d); fi; #(1/2)angle subtended by
  secant to circle out[1]:=sigma-theta: out[2]:=sigma+theta:
  >  if v<0 and u>0 then out[1]:=sigma+theta: out[2]:=sigma-theta:
  fi: end proc;
```

Two issues stem from the foregoing discussion, namely,

(a) Linkage feasibility: For the four link lengths to yield a *feasible linkage*, they must define a quadrilateron. The condition on four given side lengths to close a quadrilateron, as given in Sect. 3.3.1, is that every length be smaller than the sum of the remaining three. When four link lengths are given as candidates to define a planar four-bar linkage, these lengths must first and foremost be capable of defining a quadrilateron. If they do, the lengths are said to be *feasible*; otherwise, they are *unfeasible*.

(b) Link mobility: A link may or may not be capable of a full turn; if capable, it is called a *crank*; otherwise, a *rocker*. This gives rise to various types of linkages, depending on the type of its input and output links, namely, *double crank*, *crank-rocker*, *rocker-crank*, or *double rocker*. Double crank linkages are

[3] in Maple code.

known as *drag-link* mechanisms. This variety of linkage type leads, in turn, to what is known as *Grashof mechanisms*.

A major fundamental result in linkage theory is the *Grashof classification* of planar four-bar linkages. This classification looks at the mobility of three links with respect to the remaining one. Obviously, which of the four links is considered the "remaining one" is immaterial. A linkage is termed *Grashof* if *at least one* of its links is capable of a full turn with respect to any other link. Otherwise, the linkage is termed *non-Grashof*. Now we have the main result—for a proof, see, e.g., Waldron and Kinzel (1999)—below:

> A planar four-bar linkage is Grashof if and only if the sum of the lengths of its shortest and longest links is smaller than or equal to the sum of the two other link lengths.

In linkage synthesis, we are interested in meeting mobility conditions either on the input or on the output links, or even on both. We derive below these conditions in terms of the Freudenstein parameters.

Mobility of the Input and Output Links

The condition under which the input link is a crank is quite useful because four-bar linkages are frequently driven at a constant angular velocity, and hence, the input link might as well be capable of a full turn. To find this condition, we recall Eq. (3.73a), whose discriminant is a function not only of constants k_1, k_2 and k_3, but also of ψ and, hence, it is not only linkage- but also posture-dependent. In the discussion below, we assume that the linkage parameters are fixed, and hence, the *linkage discriminant* $\Delta(\psi)$, stemming from Eq. (3.73a), will be regarded as a function of ψ only. This is given by

$$\Delta(\psi) \equiv E^2(\psi) - D(\psi)F(\psi) \tag{3.81a}$$

Upon expansion, the above discriminant becomes

$$\Delta(\psi) \equiv -k_3^2\cos^2\psi + 2(k_1k_3 - k_2)\cos\psi + (1 - k_1^2 + k_2^2) \tag{3.81b}$$

which is, clearly, a parabola in $\cos\psi$ with concavity downward—its second derivative w.r.t. $\cos\psi$ is $-2k_3^2$. Moreover, since $\Delta(\psi)$ is a function only of $\cos\psi$, which is an *even function* of ψ, $\Delta(\psi)$ is also an even function of ψ, i.e.,

$$\Delta(\psi) = \Delta(-\psi)$$

For the input link to be a crank, then, the discriminant $\Delta(\psi)$ should attain nonnegative values in the range $-1 \le \cos\psi \le +1$. Moreover, by virtue of the parabolic shape of the $\Delta(\cos\psi)$ vs. $\cos\psi$ plot, Δ is nonnegative for *any* value of ψ if and only if $\Delta(\cos\psi) \ge 0$ when $\cos\psi = \pm 1$. It is noteworthy that, even thought $\Delta(\psi)$ is an even function of ψ, $\Delta(\cos\psi)$ is not; in general, then, $\Delta(\cos\psi)$ attains distinct values at $\cos\psi = -1$ and at $\cos\psi = +1$. Let Δ_+ and Δ_- denote the values that Δ attains when $\cos\psi$ equals $+1$ and -1, respectively, which are given below:

$$\Delta_+ = -k_3^2 + 2(k_1k_3 - k_2) + 1 - k_1^2 + k_2^2, \quad \Delta_- = -k_3^2 - 2(k_1k_3 - k_2) + 1 - k_1^2 + k_2^2$$

The necessary and sufficient conditions for a nonnegative linkage discriminant, for any value of ψ, are now derived. Note first that Δ_+ and Δ_- can be expressed as differences of squares, namely,

$$\Delta_+ = (1 - k_2)^2 - (k_1 - k_3)^2 \,, \quad \Delta_- = (1 + k_2)^2 - (k_1 + k_3)^2$$

Apparently, Δ_+ and Δ_- are nonnegative if and only if the relations below hold:

$$(k_1 - k_3)^2 - (1 - k_2)^2 \le 0 \;\text{ and }\; (k_1 + k_3)^2 - (1 + k_2)^2 \le 0 \tag{3.82}$$

Upon expressing left-hand sides of the foregoing inequalities as the products of conjugate binomials, the inequalities become

$$(k_1 - k_3 - 1 + k_2)(k_1 - k_3 + 1 - k_2) \le 0 \tag{3.83a}$$

$$(k_1 + k_3 - 1 - k_2)(k_1 + k_3 + 1 + k_2) \le 0 \tag{3.83b}$$

Each of the above inequalities holds if its two left-hand side factors have opposite signs, the first inequality thus leading to

$$k_1 - k_3 - 1 + k_2 \ge 0 \;\;\&\;\; k_1 - k_3 + 1 - k_2 \le 0 \tag{3.83c}$$

or

$$k_1 - k_3 - 1 + k_2 \le 0 \;\;\&\;\; k_1 - k_3 + 1 - k_2 \ge 0 \tag{3.83d}$$

The second inequality, likewise, leads to

$$k_1 + k_3 - 1 - k_2 \ge 0 \;\;\&\;\; k_1 + k_3 + 1 + k_2 \le 0 \tag{3.83e}$$

or

$$k_1 + k_3 - 1 - k_2 \le 0 \;\;\&\;\; k_1 + k_3 + 1 + k_2 \ge 0 \tag{3.83f}$$

Thus, the region $\mathcal{K}$ of the **k**-space that contains input cranks is the *intersection* of two subregions, that defined by the four inequalities (3.83c and 3.83d) and that defined by their counterparts (3.83e and 3.83f). Moreover, the subregion represented by each quadruplet is the *union* of the intersections of the regions defined by each pair of linear inequalities. Let $\mathcal{C}_1$ and $\mathcal{C}_2$ be the subregions defined by the first and the second inequalities (3.83c), with similar definitions for subregions $\mathcal{D}_1$, $\mathcal{D}_2$, $\mathcal{E}_1$, $\mathcal{E}_2$, $\mathcal{F}_1$ and $\mathcal{F}_2$. Let, further, $\mathcal{R}_1$ denote the subregion defined by the four inequalities (3.83c) and (3.83d), and hence,

$$\mathcal{R}_1 = (\mathcal{C}_1 \cap \mathcal{C}_2) \cup (\mathcal{D}_1 \cap \mathcal{D}_2) \tag{3.84a}$$

with a similar definition for the subregion $\mathcal{R}_2$, stemming from the four inequalities (3.83e) and (3.83f):

$$\mathcal{R}_2 = (\mathcal{E}_1 \cap \mathcal{E}_2) \cup (\mathcal{F}_1 \cap \mathcal{F}_2) \tag{3.84b}$$

Therefore, the region $\mathcal{R}_I$ containing input cranks is

$$\mathcal{R}_I = (\mathcal{R}_1 \cap \mathcal{R}_2) \tag{3.85}$$

Each inequality, furthermore, divides $\mathcal{K}$ into two halves, one on each side of the plane obtained when turning the inequality sign of each relation into an equality sign. As the reader can readily realize, inequalities (3.83c and 3.83d) lead to the same pair of planes; likewise inequalities (3.83e and 3.83f) lead to a second pair of planes, namely,

$$k_1 + k_2 - k_3 - 1 = 0\,, \quad k_1 - k_2 - k_3 + 1 = 0$$
$$k_1 - k_2 + k_3 - 1 = 0\,, \quad k_1 + k_2 + k_3 + 1 = 0$$

In summary, the two original quadratic inequalities of Eq. (3.82) represent a region of $\mathcal{K}$ bounded by four planes, as displayed in Fig. 3.11. Hence, the region containing input cranks comprises a regular tetrahedron with its centroid located at the origin of the above space, and two *open convexes*. Thus, all points within that region represent linkages whose input link is a crank.

It is noteworthy that the k_1-axis represents linkages for which $a_2,\ a_4 \to \infty$, i.e., the k_1-axis represents, actually, all PRRP linkages. However, as the reader is invited to verify, the origin does not represent a feasible linkage.

Now, in order to analyze the mobility of the output link, we simply exchange the roles of k_2 and k_3 in the foregoing results, which is apparent from the definitions of the linkage parameters $\{k_i\}_1^3$, as given in Eqs. (3.70) and the Freudenstein equation (3.10). Actually, then, the region containing output cranks can be obtained by mapping that containing input cranks by means of a linear transformation:

$$k_1 = k_1, \quad k_2 = -k_3, \quad k_3 = -k_2$$

The above transformation can be represented in matrix form as a *reflection* $\mathbf{R}$ about a plane of unit normal $[0,\ \sqrt{2}/2,\ \sqrt{2}/2]^T$, given by Angeles (2014)

$$\mathbf{R} = \begin{bmatrix} 1 & 0 & 0 \\ 0 & 0 & -1 \\ 0 & -1 & 0 \end{bmatrix}$$

As the reader can readily verify, $\mathbf{R}\mathbf{R}^T = \mathbf{1}$, with $\mathbf{1}$ denoting the 3×3 identity matrix and $\det(\mathbf{R}) = -1$, which shows that $\mathbf{R}$ is indeed a reflection.

By means of the foregoing exchange in Eqs. (3.83a) and (3.83b), the inequalities leading to an output crank are obtained as

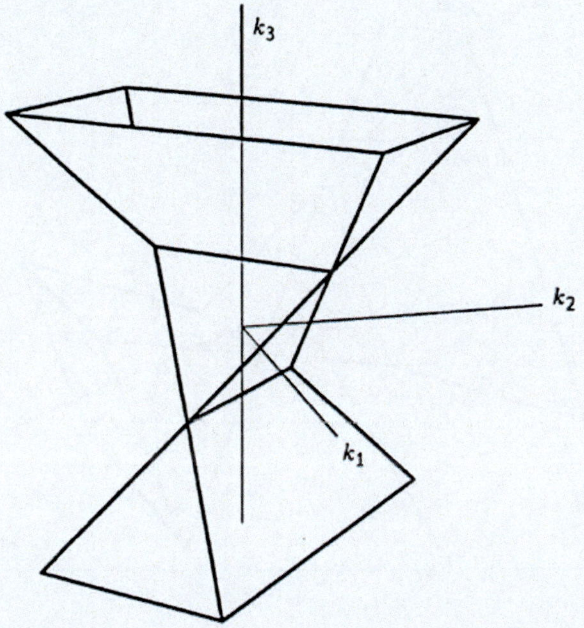

Fig. 3.11 Region comprising planar four-bar linkages with an input crank

$$(k_1 - k_2)^2 < (1 - k_3)^2 \tag{3.86a}$$

$$(k_1 + k_2)^2 < (1 + k_3)^2 \tag{3.86b}$$

The mobility region $\mathcal{R}_O$ represented by the two foregoing inequalities comprises all four-bar linkages with an *output crank*. This region is, then, the mirror image of that of Fig. 3.11 when reflected about a plane Π passing through the k_1-axis and intersecting the k_2-k_3 plane along a line passing through the origin and contained in the third and fourth quadrants of this plane. The foregoing region is represented in Fig. 3.12. Notice that this region comprises a tetrahedron identical to that of Fig. 3.11 and, hence, the tetrahedron is common to both mobility regions. Consequently, any point within this tetrahedron represents a double-crank four-bar linkage, except for the origin, of course.

Furthermore, the central tetrahedron of Figs. 3.11 and 3.12 can be shown to have axes of length $2\sqrt{2}$.

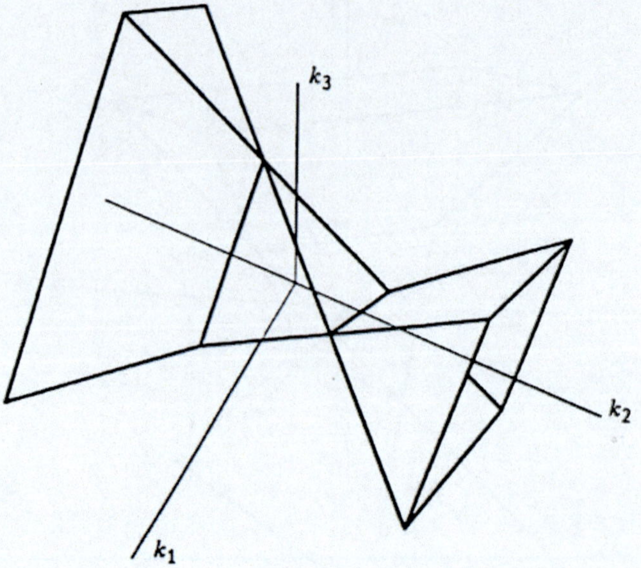

Fig. 3.12 Region comprising four-bar linkages with an output crank

All linkages outside of the two foregoing regions are either of the double-rocker type or unfeasible.

We have finished the mobility analysis for input and output links. The analysis for the coupler link, not covered in this book, is available in literature (Bai, 2017).

3.4.2 *Spherical Four-Bar Linkages*

The analysis of the spherical four-bar linkage parallels that of its planar counterpart. Indeed, upon introduction of the tan-half identities of Eq. (3.72) into the IO equation of the spherical linkage, Eq. (3.27), we obtain, again, a quadratic equation in T of the form of Eq. (3.73a), namely,

$$D(\psi)T^2 + 2E(\psi)T + F(\psi) = 0 \tag{3.87a}$$

but now with the coefficients

$$D(\psi) \equiv k_1 + (k_2 - k_3)\cos\psi + k_4 \tag{3.87b}$$

$$E(\psi) \equiv \sin\psi \tag{3.87c}$$

$$F(\psi) \equiv k_1 + (k_2 + k_3)\cos\psi - k_4 \tag{3.87d}$$

Similar to the planar case, rather than attempting a solution of the quadratic equation as such, we cast the input-output equation (3.27) in the same form as we did for the planar case:

$$\mathcal{L}: \quad A(\psi)u + B(\psi)v + C(\psi) = 0 \tag{3.88a}$$

with coefficients given below:

$$A(\psi) = k_3\cos\psi - k_4, \quad B(\psi) = \sin\psi, \quad C(\psi) = k_1 + k_2\cos\psi \tag{3.88b}$$

which is the equation of a line $\mathcal{L}$ in the u-v plane, with u and v defined as

$$u \equiv \cos\phi, \quad v \equiv \sin\phi \tag{3.88c}$$

and hence, these variables are subject to the constraint

$$\mathcal{C}: \quad u^2 + v^2 = 1 \tag{3.88d}$$

The two conjugate values of ϕ for a given value of ψ can thus be computed as the intersection of the line $\mathcal{L}$ with the circle $\mathcal{C}$, in exactly the same way as in the planar case. As in the planar case, an interesting singularity occurs when coefficients $A(\psi)$, $B(\psi)$, and $C(\psi)$ of the line equation (3.88a) all vanish, i.e., if

$$k_3\cos\psi - k_4 = 0, \quad \sin\psi = 0, \quad k_1 + k_2\cos\psi = 0 \tag{3.89}$$

The second of the foregoing equations leads to $\psi = 0$ or π. If $\psi = 0$, then the first equation implies $k_3 = k_4$, and hence,

$$\cos\alpha_1\sin\alpha_2 - \sin\alpha_1\cos\alpha_2 = 0$$

i.e.,

$$\alpha_2 = \alpha_1 \quad \text{or} \quad \alpha_2 = \alpha_1 + \pi$$

If $\alpha_2 = \alpha_1$, then the third equation in (3.89) leads to $\cos\alpha_4 = \cos\alpha_3$, and hence, $\alpha_4 = \pm\alpha_3$. If $\psi = \pi$, a similar reasoning to that introduced for the planar case leads exactly to the same result as for $\psi = 0$.

As a consequence, then, the singularity under study leads to a set of postures of the spherical linkage under which the joint axes OA and OD coincide, the coupler and the output links then being free to move as a single rigid body.

A procedure[4] that produces the input-output analysis of spherical four-bar linkages is included below.

Input-Output Analysis of Spherical Four-Bar Linkages

This procedure is based on the *graphical* solution of the input-output *equation for spherical mechanisms*: the intersection of a line $\mathcal{L}$, describing the underlying linear relation between the cosine and the sine of the output angle, and the unit circle $\mathcal{C}$ centered at the origin of the cos(output)-sin(output) plane, is determined. Line $\mathcal{L}$ depends on the linkage parameters, $k[1], k[2], k[3]$ and $k[4]$, which are entered into the procedure as array **k**, as well as on the input-angle value, which is entered as variable "input." If $\mathcal{L}$ does not intersect $\mathcal{C}$, then either the linkage is unfeasible in that its link lengths do not form a quadrilateron, or the input link is a rocker and the prescribed value of the input angle lies outside its range of motion. If $\mathcal{L}$ intersects $\mathcal{C}$, then it does normally so at two points, the procedure returning the two conjugate values of the output angle as out[1] and out[2]. In case $\mathcal{L}$ is tangent to $\mathcal{C}$, the procedure detects a deadpoint, and returns out[2] = out[1].

```
> restart:

> with(plots): with(plottools):

> sph4barIO:= proc( k, input )#Computes the intersection of one
line L
> and the unit circle centered at the origin of the
> cos(output)-sin(output) plane. The intersection points, when
they
> exist, are returned in array out, with out[1] and out[2]
denoting
> the two conjugate values of the output angle

> local dpoint, feasible, pathos, D_d, N_d, d, q, u, v, sigma,
theta;

> global out;

> dpoint:=false;#we assume that we are not in the presence of a
> deadpoint

> feasible:=true;#we assume that current psi-value is feasible
```

[4] Written in Maple.

```
>  pathos:=false;#we assume that we are not in the presence of
>  pathological case whereby linkage becomes a one-dof open chain

>  if k[3]=k[4] and input=0 then pathos:=true; print(patholo=
pathos);
>  return; fi; #Note that, if k[3] = k[4], then alpha[1]=
alpha[2]

>  D_d:=(k[3]*cos(input)-k[4])^2 + (sin(input))^2: #Maple
reserves D
>  for differential operator!

>  N_d:=-k[1] - k[2]*cos(input): q:= N_d/D_d: print(Num, Den =
N_d,
>  D_d);

>  d:=abs(N_d)/sqrt(D_d): print(distance = d);#distance of line L
to
>  origin

>  if d > 1.0 then feasible:=false; print(feas=unfeasible);
return; fi;

>  if d = 1.0 then dpoint:=true; print(dead=dpoint);  fi;

>  u:=q*(k[3]*cos(input)-k[4]): v:=q*sin(input): print(inters = u,

>  v);#coordinates of intersection of line L and its normal N
passing
>  through the origin

>  if d = 0 then sigma:=arctan(-sin(input), k[3]*cos(input)-k[4])
else
>  sigma:=arctan(v,u); fi: print(sig=sigma);#angle that normal
makes
>  with u-axis

>  theta:=arccos(d); print(th=theta); #(1/2) angle subtended by
secant
>  to circle

>  if d=0 then out[2]:=sigma; out[1]:=evalf(out[2]+Pi) else
>  out[1]:=sigma-theta; out[2]:=sigma+theta end if; print(out1=
out[1]);
>  print(out2=out[2]);
>  end;
```

Mobility of the Input and Output Links

This analysis is conducted in the space of the four Freudenstein parameters $\{ k_i \}_1^4$, with results similar to the planar case. Obviously, in this case the visualization is more challenging.

3.4.3 Spatial Four-Bar Linkages

The analysis of the spatial four-bar linkage, instantiated in Fig. 3.5, parallels that of its planar and spherical counterparts. There are, however, a few remarkable differences, as described below.

For starters, we cast the input-output equation (3.36) in the *dual form*

$$\hat{A}\hat{u} + \hat{B}\hat{v} + \hat{C} = 0 \tag{3.90a}$$

where

$$\hat{u} = u - \epsilon d_1 v, \quad \hat{v} = v + \epsilon d_1 u, \quad u \equiv \cos\phi, \quad v \equiv \sin\phi \tag{3.90b}$$

with d_1 denoting the translation of the output cylindrical pair, while ϵ is the dual unit, which, we recall, has the properties $\epsilon \neq 0$ and $\epsilon^2 = 0$. Moreover,

$$\hat{A} = A(\psi) + \epsilon A_o(\psi), \quad \hat{B} = B(\psi) + \epsilon B_o(\psi), \quad \hat{C} = C(\psi) + \epsilon C_o(\psi) \tag{3.90c}$$

whose primal parts $A(\psi)$, $B(\psi)$ and $C(\psi)$ are identical to those of the spherical linkage, as displayed in Eqs. (3.88b), their dual parts $A_o(\psi)$, $B_o(\psi)$ and $C_o(\psi)$ being obtained with the aid of computer algebra and the rules of operations with dual numbers, namely,

$$A_o = k_{3o} c\psi - k_3 d_2 s\psi - k_{4o} \tag{3.91a}$$

$$B_o = s\psi + d_2 c\psi \tag{3.91b}$$

$$C_o = k_{1o} + k_{2o} c\psi - k_2 d_2 s\psi \tag{3.91c}$$

in which the Freudenstein parameters are now dual numbers: $\hat{k}_i = k_i + \epsilon k_{io}$, for $i = 1, \ldots, 4$. Moreover, their primal part is identical to that of the spherical four-bar linkage, their dual parts being displayed in Eqs. (3.42) and reproduced below for quick reference.

$$k_{o1} = -\frac{a_1\lambda_2\lambda_4\mu_1\mu_2\mu_4 + a_2(\lambda_1\lambda_4 - \lambda_2\lambda_3)\mu_4 - a_3\mu_2\mu_3\mu_4 + a_4(\lambda_1\lambda_2 - \lambda_3\lambda_4)\mu_2}{\mu_2^2\mu_4^2}$$

$$k_{o2} = \frac{a_1\lambda_1\lambda_4\mu_4 - a_4\mu_1}{\mu_4^2}, \quad k_{o3} = -a_1\mu_1, \quad k_{o4} = \frac{a_1\lambda_1\lambda_2\mu_2 - a_2\mu_1}{\mu_2^2}$$

Once we have obtained the input-output equation in terms of dual angles, it is possible to analyze the RCCC linkage, which allows us, in turn, to compute all the joint rotations and translations. The input-output equation (3.90a) can be generally written as

$$\hat{\mathcal{L}}: \quad \hat{A}\hat{u} + \hat{B}\hat{v} + \hat{C} = 0 \tag{3.92a}$$

and

$$\hat{\mathcal{C}}: \quad \hat{u}^2 + \hat{v}^2 = 1 \tag{3.92b}$$

where

$$\hat{u} = \cos\hat{\phi}, \quad \hat{v} = \sin\hat{\phi} \tag{3.92c}$$

Equations (3.92a–3.92c) represent a *dual line* $\hat{\mathcal{L}}$ and a *dual unit circle* $\hat{\mathcal{C}}$ in the dual $\hat{u}$-$\hat{v}$ plane, respectively. Now, it is possible to decompose the equation of the "line" $\hat{\mathcal{L}}$ into two real equations, one for its primal, and one for its dual part, namely,

$$\mathcal{P}: \quad Au + Bv + C = 0 \tag{3.93a}$$

$$\mathcal{H}: \quad (A_o + Bd_1)u - Ad_1 v + C_o = 0 \tag{3.93b}$$

For the circle $\hat{\mathcal{C}}$, the dual part vanishes identically, the primal part leading to a *real circle*, namely,

$$\mathcal{C}: \quad u^2 + v^2 = 1 \tag{3.93c}$$

Equation (3.93a) represents a plane $\mathcal{P}$ parallel to the d_1-axis in the (u, v, d_1)-space, while Eq. (3.93b) represents a hyperbolic paraboloid $\mathcal{H}$ in the same space. Moreover, Eq. (3.93c) represents a cylinder $\mathcal{C}$ of unit radius and axis parallel to the d_1-axis, all foregoing items shown in Fig. 3.13a, b.

The three-dimensional interpretation of Eqs. (3.93a)–(3.93c) is illustrated in Fig. 3.13a, b, whereby line $\mathcal{L}_i$, for $i = 1, 2$, is defined by the intersection of the plane of Eq. (3.93a) with the cylinder (3.93c). Moreover, each line $\mathcal{L}_i$ intersects the paraboloid (3.93b) at one single point, as illustrated in Fig. 3.13b, and as made apparent below.

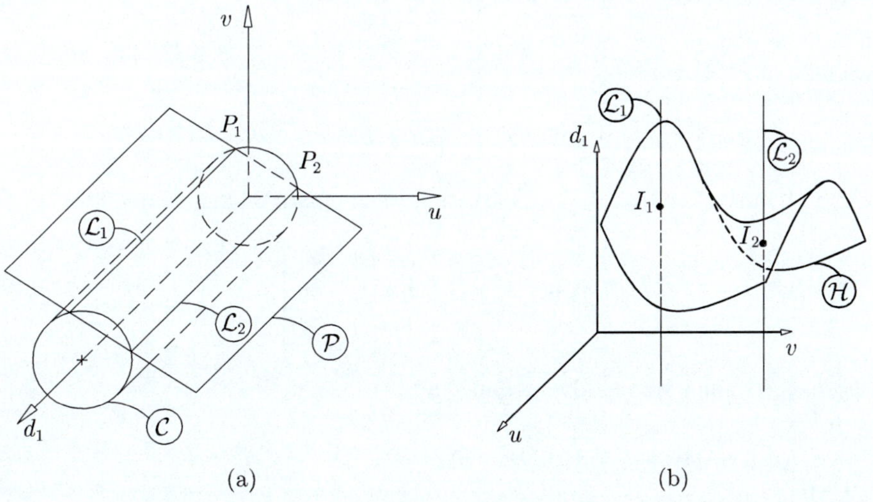

Fig. 3.13 Intersections of **a** $\mathcal{P}$ with $\mathcal{C}$; and **b** $\mathcal{L}_i$ with $\mathcal{H}$, for $i = 1, 2$

The system of equations (3.93a–3.93c) should be solved for u, v and d_1 in order to calculate the two conjugate output angles and their corresponding output translations. The intersections $\mathcal{L}_1$ and $\mathcal{L}_2$ of the plane $\mathcal{P}$ and the cylinder intersect the u-v plane at points P_1 and P_2, as shown in Fig. 3.13a, while $\mathcal{L}_1$ and $\mathcal{L}_2$ intersect the hyperbolic paraboloid $\mathcal{H}$ at points I_1 and I_2, as depicted in Fig. 3.13b. The intersection points P_1 and P_2 thus yield the two conjugate output angles ϕ_1 and ϕ_2. Once the two conjugate solutions u and v are known, via the coordinates of P_1 and P_2, the unique value of d_1 corresponding to each solution, and defining the intersection points I_1 and I_2, is determined from Eq. (3.93b), namely,

$$d_1(\psi) = \frac{A_o u + C_o}{Av - Bu}, \quad Av \neq Bu \tag{3.94}$$

where we have dispensed with the argument ψ in coefficients A, A_o, B and C_o for simplicity.

Notice that the denominator of Eq. (3.94) vanishes if $Av = Bu$; then, as can be readily verified, the numerator of d_1 in the above expression vanishes as well, and d_1 is indeterminate. In this case, the surface $\mathcal{H}$ disappears for all values of the output translations d_1 and we are left with the plane $\mathcal{P}$ and the cylinder $\mathcal{C}$, which means that d_1 is free to take any value. That is, the motion of this linkage in the plane normal to its joint axes is independent of the translations along these axes. We are here in the presence of a parametric singularity producing a degeneracy of the linkage, similar to those described for the planar and spherical linkages in Sects. 3.4.1 and 3.4.2. Under this singularity, all joint axes are parallel ($\alpha_i = 0, i = 1, \ldots, 4$) and, hence, the coupler and the output links can freely slide along their cylindrical-joint axes.

Canonical Equation of the Hyperbolic Paraboloid $\mathcal{H}$

In order to gain insight into the problem geometry, we derive below the *canonical equation* of $\mathcal{H}$. To this end, we let

$$\mathbf{x} \equiv [\, u \;\; v \;\; d_1 \,]^T, \quad Q(\mathbf{x}) \equiv A_o u + B d_1 u - A d_1 v + C_o = 0$$

where $Q(\mathbf{x})$ is taken from Eq. (3.93b), its Hessian matrix $\mathbf{H}$ then being

$$\mathbf{H} \equiv \frac{\partial^2 Q}{\partial \mathbf{x}^2} = \begin{bmatrix} 0 & 0 & B \\ 0 & 0 & -A \\ B & -A & 0 \end{bmatrix} \tag{3.95}$$

whose eigenvalues are readily computed as

$$\lambda_1 = -\sqrt{A^2 + B^2}, \quad \lambda_2 = 0\,, \quad \lambda_3 = \sqrt{A^2 + B^2}$$

The corresponding non-normalized eigenvectors $\mathbf{e}_i$, for $i = 1, 2, 3$, are

$$\mathbf{e}_1 = \begin{bmatrix} B \\ -A \\ \sqrt{A^2+B^2} \end{bmatrix}, \quad \mathbf{e}_2 = \begin{bmatrix} A \\ B \\ 0 \end{bmatrix}, \quad \mathbf{e}_3 = \begin{bmatrix} -B \\ A \\ \sqrt{A^2+B^2} \end{bmatrix}$$

and hence, the canonical equation of the surface $\mathcal{H}$ is of the form:

$$\zeta = \frac{\xi^2}{K} - \frac{\eta^2}{K}, \quad K = \frac{2A_oA}{A^2+B^2}$$

where

$$\xi = \frac{-\sqrt{2}}{2\sqrt{A^2+B^2}}\left(Bu + Av + d_1 + \frac{A_oB}{4A_oA}\right)$$

$$\eta = \frac{\sqrt{2}}{4\sqrt{A^2+B^2}}\left(Bu - Av + d_1 + \frac{A_oB}{A_oA}\right)$$

$$\zeta = \frac{1}{\sqrt{A^2+B^2}}\left[Au + Bv + \frac{(A^2+B^2)C_oA}{A_oA}\right]$$

which proves that $\mathcal{H}$ is indeed a hyperbolic paraboloid.

The Case of d_1 Acting as Input

We include here a case that has been overlooked in the literature. In this case we regard the translational displacement of the output C joint of a RCCC linkage as input, the two outputs being angles ψ and ϕ. This problem can occur when designing an analogue temperature sensor in which temperature is measured via the expansion of a rod whose axis is offset with respect to the axis of a dial, the two axes lying at an arbitrary angle.

The problem no longer leads to a quadratic equation, but rather to a system of one quartic and one quadratic equation in two variables, as described presently.

Equations (3.93a and 3.93b) are both linear in u and v, which allows us to solve for these variables in terms of d_1, namely,

$$u = u(p,q) = \frac{-BC_o - CAd_1}{BA_o + B^2d_1 + A^2d_1} \tag{3.96a}$$

$$v = v(p,q) = \frac{-CA_o - AC_o + CBd_1}{BA_o + B^2d_1 + A^2d_1} \tag{3.96b}$$

where, in light of Eqs. (3.91a), with $p = \cos\psi$ and $q = \sin\psi$, u and v become functions of p and q. The latter, moreover, are subject to

$$p^2 + q^2 = 1 \tag{3.97}$$

Substituting the values of u and v given above into Eq. (3.93c) produces an equation free of u and v or, correspondingly, free of ϕ, namely,

Table 3.2 D-H parameters of a RCCC mechanism

Link	1	2	3	4
a_i [in]	5	2	4	3
α_i [deg]	60	30	55	45
d_i [in]	0	Variable	Variable	Variable

$$f(p, q) = 0 \tag{3.98}$$

From Eq. (3.71b) and Eqs. (3.91a)–(3.91c), both u and v, as given by Eqs. (3.96a) and (3.96b), are rational functions in these variables, with both numerator and denominator quadratic in p and q. Hence, u^2 and v^2 are rational functions with both numerator and denominator quartic in p and q. Therefore, $f(p, q) = 0$ leads, after clearing denominators, to a quartic equation in p and q.

The system of polynomial equations (3.97) and (3.98) apparently has a Bezout number, introduced in Sect. 1.3, of $4 \times 2 = 8$.

Numerical Examples

The foregoing procedure is validated with two numerical examples. All numerical and symbolic calculations were conducted with the aid of computer algebra.

Example 1: The Yang-and-Freudenstein Linkage

The first example is taken from Yang and Freudenstein (1964), with the data given in Table 3.2. The output displacements, which vary with the input angle, are recorded in Table 3.3. For conciseness, we list only the results for $0 \le \psi \le \pi$. Our results match those reported by Yang and Freudenstein, considering the difference of input and output angles in both works, as explained in Sect. 3.2.4. It is noteworthy that only two displacement equations need to be solved in our method, a gain in simplicity when compared with the system of six equations in six unknowns formulated by Yang and Freudenstein, within a purely numerical approach.

Example 2: Prescribing d_1 as Input

In the second example, we try to find the rotations, ψ and ϕ, for a given d_1, and the given dimensions of a RCCC linkage. The dimensions are the same as those in Example 1, with $d_1 = 1.0$. In this example, Eq. (3.98) takes the form:

$$A_0 p^4 + A_1(q)p^3 + A_2(q)p^2 + A_3(q)p + A_4(q) = 0 \tag{3.99}$$

whose coefficients $A_i(q)$, for $i = 0, \ldots, 4$, are given below:

$$\begin{aligned}
A_0 &= 0.09209746694\\
A_1(q) &= -0.06765823468q - 0.0073324502\\
A_2(q) &= -0.1754806581q^2 + 0.01487658368q - 0.1902460942\\
A_3(q) &= 0.1353164694q^3 + 0.1202907568q^2 + 0.2424947249q + 0.04203177757\\
A_4(q) &= -0.015625q^4 - 0.0811898817q^3 - 0.020697377q^2 - 0.1362382267q\\
&\quad + 0.0484753242
\end{aligned}$$

Table 3.3 RCCC displacements

	Branch 1		Branch 2	
ψ[deg]	ϕ[deg]	d_1 [in]	ϕ[deg]	d_1 [in]
0	83.70015289	−0.1731633183	−83.70015289	0.1731633183
20	68.59658457	0.01107737578	−105.3298310	0.8429100445
40	64.21379652	−0.5291731100	235.9479009	1.085719194
60	67.55907283	−1.262205018	223.0109192	0.9378806915
80	75.72376603	−1.888758476	214.5328380	0.6631677103
100	87.21970033	−2.259417488	209.1315343	0.3676536240
120	101.1949772	−2.248309766	206.1460158	0.08437533590
140	116.6745934	−1.770565950	205.6297490	−0.1502382358
160	131.8997404	−0.9205435228	208.4003706	−0.2203697101
180	144.2093802	−0.1150813726	215.7906198	0.1150813650

Table 3.4 Possible values of ψ and ϕ

	$[p, q]$	ψ[deg]	ϕ[deg]
1	[0.6047587377, −.7964087325]	−52.78	{−65.68, −227.07}
2	[−.9289796338, −.3701308418]	−158.27	{−130.66, −207.99}
3	[0.5819053587, 0.8132565115]	54.41	{66.04, 226.10}
4	[0.8869350365, 0.4618941881]	27.50	{65.79, −113.02}

Equation (3.99) represents a curve in the p-q plane, whose intersections with the circle of Eq. (3.97) yield all real roots of the system at hand. Notice that all such roots are bound to lie in the above circle. The four real solutions of the foregoing system are given by the four intersections depicted in Fig. 3.14. The solutions are listed in Table 3.4, including the corresponding angles of rotation.[5]

Mobility of the Input and Output Links

In this case, the mobility analysis applies *only to the input* ψ *and the output* ϕ, as this analysis allows us to decide whether a joint is fully rotatable—can sweep an angle of 2π—or not. This analysis thus reduces to that of the spherical mechanism whose IO equation is the primal part of the dual equation of this linkage.

[5] In this table only p and q are given with 10 digits; all other values are given with only four, for the sake of economy of space.

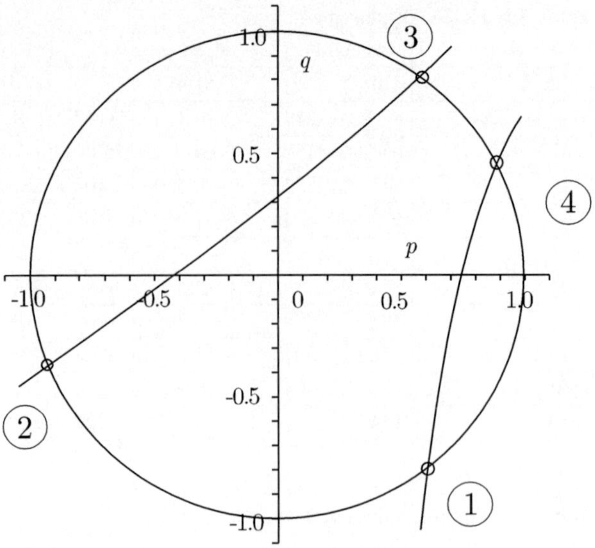

Fig. 3.14 The case of an input translation

3.5 Approximate Synthesis

Regardless of the type of four-bar linkage, $\mathbf{k}$ ($\hat{\mathbf{k}}$) is a n-dimensional real (dual) vector; for planar linkages, $n = 3$, for spherical, $n = 4$. For spatial linkages, n is still four, but now, the parameters being dual, the total number of real parameters is eight. In general, for m input-output pairs—in the spatial case, the output is dual!—to meet with $m > 3$ in the planar case, and $m > 4$ in the spherical and spatial cases, no set of Freudenstein parameters can verify all synthesis equations, thereby ending up with an error vector $\mathbf{e}$:

$$\mathbf{e} \equiv \mathbf{b} - \mathbf{S}\mathbf{k} \tag{3.100a}$$

which, in the case of RCCC linkages, becomes dual, i.e.,

$$\hat{\mathbf{e}} \equiv \hat{\mathbf{b}} - \hat{\mathbf{S}}\hat{\mathbf{k}} \tag{3.100b}$$

The foregoing error vector, in its two versions, primal and dual, is termed the *design-error vector*. A positive scalar derived from this vector will be termed the *design error*.

The *design error* e_d adopted here is the rms value of the components of vector $\mathbf{e}$, namely,

$$e_d \equiv \sqrt{\frac{1}{m}\sum_1^m e_i^2} \tag{3.101a}$$

where e_i is the ith component of vector $\mathbf{e}$, i.e., the *residual* of the ith synthesis equation. Hence, the design error is proportional to the Euclidean norm of the design-error vector:

$$e_d \equiv \sqrt{\frac{1}{m}}\,\|\mathbf{e}\| \tag{3.101b}$$

It is apparent that, for fixed m, if we minimize $\|\mathbf{e}\|$, we minimize e_d. In the case of the spatial four-bar linkage, of course, e_d is defined as

$$\hat{e}_d = \sqrt{\frac{1}{m}}\,\|\hat{\mathbf{e}}\| \tag{3.101c}$$

where, from Eq. (A.9e),

$$\|\hat{\mathbf{e}}\| = \sqrt{\hat{\mathbf{e}}^T\hat{\mathbf{e}}}, \quad \hat{\mathbf{e}}^T\hat{\mathbf{e}} = \|\mathbf{e}\|^2 + \epsilon 2\mathbf{e}^T\mathbf{e}_o \tag{3.101d}$$

The value $\mathbf{k}_0$ of $\mathbf{k}$ that minimizes $\|\mathbf{e}\|$, as derived in Sect. 1.4, is applicable to the planar and spherical cases; it is given in Eq. (1.41)[6]—The value $\hat{\mathbf{k}}_0$ that minimizes $\hat{e}_d$ is discussed in Sect. 3.5.3. In the planar and spherical cases, this equation leads to

$$\mathbf{k}_0 = \mathbf{S}^I\mathbf{b} \tag{3.102a}$$

which is the *least-square approximation* of the given overdetermined system of linear equations, $\mathbf{S}^I$ being the *left Moore-Penrose generalized inverse* of $\mathbf{S}$, as introduced in Eq. (1.42), and is given by

$$\mathbf{S}^I = (\mathbf{S}\mathbf{S}^T)^{-1}\mathbf{S}^T \tag{3.102b}$$

Hence,

$$\mathbf{e}_0 \equiv \mathbf{b} - \mathbf{S}\mathbf{k}_0 \tag{3.103}$$

is the *least-square error vector*. Correspondingly,

$$e_{d0} \equiv \sqrt{\frac{1}{m}}\,\|\mathbf{e}_0\| \tag{3.104}$$

is the *least-square design error* of the approximation to the overdetermined system of synthesis equations.

Remark 3.5.1 Expression (3.102a) for $\mathbf{k}_0$ can be derived upon multiplying both sides of Eq. (3.12) by $\mathbf{S}^T$:

$$(\mathbf{S}^T\mathbf{S})\mathbf{k} = \mathbf{S}^T\mathbf{b} \tag{3.105}$$

where $\mathbf{S}^T\mathbf{S}$ is a $n \times n$ matrix. If this matrix is nonsingular, then

[6] $\mathbf{k}_0$ shouldn't be mistaken by $\mathbf{k}_o$, the dual part of $\hat{\mathbf{k}}$.

$$\mathbf{k} \equiv \mathbf{k}_0 = (\mathbf{S}^T\mathbf{S})^{-1}\mathbf{S}^T\mathbf{b}$$

Remark 3.5.2 The least-square approximation $\mathbf{k}_0$ can be thought of as being derived upon "inverting" the rectangular **S** matrix in the original overdetermined system, Eq. (3.12), with the "inverse" of **S** understood in the generalized sense.

Remark 3.5.3 $\mathbf{k}_0$ minimizes the Euclidean norm of **e**, which is proportional to the design error.

Remark 3.5.4 The least-square error of the approximation of the overdetermined system of synthesis equations *does not* measure the *positioning error*, a.k.a. the *structural error*, but rather the design error **e** defined above. The structural error produced by the synthesized linkage must be measured with respect to the task, not with respect to the synthesis equations! That is, if we let $\overline{\phi}_i$ denote the *prescribed* value of the output angle, corresponding to the ψ_i value, with ϕ_i denoting the *generated* value of the output angle, then the structural error is the vector **s** defined below:

$$\mathbf{s} \equiv [\phi_1 - \overline{\phi}_1,\ \phi_2 - \overline{\phi}_2,\ \cdots,\ \phi_m - \overline{\phi}_m]^T \tag{3.106}$$

Computing the least-square approximation $\mathbf{k}_0$ *verbatim* as appearing in Eqs. (3.102a) and (3.102b) *is not advisable* because of Remark 1.4.3 and the ensuing discussion. That is, if $\kappa(\mathbf{S})$ is moderately large, say, of the order of 1000, $\kappa(\mathbf{S}^T\mathbf{S})$ is inadmissibly large, of the order of 10^6.

Numerical methods for the solution of Eq. (3.12) in the presence of a rectangular **S** are available in the literature (Golub and Van Loan, 1983), as outlined in Sect. 1.4.5 and implemented in scientific software. The two methods outlined therein fall into what is called the *QR decomposition*: **S** is factored into an orthogonal matrix **Q** and an upper-triangular matrix **R**.

Maple uses Householder reflections to find numerically the least-square approximation of an overdetermined system of linear equations; it uses Gram-Schmidt orthogonalization to do the same if data are given *symbolically*.

In any event, the original system (3.12) is transformed into the form

$$\mathbf{T}\mathbf{k} = \mathbf{c} \tag{3.107}$$

where **T** and **c** are the transforms of **S** and **b** of Eq. (3.100a), respectively, with **T** of the form

$$\mathbf{T} = \begin{bmatrix} \mathbf{U} \\ \mathbf{O} \end{bmatrix} \tag{3.108}$$

while **U** and **O** are

$$\mathbf{U} = \begin{bmatrix} u_{11} & u_{12} & \cdots & u_{1n} \\ 0 & u_{22} & \cdots & u_{2n} \\ \vdots & \vdots & \ddots & \vdots \\ 0 & 0 & \cdots & u_{nn} \end{bmatrix}, \quad \mathbf{O}: \quad (m-n) \times n \text{ zero matrix} \tag{3.109}$$

In order to solve Eq. (3.107) for $\mathbf{k}$, we partition vector $\mathbf{c}$ into a n-dimensional upper part $\mathbf{c}_U$ and a $(m-n)$-dimensional lower part $\mathbf{c}_L$:

$$\mathbf{c} = \begin{bmatrix} \mathbf{c}_U \\ \mathbf{c}_L \end{bmatrix} \tag{3.110}$$

where, in general, $\mathbf{c}_L \neq \mathbf{0} \in \mathbb{R}^{m-n}$.

System (3.107) thus takes the form

$$\begin{bmatrix} \mathbf{U} \\ \mathbf{O} \end{bmatrix} \mathbf{k} = \begin{bmatrix} \mathbf{c}_U \\ \mathbf{c}_L \end{bmatrix} \quad \Rightarrow \quad \begin{cases} \mathbf{Uk} = \mathbf{c}_U \\ \mathbf{Ok} = \mathbf{c}_L \neq \mathbf{0} \end{cases} \tag{3.111}$$

Remark 3.5.5 If $\mathbf{S}$ is of full rank, then so is $\mathbf{T}$ and hence, $\mathbf{U}$ is nonsingular.

Remark 3.5.6 If $\mathbf{U}$ is nonsingular, then none of its diagonal entries vanishes, for $\det(\mathbf{U}) = u_{11}u_{22}\cdots u_{nn}$.

Remark 3.5.7 If $\mathbf{U}$ is nonsingular, then $k_1, k_2, \ldots, k_n$ can be computed from the first of Eqs. (3.111) by *backward substitution*.

Remark 3.5.8 The second of Eqs. (3.111) is a contradiction: its RHS is zero, but its LHS is not! Hence, $\mathbf{c}_L$ is the error vector—although not in the original vector basis, but in a new, orthonormal basis—the error in the approximation of the synthesis equations thus being

$$e_{d0} = \sqrt{\frac{1}{m}}\|\mathbf{c}_L\| \tag{3.112}$$

3.5.1 The Approximate Synthesis of Planar Four-Bar Linkages

For planar linkages the procedure is straightforward, as illustrated with the example below.

Example 3.5.1 (Approximate synthesis of the gripper mechanism) The mechanism of Fig. 3.6 is to be synthesized, but now with a large number of IO values. For comparison purposes, the data points used by Dudiţă et al. (1989) are used here, which are prescribed by equally spacing 61 IO values between $\psi_1 = 30°$, $\phi = 240°$ and $\psi_{61} = 60°$, $\phi_{61} = 210°$, as depicted in Fig. 3.15.

Solution: The 61×3 synthesis matrix $\mathbf{S}$ and the 61-dimensional vector $\mathbf{b}$ are not displayed for the sake of economy of space. Details of the solution are available in

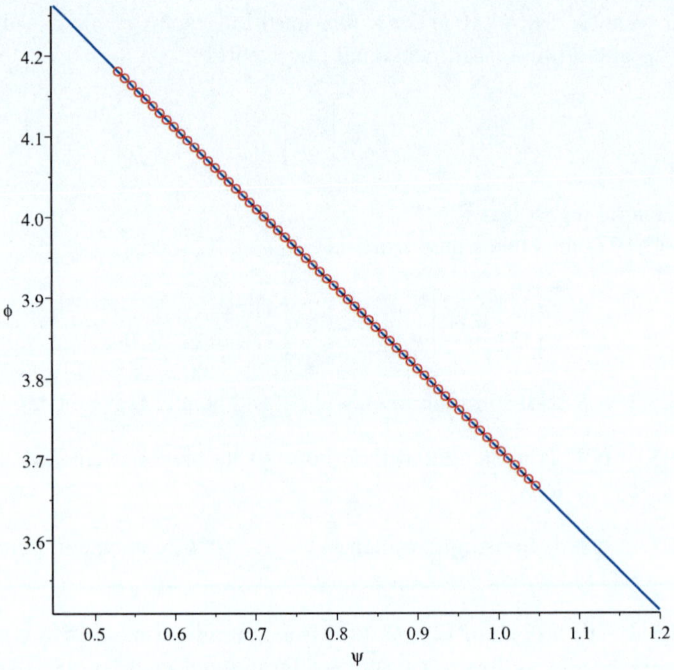

Fig. 3.15 The 61 points prescribed in the ϕ-versus-ψ plane

the code written for the purpose at hand.[7] The least-square approximation was computed using Householder reflections, which yielded, using 16 digits for comparison purposes with the results reported by Dudiţă et al. (1989):

$$\mathbf{k}_0 = [2.9417068638, 2.7871366821, 2.7869959265]^T$$

with corresponding link parameters

$$a_1 = 1.0, \quad a_2 = 0.3587911588, \quad a_3 = 0.7071482506, \quad a_4 = 0.3588092794$$

in units of length. In the foregoing reference, the authors computed the least-square approximation using the *normal equations*, which produced

$$\mathbf{k}_D = [2.9398767070, 2.7857633820, 2.7857633820]^T$$

thereby obtaining the link lengths[8] below:

[7] See `Dudita2.mw`.

[8] Dudiţă et al. adjusted the values of the second and the third components of the $\mathbf{k}_D$ array to be identical.

$$a_1 = 1.0, \quad a_2 = 0.3589680324, \quad a_3 = 0.7071510069, \quad a_4 = a_2$$

in units of length. The values of both $\mathbf{k}$ and $\mathbf{k}_D$ coincide up to the first two digits, those of the link lengths up to the first three digits, which is a fair matching, a consequence of the moderate condition number of the synthesis matrix, namely $\kappa = 195$, as computed in `Dudita2.mw`. However, the *normality conditions* were met, with the values provided in the foregoing reference, with an error of $O(-4)$; the same conditions were met with an error of $O(-14)$ in the code mentioned above, which uses HHR.

To the naked eye, the synthesized linkage doesn't appear different from that in Fig. 3.7.

3.5.2 The Approximate Synthesis of Spherical Linkages

This case parallels that of planar linkages, with the provision that, as in the case of exact synthesis of spherical linkages, nothing guarantees that the computed least-square approximation complies with the two conditions (3.33). The first of these, $|k_3| \le 1$, can be enforced in the least-square solution by adding one more equation, $k_3 = 0$, to the synthesis equations. Compliance with this condition, however, will invariably lead to a larger value of e_{d0}. Enforcing the second condition of Eq. (3.33) is less straightforward, as it requires techniques for solving problems of *constrained least squares* with nonlinear equality constraints, which fall outside of the scope of this book, and will not be further discussed. The reader is referred to the literature on engineering optimization whenever confronted with this problem.

Adjoining the above equation, $k_3 = 0$, to the synthesis equation, then, leads to the *augmented synthesis equations*

$$\mathbf{S}_a\mathbf{k} = \mathbf{b}_a \tag{3.113a}$$

where

$$\mathbf{S}_a = \begin{bmatrix} \mathbf{S} \\ \mathbf{u}^T \end{bmatrix}, \quad \mathbf{b}_a = \begin{bmatrix} \mathbf{b} \\ 0 \end{bmatrix} \tag{3.113b}$$

with $\mathbf{u} = [0,\ 0,\ 1,\ 0]^T$, and hence, $\mathbf{S}_a$ now becomes of $(m+1) \times 4$, while $\mathbf{b}_a$ is now $(m+1)$-dimensional.

Furthermore, least-square approximations allow for more flexibility, if we introduce *weights* in Eq. (3.113a), by means of a $(m+1) \times (m+1)$ constant matrix $\mathbf{V}_a$:

$$\mathbf{V}_a\mathbf{S}_a\mathbf{k} = \mathbf{V}_a\mathbf{b}_a \tag{3.114a}$$

with

$$\mathbf{V}_a = \begin{bmatrix} \mathbf{V} & \mathbf{0}_m \\ \mathbf{0}_m{}^T & v \end{bmatrix} \tag{3.114b}$$

in which $\mathbf{V}$ is a $m \times m$ block, $\mathbf{0}_m$ is the m-dimensional zero matrix, and v is a scalar. Both $\mathbf{V}$ and v are assigned by the user under the only constraint of avoiding the introduction of large roundoff-error amplification. We will describe presently how to prescribe $\mathbf{V}$ and v.

Notice that the least-square approximation $\mathbf{k}_0$ of Eq. (3.114a) now becomes, symbolically,

$$\begin{aligned}\mathbf{k}_0 &= [(\mathbf{V}_a\mathbf{S}_a)^T(\mathbf{V}_a\mathbf{S}_a)]^{-1}(\mathbf{V}_a\mathbf{S}_a{}^T)\mathbf{V}_a\mathbf{b}_a \\ &= (\mathbf{S}_a{}^T\mathbf{W}_a\mathbf{S}_a)^{-1}\mathbf{S}_a{}^T\mathbf{W}_a\mathbf{b}_a, \quad \mathbf{W}_a \equiv \mathbf{V}_a{}^T\mathbf{V}_a\end{aligned} \tag{3.115}$$

in which the symmetric and positive-definite $\mathbf{W}_a$ is termed a *weighting matrix*.

Also notice that

$$\mathbf{W}_a = \begin{bmatrix}\mathbf{V}^T & \mathbf{0}_m \\ \mathbf{0}_m{}^T & v_{m+1}\end{bmatrix}\begin{bmatrix}\mathbf{V} & \mathbf{0}_m \\ \mathbf{0}_m{}^T & v_{m+1}\end{bmatrix} = \begin{bmatrix}\mathbf{W} & \mathbf{0}_m \\ \mathbf{0}_m{}^T & w_{m+1}\end{bmatrix} \tag{3.116a}$$

with

$$\mathbf{W} = \mathbf{V}^T\mathbf{V}, \quad w_{m+1} \equiv v_{m+1}^2 \tag{3.116b}$$

Since no constraint is imposed on $\mathbf{V}$, besides robustness to round-off error amplification, $\mathbf{V}$ can be freely chosen as *symmetric and positive-definite*, and hence, nonsingular, i.e.,

$$\mathbf{V}^2 = \mathbf{W} \quad \Rightarrow \mathbf{V} = \sqrt{\mathbf{W}} \tag{3.117a}$$

where $\sqrt{\mathbf{W}}$ denotes the *positive-definite square root* of $\mathbf{W}$. Now, the matrices simplest to square-rooting are diagonal, $\mathbf{W}$ then being chosen as

$$\mathbf{W} = \operatorname{diag}(w_1, w_2, \ldots, w_m), \quad w_i > 0, \quad i = 1, \ldots, m \tag{3.117b}$$

Thus, the error vector in the approximation of Eqs. (3.114a) is

$$\mathbf{e}_a = \mathbf{V}_a(\mathbf{b}_a - \mathbf{S}_a\mathbf{k}) = \begin{bmatrix}\mathbf{V} & \mathbf{0}_m \\ \mathbf{0}_m{}^T & v_{m+1}\end{bmatrix}\begin{bmatrix}\mathbf{b} - \mathbf{Sk} \\ k_3\end{bmatrix} \tag{3.118a}$$

whose Euclidean norm is

$$\begin{aligned}\|\mathbf{e}_a\|^2 &= \begin{bmatrix}\mathbf{b}^T - \mathbf{k}^T\mathbf{S}^T & k_3\end{bmatrix}\begin{bmatrix}\mathbf{V}^2 & \mathbf{0}_m \\ \mathbf{0}_m{}^T & v_{m+1}^2\end{bmatrix}\begin{bmatrix}\mathbf{b} - \mathbf{Sk} \\ k_3\end{bmatrix} \\ &= (\mathbf{b}^T - \mathbf{k}^T\mathbf{S}^T)\mathbf{W}(\mathbf{b} - \mathbf{Sk}) + w_{m+1}k_3^2 \\ &= \sum_{i=1}^{m} w_i e_i^2 + w_{m+1}k_3^2\end{aligned} \tag{3.118b}$$

thus yielding a *weighted error-norm*. In order to avoid large roundoff-error amplification, we choose the weighting factors $\{ w_i \}_1^{m+1}$ as

$$\sum_{i=1}^{m+1} w_i = 1, \quad 0 \leq w_i \leq 1, \quad i = 1, \ldots, m+1 \tag{3.119}$$

so that $\|\mathbf{e}_a\|^2$ becomes a *convex combination*[9] of all $m+1$ errors. If no preference is given to the set $\{ e_i \}_1^m$, then the first m weights can be chosen all equal, while w_{m+1} is to be chosen so as to enforce $|k_3|$ to be smaller than unity but, if w_{m+1} is chosen unnecessarily large, then $|k_3|$ will be "too small" at the expense of a "large" design error. The best compromise is to be found *by trial and error*.

An application of the approximate synthesis of spherical four-bar linkages to the replacement of a bevel-gear transmission with input and output axes at right angles was reported by Alizadeh et al. (2013).

3.5.3 The Approximate Synthesis of Spatial Linkages

This subsubsection is largely based on a previous publication (Angeles, 2012). The synthesis equations (3.50) for the spatial four-bar linkage are reproduced below for quick reference:

$$\mathbf{S}\mathbf{k} = \mathbf{b} \tag{3.120a}$$

$$\mathbf{S}\mathbf{k}_o = \mathbf{b}_o - \mathbf{S}_o\mathbf{k} \tag{3.120b}$$

which can be cast in the standard form (1.28) of an overdetermined system of linear equations, in this case of $2m$ equations in $2 \times 4 = 8$ unknowns, the four components of each of $\mathbf{k}$ and of $\mathbf{k}_o$. Indeed, assembling the above equations into one single system yields

$$\underbrace{\begin{bmatrix} \mathbf{S} & \mathbf{O} \\ \mathbf{S}_o & \mathbf{S} \end{bmatrix}}_{\mathbf{A}} \underbrace{\begin{bmatrix} \mathbf{k} \\ \mathbf{k}_o \end{bmatrix}}_{\mathbf{x}} = \underbrace{\begin{bmatrix} \mathbf{b} \\ \mathbf{b}_o \end{bmatrix}}_{\mathbf{r}} \tag{3.121}$$

whose matrix $\mathbf{A}$ has four $m \times 4$ blocks, while $\mathbf{x}$ is the unknown eight-dimensional *design vector*. One could think of submitting Eq. (3.121) to a linear least-square solver and, sure enough, obtain a least-square solution $\mathbf{x}_0$ through Eq. (1.41). Problem is, this solution would be *meaningless* because the error $\mathbf{e} \equiv \mathbf{r} - \mathbf{A}\mathbf{x}$ **does not admit a norm**. The reason is that the first m components of $\mathbf{e}$ are dimensionless, as they refer to the spherical linkage associated with the spatial linkage at hand, while the last m bear units of length, as they refer to sliding errors. This approach is thus ruled out. Instead, the synthesis equations in dual form, Eq. (3.45), with the definitions appearing in Eq. (3.46), are recalled, as reproduced below for quick reference:

$$\hat{\mathbf{S}}\hat{\mathbf{k}} = \hat{\mathbf{b}} \tag{3.122}$$

---

[9] www.mathworld.wolfram.com.ConvexCombination.html.

which, for $m > 4$, cannot be satisfied exactly; the *dual error* incurred is

$$\hat{\mathbf{e}} = \hat{\mathbf{b}} - \hat{\mathbf{S}}\hat{\mathbf{k}} \tag{3.123}$$

Equation (3.122) can be shown to admit the *dual least-square solution*

$$\hat{\mathbf{k}}_0 = \hat{\mathbf{S}}^I\hat{\mathbf{b}}, \quad \hat{\mathbf{S}}^I = (\hat{\mathbf{S}}^T\hat{\mathbf{S}})^{-1}\hat{\mathbf{S}}^T \tag{3.124}$$

where

$$\hat{\mathbf{S}}^T\hat{\mathbf{S}} = \mathbf{S}^T\mathbf{S} + \epsilon(\mathbf{S}^T\mathbf{S}_o + \mathbf{S}_o^T\mathbf{S}) \tag{3.125}$$

whose inverse is readily computed using Eq. (A.13) of the Appendix:

$$(\hat{\mathbf{S}}^T\hat{\mathbf{S}})^{-1} = (\mathbf{S}^T\mathbf{S})^{-1} - \epsilon(\mathbf{S}^T\mathbf{S})^{-1}(\mathbf{S}^T\mathbf{S}_o + \mathbf{S}_o^T\mathbf{S})(\mathbf{S}^T\mathbf{S})^{-1} \tag{3.126}$$

and hence,

$$\hat{\mathbf{S}}^I = \mathbf{S}^I + \epsilon[-\mathbf{S}^I\mathbf{S}_o\mathbf{S}^I + \underbrace{(\mathbf{S}^T\mathbf{S})^{-1}\mathbf{S}_o^T - (\mathbf{S}^T\mathbf{S})^{-1}\mathbf{S}_o^T\mathbf{S}\mathbf{S}^I}_{\Delta}] \tag{3.127}$$

Were it not for the Δ term in the above expression, Eq. (3.127) would mimic faithfully the expression for the dual inverse appearing in Eq. (A.13). It will become apparent that this term can be dropped from the above expression, thereby a) simplifying the expression of interest and b) leading to a *minimum-size linkage*.

Upon substitution of expression (3.127) into Eq. (3.124), and expansion of the expression thus resulting, the least-square solution $\hat{\mathbf{k}}_0$ is obtained as

$$\hat{\mathbf{k}}_0 = \underbrace{(\mathbf{S}^T\mathbf{S})^{-1}\mathbf{S}^T\mathbf{b}}_{\mathbf{k}_0} + \epsilon\underbrace{(\mathbf{S}^T\mathbf{S})^{-1}[\mathbf{S}_o^T\mathbf{b} + \mathbf{S}^T\mathbf{b}_o - (\mathbf{S}^T\mathbf{S}_o + \mathbf{S}_o^T\mathbf{S})(\mathbf{S}^T\mathbf{S})^{-1}\mathbf{S}^T\mathbf{b}]}_{\mathbf{k}_{o0}} \tag{3.128}$$

While the above expressions for the least-square solution of both the primal part of $\hat{\mathbf{k}}$, $\mathbf{k}_0$, and its dual counterpart $\mathbf{k}_{o0}$ are theoretically sound, they are not appropriate for computations *verbatim*, given the large amount of floating-point operations involved, and their need of the inverse of $\mathbf{S}^T\mathbf{S}$. As pointed out in Remark 1.4.3, it is not advisable to compute verbatim that inverse because of the likely amplification of the condition number of the matrix product. It will be made apparent in the sequel that a terser solution $\mathbf{k}_{o0}$ can be obtained.

Indeed, if first the least-square solution $\mathbf{k}_0$ for the primal part of $\hat{\mathbf{k}}$ is computed from Eq. (3.120a), using the left Moore-Penrose generalized inverse $\mathbf{S}^I$, and then this expression is substituted into Eq. (3.120b), the least-square solution $\mathbf{k}_{o0}$ is derived as

$$\mathbf{k}_{o0} = \mathbf{S}^I(\mathbf{b}_o - \mathbf{S}_o\mathbf{S}^I\mathbf{b}) \tag{3.129}$$

which is much terser than its counterpart expression in Eq. (3.128). The difference between the two expressions can be explained based on the observation that the dual generalized inverse $\hat{\mathbf{S}}^I$ *is not unique*, contrary to its real counterpart. This fact is made apparent below.

Paraphrasing the derivation of the expression (A.13) for the dual inverse, let $\hat{\mathbf{B}} = \mathbf{B} + \epsilon\mathbf{B}_o$ be the generalized inverse of a $m \times n$ dual matrix $\hat{\mathbf{A}} = \mathbf{A} + \epsilon\mathbf{A}_o$, with $m > n$. As $\hat{\mathbf{A}}$ has been assumed of $m \times n$, $\hat{\mathbf{B}}$ is bound to be of $n \times m$.

Then,

$$\hat{\mathbf{B}}\hat{\mathbf{A}} = \mathbf{1}_n \tag{3.130}$$

with $\mathbf{1}_n$ denoting the $n \times n$ identity matrix. Upon expansion of the left-hand side of the above equation, two real equations are obtained, one for the primal, one for the dual part:

$$\mathbf{B}\mathbf{A} = \mathbf{1}_n, \quad \mathbf{B}_o\mathbf{A} + \mathbf{B}\mathbf{A}_o = \mathbf{O}_n \tag{3.131}$$

the first equation leading to the not so unexpected result $\mathbf{B} = \mathbf{A}^I$, which, when substituted into the second equation, yields a matrix equation for $\mathbf{B}_o$:

$$\mathbf{B}_o\mathbf{A} = -\mathbf{A}^I\mathbf{A}_o$$

A more suitable form of the above equation is obtained, with the unknown $\mathbf{B}_o$ as the right-hand factor of the left-hand side upon transposing the two sides of the above equation, namely,

$$\mathbf{A}^T\mathbf{B}_o^T = -\mathbf{A}_o^T(\mathbf{A}^I)^T \equiv -\mathbf{A}_o^T\mathbf{A}(\mathbf{A}^T\mathbf{A})^{-1}$$

which is a system of n^2 equations in $m \times n > n^2$ unknowns. The system is, thus, underdetermined, thereby admitting infinitely many solutions. The conclusion is, then, that the dual left generalized inverse is not unique. Among all that many solutions, one of minimum *Frobenius norm* can be obtained if one resorts to the *right Moore-Penrose generalized inverse* of $\mathbf{A}^T$, denoted $(\mathbf{A}^T)^\dagger$ (Nash and Sofer, 1996):

$$(\mathbf{A}^T)^\dagger = \mathbf{A}(\mathbf{A}\mathbf{A}^T)^{-1} \tag{3.132}$$

After some obvious manipulations,

$$\mathbf{B}_o = -\mathbf{A}^I\mathbf{A}_o\mathbf{A}^I \tag{3.133}$$

Therefore, the minimum-Frobenius-norm $\hat{\mathbf{A}}^I$ is

$$\hat{\mathbf{A}}^I = \mathbf{A}^I - \epsilon\mathbf{A}^I\mathbf{A}_o\mathbf{A}^I \tag{3.134}$$

The reader is invited to show that, if the foregoing formula is applied to compute the least-square solution $\hat{\mathbf{k}}_0$, the expression below is obtained:

$$\hat{\mathbf{k}}_0 = \mathbf{S}^I\mathbf{b} + \epsilon\mathbf{S}^I(\mathbf{b}_o - \mathbf{S}_o\mathbf{S}^I\mathbf{b}) \tag{3.135}$$

whose dual part is exactly the one obtained in Eq. (3.129). By the same token, the reader is invited to prove that Δ^T, with Δ as appearing in Eq. (3.127), is *an orthogonal complement* (OC) of $\mathbf{S}^T$, which is the reason why this term was *filtered out* in Eq. (3.129). Below we expand on the OC concept.

Given a $m \times n$ matrix $\mathbf{M}$, with $m < n$, i.e., with more columns than rows—for simplicity, its m rows will be assumed linearly independent—its ith row can be regarded as a vector $\mathbf{m}_i \in \mathbb{R}^n$. Since $\mathbf{M}$ has m such vectors, it is possible to find $n - m$ linearly independent vectors $\{\mathbf{p}_k\}_1^{n-m}$ orthogonal to the m rows—picture this with $m = 2$ and $n = 3$. If the vectors of this set are arrayed as the columns of a $n \times (n - m)$ matrix $\mathbf{P}$, then

$$\mathbf{MP} = \mathbf{O}_{mn'} \tag{3.136}$$

where $\mathbf{O}_{mn'}$ denotes the $m \times (n - m)$ zero matrix. Every matrix $\mathbf{P}$ that verifies Eq. (3.136) is termed an orthogonal complement of $\mathbf{M}$. It follows that *the OC is not unique*. Indeed, a rearrangement of the columns of $\mathbf{P}$ yields another OC of $\mathbf{M}$. Likewise, any multiple of a given $\mathbf{P}$ is also an OC of $\mathbf{M}$.

In summary, then, the approximate synthesis of a RCCC linkage proceeds sequentially:

1. Decouple the synthesis problem into two subproblems: one leading to the optimum Freudenstein parameters of the spherical linkage associated with the spatial linkage of interest, Eq. (3.120a), the other with the optimum dual counterparts of the foregoing parameters, Eq. (3.120b).
2. Apply Householder reflections to the primal part $\mathbf{S}$ of the dual synthesis matrix, thereby obtaining a $m \times m$ orthogonal matrix $\mathbf{H}$ and a $m \times 4$ matrix $\mathbf{E}$, with an upper-triangular 4×4 block occupying its first four rows and a $(m - 4) \times 4$ block of zeros. Apply the same reflections to $\mathbf{b}$ of the right-hand side of Eq. (3.120a) and obtain $\mathbf{k}_0$ by forward substitution on the first four equations.
3. Substitute $\mathbf{k}$ into Eq. (3.120b) with $\mathbf{k}_0$ and then apply the same Householder reflections to the right-hand side of the same equations.
4. Compute $\mathbf{k}_{o0}$ from the transformed Eqs. (3.120b) by backward substitution over the same 4×4 block.
5. Compute the skew angles $\{\alpha_i\}_1^4$ from Eqs. (3.32a)–(3.32d) by nonlinear-equation solving.
6. Compute the distances $\{a_i\}_1^4$ from Eqs. (3.42) by linear-equation solving.
7. Done!

3.6 Linkage Performance Evaluation

3.6.1 *Planar Linkages: Transmission Angle and Transmission Quality*

A variable of merit that is used to assess linkage performance is the *transmission angle* μ, illustrated in Fig. 3.1. The transmission angle is thus defined as *the angle between line segments* $\overline{CB}$ *and* $\overline{CD}$ *of the output and the coupler links*.

The relevance of this angle is apparent from a kinetostatic analysis[10]: in Fig. 3.16, the *internal forces of constraint* are indicated as F_{ij}, to denote the force exerted by the ith link on the jth link, using a standard notation. Therefore, the force transmitted by the output link to the frame has a magnitude $|F_{41}|$ given by

$$|F_{41}| = |F_{14}| = |F_{34}| \tag{3.137}$$

where, from the static equilibrium of the coupler and the input links,

$$|F_{34}| = |F_{32}| = \left|\frac{\tau_\psi}{a_2 \sin(\psi - \theta)}\right| \tag{3.138}$$

and τ_ψ is the *applied torque* that *balances statically* the *load torque* τ_ϕ.

The magnitude of the radial component of F_{14}, denoted $|F_{14}|_r$, is derived upon substitution of Eq. (3.138) into Eq. (3.137), thus obtaining

$$|F_{14}|_r \equiv |F_{14} \cos \mu| = \left|\frac{\tau_\psi}{a_2 \sin(\psi - \theta)} \cos \mu\right| \tag{3.139}$$

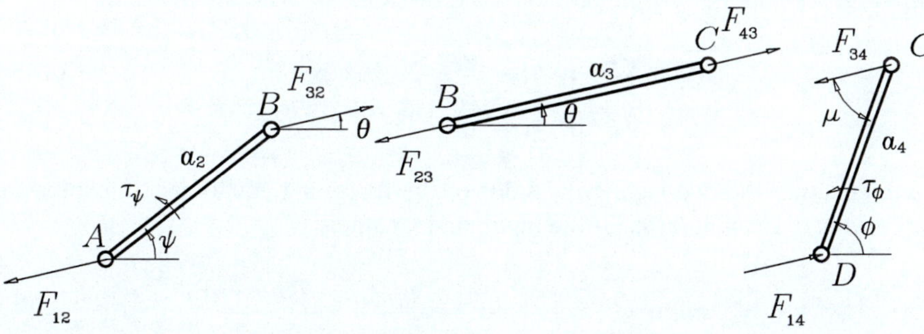

Fig. 3.16 A static analysis of the four-bar linkage

[10] This is an analysis of forces and moments of *a mechanical system in motion under static, conservative conditions*, and hence, with friction forces neglected.

from which it is apparent that $|F_{14}|_r$ is proportional to both the magnitude of the applied torque and the cosine of the transmission angle. Since this is a nonworking force, one is interested in keeping it as low as possible. However, it cannot be made zero by simply making zero the applied torque because, then, no useful force would be transmitted! Thus, the only possible way of keeping that force as small as possible is by keeping $|\cos\mu|$ as small as possible, i.e., by keeping μ as close as possible to $\pm 90°$.

The transmission angle is posture-dependent, of course; hence, it cannot be maintained at a fixed value for all the linkage postures. In practice, a minimum allowable value on the transmission angle or, rather, on its absolute value, is prescribed. This is commonly accepted as 45°, i.e., a specification when designing four-bar linkages is

$$|\mu| \geq 45° \tag{3.140}$$

If one is interested in a global evaluation of the performance of a four-bar linkage throughout its full range of motion, namely, $\psi_1 \leq \psi \leq \psi_2$, then a merit function of the linkage that takes into account *all possible postures* is needed. This quantity can be fairly termed the *transmission quality* of the linkage, defined here as the *root-mean-square* (rms) value of $\sin\mu$:

$$Q \equiv \sqrt{\frac{1}{\Delta\psi}\int_{\psi_1}^{\psi_2} \sin^2 \mu d\psi}, \quad \Delta\psi \equiv \psi_2 - \psi_1 \tag{3.141}$$

From the foregoing definition,

$$0 < Q < 1 \tag{3.142}$$

Evaluating Q as given above is rather difficult because an expression for $\sin\mu$ is not readily available. However, an expression for $\cos\mu$ can be readily available. Indeed, from Fig. 3.1 and the "cosine law", two expressions for $\overline{BD}^2$ are available:

$$\overline{BD}^2 = a_3^2 + a_4^2 - 2a_3a_4\cos\mu \tag{3.143a}$$

$$\overline{BD}^2 = a_1^2 + a_2^2 - 2a_1a_2\cos\psi \tag{3.143b}$$

Upon equating the two right-hand sides of the foregoing equations, an expression for $\cos\mu$ is derived in terms of the input angle, namely

$$\cos\mu = \frac{a_3^2 + a_4^2 - a_1^2 - a_2^2 + 2a_1a_2\cos\psi}{2a_3a_4} \tag{3.144}$$

If now relations (3.70) are recalled, an expression for $\cos\mu$ in terms of the linkage parameters $\{k_i\}_1^3$ is obtained:

$$\cos\mu = \operatorname{sgn}(k_2k_3)(c_1 + c_2\cos\psi) \tag{3.145a}$$

where coefficients c_1 and c_2 are defined as

$$c_1 \equiv \frac{k_2 - k_1 k_3}{\sqrt{D}}, \quad c_2 = \frac{k_3^2}{\sqrt{D}}, \quad D \equiv k_2^2 + k_3^2 + k_2^2 k_3^2 - 2k_1 k_2 k_3 \tag{3.145b}$$

Now the transmission quality Q can be written as $Q = \sqrt{1 - \delta^2}$ where δ is the square root of the integral of $\cos^2 \mu$ over the full mobility interval of the input link, i.e.,

$$\delta \equiv \sqrt{\frac{1}{\Delta\psi} \int_{\psi_1}^{\psi_2} \cos^2 \mu d\psi}, \quad \Delta\psi \equiv \psi_2 - \psi_1 \tag{3.146}$$

and, by virtue of the relation between the transmission quality Q and the transmission defect δ, we have the result

$$Q^2 + \delta^2 = 1 \tag{3.147}$$

It is reasonable to call δ the *transmission defect* of the linkage. Hence, maximizing Q is equivalent to minimizing δ, whose square can be readily obtained as

$$\delta^2 \equiv \frac{1}{\Delta\psi} \left[c_1^2 \Delta\psi + 2c_1 c_2 (\sin\psi_2 - \sin\psi_1) + \frac{1}{2} c_2^2 \Delta\psi + \frac{c_2^2}{4} (\sin 2\psi_2 - \sin 2\psi_1) \right] \tag{3.148}$$

If, in particular, the input link is a *crank*, then,

$$\delta^2 = c_1^2 + \frac{1}{2} c_2^2 \tag{3.149}$$

In synthesizing a four-bar linkage for function generation, the location of the zeros of the dials of the ψ and ϕ values is usually *immaterial*. What matters is the *incremental values* of these angles from those zeros. We can thus introduce parameters α and β denoting the location of the zeros on the ψ and the ϕ dials, respectively, so that now

$$\psi_i = \alpha + \Delta\psi_i, \quad \phi_i = \beta + \Delta\phi_i, \quad \text{for} \quad i = 1, 2, \ldots, m \tag{3.150}$$

We can therefore regard the least-square approximation $\mathbf{k}_0$ as a function of α and β, i.e.,

$$\mathbf{k}_0 = \mathbf{k}_0(\alpha, \beta) \tag{3.151}$$

It is apparent, then, that the two new parameters can be used to optimize the linkage performance, e.g., upon minimizing its defect δ.

As it turns out, the transmission angle plays an important role not only in the force-transmission characteristics of the linkage, but also in the *sensitivity* of its positioning accuracy to changes in the nondimensional parameters **k**. Indeed, if we make abstraction of the parameters α and β, for simplicity, we can calculate the sensitivity of the synthesized angle ϕ_i to changes in **k** from the input-output equation (3.11) written for the m prescribed input-output pairs. We display below the ith component of this vector equation:

$$F_i(\psi_i, \phi_i, \mathbf{k}) = k_1 + k_2 \cos\phi_i - k_3 \cos\psi_i - \cos(\psi_i - \phi_i) = 0, \; i = 1, 2, \ldots, m \tag{3.152}$$

where ϕ_i is one of the two values of ϕ that verify the above equation for $\psi = \psi_i$, namely, the one lying closest to $\overline{\phi}_i$, namely, the mean value of $\{\phi_i\}_1^m$, as introduced in Eq. (3.106). The sensitivity of interest is, apparently, $\partial\phi_i/\partial\mathbf{k}$, which is computed below: this partial derivative is computed upon differentiation of all sides of Eq. (3.152) w.r.t. **k**, namely,

$$\frac{\mathrm{d}F_i}{\mathrm{d}\mathbf{k}} = \frac{\partial F_i}{\partial \phi_i}\frac{\partial \phi_i}{\partial \mathbf{k}} + \frac{\partial F_i}{\partial \mathbf{k}} = \mathbf{0}$$

Hence,

$$\frac{\partial \phi_i}{\partial \mathbf{k}} = -\frac{\partial F_i/\partial \mathbf{k}}{\partial F_i/\partial \phi_i} \tag{3.153}$$

Now, we calculate $\partial F_i/\partial\phi_i$, as long as $\partial F_i/\partial\phi_i \neq 0$, from Eq. (3.152):

$$\frac{\partial F_i}{\partial \phi_i} = -k_2 \sin\phi_i - \sin(\psi_i - \phi_i) = -\frac{a_1 \sin\phi_i - a_2 \sin(\phi_i - \psi_i)}{a_2} \tag{3.154}$$

A pertinent relation among the variables and parameters involved in Eq. (3.154) is displayed in Fig. 3.17. From this figure,

$$a_1 \sin\phi_i - a_2 \sin(\phi_i - \psi_i) = a_3 \sin\mu_i \tag{3.155}$$

Upon substitution of Eq. (3.155) into Eq. (3.154), we obtain

$$\frac{\partial F_i}{\partial \phi_i} = -\frac{a_3}{a_2} \sin\mu_i \tag{3.156a}$$

which, when substituted into Eq. (3.153), yields

$$\frac{\partial \phi_i}{\partial \mathbf{k}} = \frac{a_2}{a_3 \sin\mu_i}\frac{\partial F_i}{\partial \mathbf{k}} \tag{3.156b}$$

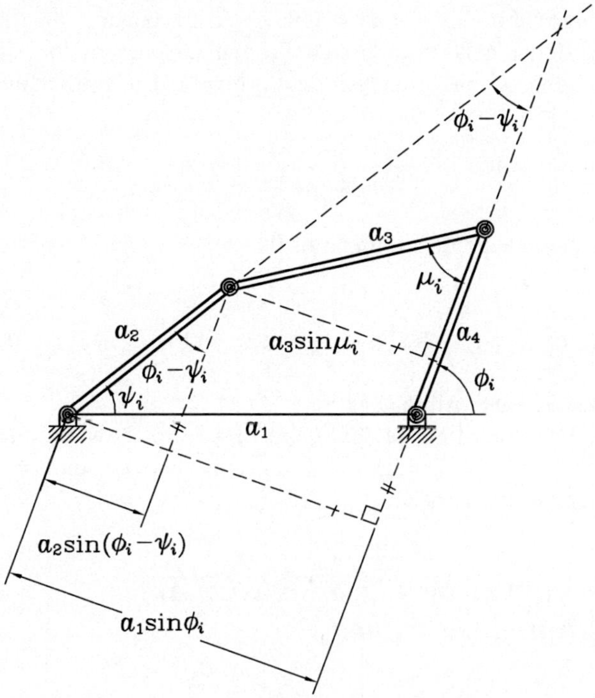

Fig. 3.17 Relation between the transmission angle and the parameters and variables of a four-bar linkage

Furthermore,

$$\frac{\partial F_i}{\partial \mathbf{k}} = \begin{bmatrix} 1 \\ \cos\phi_i \\ -\cos\psi_i \end{bmatrix} \tag{3.156c}$$

and hence,

$$\frac{\partial \phi_i}{\partial \mathbf{k}} = \frac{a_2}{a_3 \sin\mu_i} \begin{bmatrix} 1 \\ \cos\phi_i \\ -\cos\psi_i \end{bmatrix} \tag{3.156d}$$

It is now apparent that the larger $|\sin\mu_i|$, the less sensitive the positioning accuracy of the linkage is to changes in the linkage dimensions.

An interesting relation between the linkage discriminant defined in Eq. (3.81a) and the transmission angle is now derived. From the expression for $\cos\mu$ obtained in Eqs. (3.145a), an expression for $\sin^2\mu$ is readily obtained, in terms of the Freudenstein parameters, as

$$\sin^2\mu = \frac{k_3^2}{k_2^2 + k_3^2 + k_2^2 k_3^2 - 2k_1k_2k_3}\Delta(\psi) \tag{3.157a}$$

where $\Delta(\psi)$ is the linkage discriminant of Eq. (3.81b), reproduced below for quick reference:

$$\Delta(\psi) \equiv -k_3^2\cos^2\psi + 2(k_1k_3 - k_2)\cos\psi + (1 - k_1^2 + k_2^2) \geq 0 \tag{3.157b}$$

which is nonnegative at feasible postures.

Apparently, for a given linkage, the square of the sine of the transmission angle is proportional to the discriminant. Hence, both vanish at dead points of the input link, which occur when this is a rocker.

3.6.2 Spherical Linkages: Transmission Angle and Transmission Quality

Spherical linkages are elusive to a kinetostatic analysis because they are *hyperstatic*, in that the number of static equations available under the condition that all their axes intersect at one common point be smaller than the number of reaction forces and moments to be found. Rather than deriving the transmission angle for this kind of linkages by means of a kinetostatic analysis, as in the case of planar linkages, we derive it by establishing a correspondence between its geometry and that of the planar linkage.

Referring to Fig. 3.18, the transmission angle μ in this case is defined as the dihedral angle between the planes of the circular arcs of the coupler and the output links. By this means, an identity is obtained immediately:

$$\cos\mu = -\cos\theta_4 \tag{3.158}$$

where θ_4 is the joint rotation defined following DH convention.

An expression for $\cos\mu$ in terms of the input angle ψ can be found, as in the planar case, using trigonometry. Obviously, in the case at hand, spherical trigonometry is the tool to use, which then yields (McCarthy and Soh, 2011):

$$\cos\mu = \frac{c\alpha_3 c\alpha_4 - c\alpha_1 c\alpha_2 - s\alpha_1 s\alpha_2\cos\psi}{s\alpha_3 s\alpha_4} \tag{3.159}$$

The mechanical significance of the transmission angle is the same as in the planar case: the closer μ is to $\pm 90°$, the smaller the radial component of the force transmitted

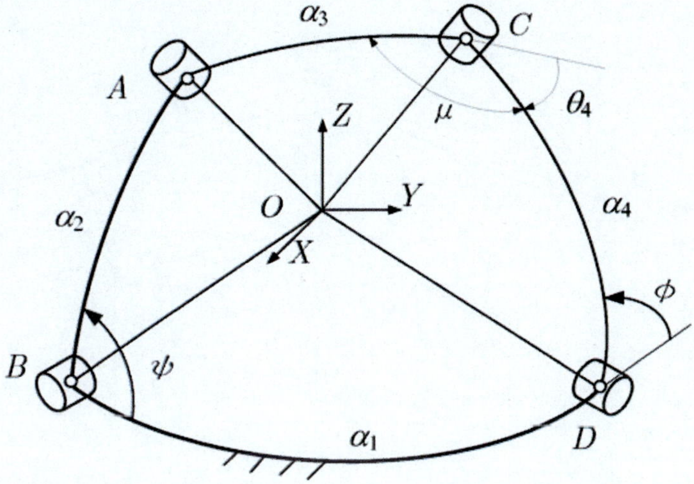

Fig. 3.18 Relation between the transmission angle and the parameters and variables of a spherical four-bar linkage

by the output link to the frame, and hence, the higher the quality of the force-transmission from the input to the output links. That is, for an acceptable performance, the dihedral angle between the planes of the circular arcs of the coupler and the output links should be such that the two planes are as far from each other as possible, which happens when the angle is $\pm 90°$, i.e., when $\cos\mu = 0$.

The transmission quality is, therefore, defined exactly as in the planar case by Eq. (3.141).

3.6.3 Spatial Linkages: Transmission Angle and Transmission Quality

For quick reference, the spatial four-bar linkage of Fig. 3.5 is reproduced here as Fig. 3.19.

Now, the simplest way of determining the transmission angle is by *dualization* of the expression in Eq. (3.158), namely, by *putting hats* on both μ and θ_4 of this expression, which yields

$$\cos\hat{\mu} = -\cos\hat{\theta}_4 \tag{3.160a}$$

For quick reference, $\hat{\theta}_4$ is the dual angle between X_4 and X_1, measured positive in the direction of X_4. Similarly, an expression for $\cos\hat{\mu}$ in terms of the input angle ψ can be found upon *dualizing* both sides of Eq. (3.159):

$$\cos\hat{\mu} = \frac{c\hat{\alpha}_3 c\hat{\alpha}_4 - c\hat{\alpha}_1 c\hat{\alpha}_2 - s\hat{\alpha}_1 s\hat{\alpha}_2 \cos\hat{\psi}}{s\hat{\alpha}_3 s\hat{\alpha}_4} \tag{3.160b}$$

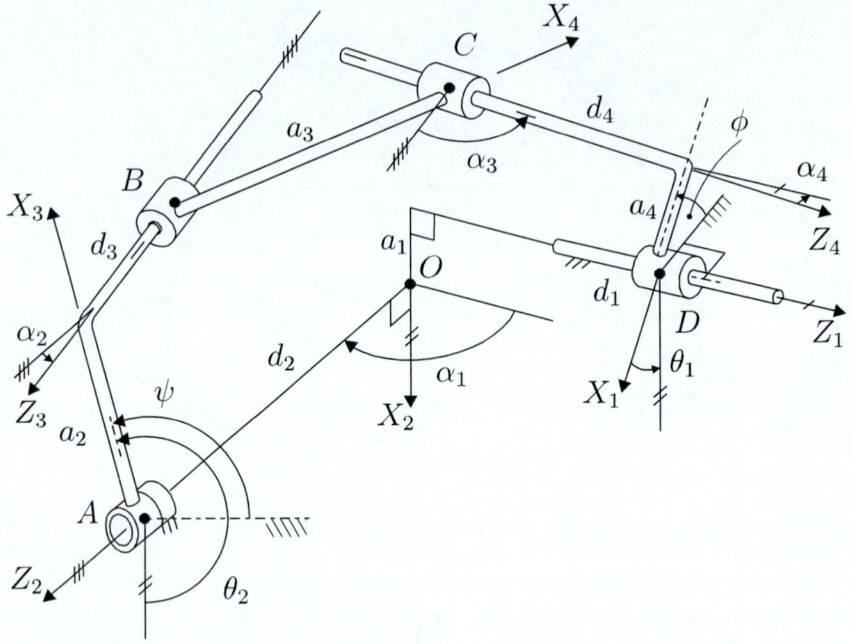

Fig. 3.19 A RCCC linkage for function generation (reproduced from Fig. 3.5)

In the foregoing equation, according with the definition of dual angle given in Eq. (A.5),

$$\hat{\mu} = \theta_4 + \epsilon d_4 \tag{3.161a}$$

with the two terms in the RHS of the above equation defined positive about and along the positive direction of Z_4, respectively. The geometric interpretation of the above expression is straightforward: while θ_4 is the angle between X_4 and X_1, d_4 is the signed distance between X_4 and X_1, the sign being determined by rule 5 of the DH notation. By the same token,

$$\hat{\alpha}_i = \alpha_i + \epsilon a_i \tag{3.161b}$$

which carries two constant quantities, while

$$\hat{\psi} = \psi + \epsilon d_1 \tag{3.161c}$$

carries two variable quantities, the input angle ψ, and the sliding d_1, with units of length. Moreover, the transmission quality is now a dual quantity, namely,

$$\hat{Q} \equiv \sqrt{\frac{1}{\Delta\psi}\int_{\psi_1}^{\psi_2} \sin^2\theta_4 d\psi} + \epsilon\, 2\sqrt{\frac{1}{\Delta\psi}\int_{\psi_1}^{\psi_2} d_4 \cos\theta_4 \sin\theta_4 d\psi}, \quad \Delta\psi \equiv \psi_2 - \psi_1 \tag{3.162}$$

Furthermore, θ_4 is the real angle between axes X_4 and X_1 in the spherical case, in which the two axes are concurrent. In the spatial case, $\hat{\theta}_4$ is the *dual angle* between two skew lines, axes X_4 and X_1. This angle, as discussed in Appendix A, comprises a primal part and a dual part, the former being the real angle between the two lines, as described above. The dual part is the signed distance between X_4 and X_1, which is positive when Z_4 points in the direction from X_4 to X_1. Thus, in the same way that the dot product of two unit vectors provides the cosine of the angle between the two vectors, the dot product of two *dual unit vectors*—a dual vector is "of unit magnitude" when its primal part is a real unit vector—provides the cosine of the (dual) angle between two lines. The dual angle in question involves both the angle between the two lines and their signed distance.

In the planar case, $\cos\mu$ determines "how far" the axis of the coupler link is from that of the output link, while keeping one common point, C, at any linkage posture: the larger $|\cos\mu|$, the "closer" the axes. In the spherical case, $\cos\mu$ determines "how far" the plane of the circular arc of the coupler link is from its counterpart of the output link, while maintaining one common line, Z_4, at any given linkage posture: again, the larger $|\cos\mu|$, the "closer" the planes. In order to provide a larger *lever arm* to the transmitted force, the foregoing "distances" should be kept as large as possible, aiming at values of $|\cos\mu|$ closer to zero than to unity.

The generalization to the spatial case then follows: $\cos\hat{\mu}$ determines "how far" X_4, while intersecting Z_4, is from X_1 at any given *linkage posture*. When the two axes coincide, the worst-case scenario, the full *wrench*—force and moment—transmitted by the coupler link to the output link goes into the linkage support, and no part of it is used to counter the load applied onto the output link. The effect of the transmitted force on the output link, under these conditions, is the same as that of a force applied onto a door along a line of action that passes through the hinge axis and a moment vector lying in a plane normal to this axis. Force and moment, in this case, are incapable of turning the door.

In summary, maximizing the dual transmission quality of the spatial four-bar linkage is equivalent to keeping the axis of the coupler link as "far away" as possible from axis X_1 of the DH notation, to increase the dual lever arm of the wrench—combination of force and moment—driving the output link directly, as transmitted by the coupler link.

3.7 Design Error Versus Structural Error

In this section we establish the relation between the design error and the structural error. In doing this, we build upon the analysis proposed by Tinubu and Gupta (1984).

The structural error was introduced in Eq. (3.106). If now ϕ and $\overline{\phi}$ denote the m-dimensional vectors of *generated and prescribed output values*, then the structural-error vector $\mathbf{s}$ can be expressed as

$$\mathbf{s} \equiv \boldsymbol{\phi} - \overline{\boldsymbol{\phi}} \tag{3.163}$$

where, it is recalled, ϕ_i denotes the *generated* value, $\overline{\phi}_i$ the *prescribed* value of the output angle for a given value ψ_i of the input angle. In the ensuing discussion we assume that the synthesis equations are cast in the general form

$$\mathbf{Sk} = \mathbf{b} \tag{3.164}$$

regardless of the type of linkage, planar, spherical or spatial. However, one should keep in mind that, in the spatial case, $\mathbf{S}$, $\mathbf{b}$ and $\mathbf{k}$ become all dual quantities: $\hat{\mathbf{S}}$, $\hat{\mathbf{b}}$ and $\hat{\mathbf{k}}$. In this context, $\mathbf{S}$ is a $m \times n$ matrix, while $\mathbf{k}$ and $\mathbf{b}$, or their dual counterparts, as the case may be, are n- and m-dimensional vectors, respectively. Again, $n =$ 3 in the planar case, 4 in the spherical and spatial cases.[11] In the case of spatial linkages, a second equation of the same structure as that of Eq. (3.164), involving a second vector of Freudenstein parameters—the dual part of the dual vector $\hat{\mathbf{k}}$—occurs, as per Eqs. (3.120a) and (3.120b). The difference $\mathbf{e} \equiv \mathbf{b} - \mathbf{Sk}$—or its dual counterpart for that matter—is to be highlighted: minimizing a norm of $\mathbf{e}$ is not equivalent to minimizing the same norm of $\mathbf{s}$. Indeed, while $\mathbf{e}$ denotes the error in meeting the synthesis equations, whose components involve trigonometric functions of the input and output angles at the m prescribed pairs, $\mathbf{s}$ denotes *the error in meeting the prescribed positions*, which is what really matters. Unfortunately, however, a relation of the form $\mathbf{g} \equiv \mathbf{g}(\mathbf{e}, \mathbf{s}) = \mathbf{0}$ between the two errors is elusive. Nevertheless, a *differential* relation between the two errors can be readily obtained, as done below, *and would suffice*.[12]

In minimizing the structural error, we aim at a minimum of the rms value of the components of vector $\mathbf{s}$ by properly choosing $\mathbf{k}$:

$$z(\mathbf{k}) \equiv \frac{1}{2m}\|\mathbf{s}\|^2 \quad \rightarrow \quad \min_{\mathbf{k}} \tag{3.165}$$

where $\|\mathbf{s}\|$ is the Euclidean norm of the structural error $\mathbf{s}$.

Function $z(\mathbf{k})$ attains a *stationary* value with respect to $\mathbf{k}$ when its *gradient* with respect to $\mathbf{k}$ vanishes, i.e., upon application of the *chain rule*,

$$\nabla z \equiv \frac{\partial z}{\partial \mathbf{k}} = \left(\frac{\partial \mathbf{s}}{\partial \mathbf{k}}\right)^T \frac{\partial z}{\partial \mathbf{s}} = \mathbf{0}_n \tag{3.166}$$

with $\mathbf{0}_n$, as usual, denoting the n-dimensional zero vector. The above equation is the *normality condition* of the minimization problem at hand. Apparently, from the definition of z in Eq. (3.165),

$$\frac{\partial z}{\partial \mathbf{s}} = \frac{1}{m}\mathbf{s} \tag{3.167}$$

[11] Read the opening paragraph of Sect. 3.5.

[12] Quoting Robert Frost (1874–1963): https://www.poetryfoundation.org/poems/44263/fire-and-ice.

Now, in order to compute $\partial\mathbf{s}/\partial\mathbf{k}$, we recall the definition of $\mathbf{s}$, Eq. (3.163), which leads to

$$\frac{\partial\mathbf{s}}{\partial\mathbf{k}} = \frac{\partial\boldsymbol{\phi}}{\partial\mathbf{k}} \tag{3.168}$$

as $\overline{\boldsymbol{\phi}}$ is a constant vector of prescribed output values. Moreover, the ith row of matrix $\partial\boldsymbol{\phi}/\partial\mathbf{k}$, for the *planar case*, is displayed in Eq. (3.156d).

Now, in order to compute $\partial\boldsymbol{\phi}/\partial\mathbf{k}$, we need an equation relating the array $\boldsymbol{\phi}$ of *generated values* of the output angle with vector $\mathbf{k}$. One candidate would be the m synthesis equations (3.164), which define the *design error* $\mathbf{e}$:

$$\mathbf{e} \equiv \mathbf{b} - \mathbf{S}\mathbf{k} \tag{3.169}$$

The above expression is, in general, different from zero, when evaluated at the *prescribed* values $\overline{\phi}_i$ of the output angle, for $i = 1, \ldots, m$, and hence, does not define an implicit equation in $\boldsymbol{\phi}$ and $\mathbf{k}$. As a matter of fact, the problem of approximate synthesis consists in minimizing the Euclidean norm—or any norm, for that matter—of the, usually, nonzero vector $\mathbf{e}$.

However, when vector $\mathbf{e}$ of Eq. (3.169) is evaluated at the *generated* values ϕ_i of the output angle, for $i = 1, \ldots, m$, the vector does vanish. Indeed, the ith component of $\mathbf{e}$, as defined in Eq. (3.169), is nothing but the input-output function $F(\psi, \phi) = 0$ evaluated at ψ_i for a given linkage defined by $\mathbf{k}$. In our case, $\mathbf{k}$ is the *current value*, within an iterative process—to be formulated in Sect. 3.7.1—of the unknown vector of linkage parameters, i.e., the Freudenstein parameters. Upon solving the input-output equation for ϕ, two values of ϕ_i are obtained, as found in Sect. 3.4, and hence, the function does vanish at these two values. We will assume that, of these two values, ϕ_i is chosen as the one closer to $\overline{\phi}_i$.

We thus have

$$F(\psi_i, \phi_i) \equiv b_i - \mathbf{s}_i^T\mathbf{k} = 0 \tag{3.170}$$

in which $\mathbf{s}_i^T$ denotes the ith row of $\mathbf{S}$ and b_i, the ith component of $\mathbf{b}$.

To avoid confusion, let us distinguish between the design error $\mathbf{e}$ when evaluated at $\boldsymbol{\phi}$ and when evaluated at $\overline{\boldsymbol{\phi}}$, by denoting the latter by $\overline{\mathbf{e}}$, i.e.,

$$\overline{\mathbf{e}} \equiv \mathbf{e}(\overline{\boldsymbol{\phi}}, \mathbf{k}) = \overline{\mathbf{b}} - \overline{\mathbf{S}}\mathbf{k} \neq \mathbf{0} \tag{3.171}$$

where $\overline{\mathbf{S}}$ and $\overline{\mathbf{b}}$ denote $\mathbf{S}$ and $\mathbf{b}$, respectively, when evaluated at the prescribed values of the input angle $\{\,\psi_i\,\}_1^m$ and at the prescribed $\overline{\boldsymbol{\phi}}$.

Moreover, when we evaluate $\mathbf{e}$ at the *generated* value $\boldsymbol{\phi}$, we obtain

$$\mathbf{e} \equiv \mathbf{e}(\boldsymbol{\phi}, \mathbf{k}) = \mathbf{b} - \mathbf{S}\mathbf{k} = \mathbf{0}_m \tag{3.172}$$

which is an *implicit* vector function of $\boldsymbol{\phi}$ and $\mathbf{k}$, and hence, allows for the evaluation of $\partial\boldsymbol{\phi}/\partial\mathbf{k}$. Upon differentiation of Eq. (3.172) with respect to $\mathbf{k}$, and zeroing of the expression thus resulting, we obtain

$$\frac{d\mathbf{e}}{d\mathbf{k}} = \frac{\partial \mathbf{e}}{\partial \mathbf{k}} + \frac{\partial \mathbf{e}}{\partial \boldsymbol{\phi}} \frac{\partial \boldsymbol{\phi}}{\partial \mathbf{k}} = \mathbf{O}_{mn} \tag{3.173}$$

where $\mathbf{O}_{mn}$ is the $m \times n$ zero matrix. Moreover, the $m \times m$ matrix $\partial \mathbf{e}/\partial \boldsymbol{\phi}$ is computed from the input-output equation (3.170), or its dual counterpart, as the case may be. Since e_k is influenced only by ϕ_k, and not by ϕ_j, for $j \neq k$, $\partial \mathbf{e}/\partial \boldsymbol{\phi}$ is diagonal, i.e.,

$$\frac{\partial \mathbf{e}}{\partial \boldsymbol{\phi}} = \operatorname{diag}\left[\frac{\partial e_1}{\partial \phi_1}, \frac{\partial e_2}{\partial \phi_2}, \dots, \frac{\partial e_m}{\partial \phi_m}\right] \equiv \mathbf{D} \tag{3.174a}$$

Under the assumption that none of the diagonal entries of $\mathbf{D}$ vanishes, this matrix is nonsingular, and hence, the matrix $\partial \boldsymbol{\phi}/\partial \mathbf{k}$ sought can be solved for from Eq. (3.173). Furthermore, it is apparent from Eq. (3.172) that $\partial \mathbf{e}/\partial \mathbf{k}$ is nothing but the negative of the synthesis matrix $\mathbf{S}$, evaluated at the generated values of the output angle, i.e.,

$$\frac{\partial \mathbf{e}}{\partial \mathbf{k}} = -\mathbf{S} \tag{3.174b}$$

Hence, $\partial \boldsymbol{\phi}/\partial \mathbf{k}$, as computed from Eq. (3.173), is

$$\frac{\partial \boldsymbol{\phi}}{\partial \mathbf{k}} \equiv \frac{\partial \mathbf{s}}{\partial \mathbf{k}} = \mathbf{D}^{-1}\mathbf{S} \tag{3.175}$$

Therefore, the *normality condition* (3.166) leads to

$$\mathbf{S}^T\mathbf{D}^{-1}\mathbf{s} = \mathbf{0}_n \tag{3.176}$$

where $\mathbf{0}_n$ denotes the n-dimensional zero vector. The normality condition thus states that, for $\mathbf{k}$ to produce a stationary value of the positioning error—proportional to the rms value of the structural error $\mathbf{s}$—the structural error $\mathbf{s}$ must lie in the null space of the matrix product $\mathbf{S}^T\mathbf{D}^{-1}$. That is, the structural error of minimum norm need not vanish and, in general, it won't, but must verify Eq. (3.176).

Now, contrary to the minimization of the design error, the minimization of the positioning error leads to a *nonlinear least-square problem*, which must be solved *iteratively*, as described in Sect. 3.7.1.

3.7.1 Minimizing the Structural Error

The approach followed here is similar to the Newton-Gauss method used to solve nonlinear least-square problems, as outlined in Sect. 1.6.1: for starters, a sequence $\mathbf{s}^0, \mathbf{s}^1, \dots, \mathbf{s}^i, \mathbf{s}^{i+1}$ of structural-error vector values is generated, which, upon convergence, should verify the normality condition. For a given $\mathbf{s}^i$, an improved vector value $\mathbf{s}^{i+1}$ is obtained from the *first-order approximation* of $\mathbf{s}$:

$$\mathbf{s}^{i+1} \approx \mathbf{s}^i + \left.\frac{\partial \mathbf{s}}{\partial \mathbf{k}}\right|_{\mathbf{k}=\mathbf{k}^i} \Delta\mathbf{k}^i = \mathbf{s}^i + \mathbf{D}_i^{-1}\mathbf{S}_i\Delta\mathbf{k}^i \tag{3.177}$$

where $\mathbf{D}_i \equiv \mathbf{D}|_{\mathbf{k}=\mathbf{k}^i}$ and $\mathbf{S}_i \equiv \mathbf{S}|_{\mathbf{k}=\mathbf{k}^i}$. Hence, if we assume that Eq. (3.177) is *exactly* verified by the current $\Delta\mathbf{k}^i$, then

$$\mathbf{D}_i^{-1}\mathbf{S}_i\Delta\mathbf{k}^i = \mathbf{s}^{i+1} - \mathbf{s}^i \tag{3.178}$$

Upon solving for $\Delta\mathbf{k}^i$, the above equation allows the updating of $\mathbf{k}$ as $\mathbf{k}^{i+1} = \mathbf{k}^i + \Delta\mathbf{k}^i$. However, in Eq. (3.178) we don't know $\mathbf{s}^{i+1}$. Moreover, upon convergence, $\mathbf{s}$ needn't vanish, and, most likely, it won't. We can thus assume that $\mathbf{s}^{i+1} \neq \mathbf{0}_m$, but, if $\mathbf{k}^{i+1}$ is an improvement over $\mathbf{k}^i$, then the corresponding structural error $\mathbf{s}^{i+1}$ will be "close" to verifying the normality condition (3.176). Accordingly, let us assume that $\mathbf{s}^{i+1}$ does verify the normality condition, with $\mathbf{S}$ and $\mathbf{D}$ evaluated at $\mathbf{k} = \mathbf{k}^i$, as we cannot evaluate them at $\mathbf{k}^{i+1}$. Further, let us multiply both sides of Eq. (3.178) from the left by $\mathbf{S}_i^T\mathbf{D}_i^{-1}$, which yields

$$\mathbf{S}_i^T\mathbf{D}_i^{-1}\mathbf{D}_i^{-1}\mathbf{S}_i\Delta\mathbf{k}^i = -\mathbf{S}_i^T\mathbf{D}_i^{-1}\mathbf{s}^i \tag{3.179}$$

where the term linear in $\mathbf{s}^{i+1}$ has dropped because it has been assumed to verify the normality conditions. In Eq. (3.179) the coefficient of $\Delta\mathbf{k}^i$ is a square $n \times n$ matrix—with n being the dimension of vector $\mathbf{k}$—which allows for the computation of $\Delta\mathbf{k}^i$ in the form

$$\Delta\mathbf{k}^i = -(\mathbf{S}_i^T\mathbf{D}_i^{-2}\mathbf{S}_i)^{-1}\mathbf{S}_i^T\mathbf{D}_i^{-1}\mathbf{s}^i \tag{3.180}$$

thereby showing that the correction $\Delta\mathbf{k}^i$ can be computed with the numerical values available at the ith iteration. In fact, the expression for $\Delta\mathbf{k}^i$ given in Eq. (3.180) should be regarded as a *formula*, not as an algorithm. Indeed, the verbatim inversion of the matrix in parentheses in the foregoing equation is *to be avoided* due to its high *condition number*.[13] As a matter of fact, the condition number, in either the Euclidean or the Frobenius norm, of the same $n \times n$ matrix is exactly the square of the same norm of the $m \times n$ matrix $\mathbf{D}_i^{-1}\mathbf{S}_i$. Hence, a formulation is sought that will allow the computation of $\Delta\mathbf{k}^i$ from a system of equations involving the foregoing rectangular matrix. If we recall Sect. 1.4.5, the right-hand side of Eq. (3.180) is the *least-square approximation* of the *overdetermined system*

$$(\mathbf{D}_i^{-1}\mathbf{S}_i)\Delta\mathbf{k}^i = -\mathbf{s}^i \tag{3.181}$$

which is identical to Eq. (3.178) when the therm $\mathbf{s}^{i+1}$ is dropped. Notice, however, that this term couldn't simply be dropped from the above-mentioned equation on the basis that the said term vanishes, because the structural error is not expected to vanish at the optimum solution. The computation of $\Delta\mathbf{k}^i$ from Eq. (3.181) now should be pursued via an orthogonalization procedure, as studied in Sect. 1.4.5. With

[13] See the definition of this concept in Sect. 1.4.4.

$\Delta\mathbf{k}^i$ calculated, the ith iteration is complete, as a new, improved value $\mathbf{k}^{i+1}$ of the design parameter vector $\mathbf{k}$ is available. Now the new structural-error vector value $\mathbf{s}^{i+1}$ can be computed, and then the normality condition verified. If the condition is not verified, a new iteration is in order; if the same condition is verified, then the procedure stops. An alternative convergence criterion, equivalent to the latter, is to verify whether $\|\Delta\mathbf{k}^i\| < \epsilon$, for a prescribed tolerance ϵ. The equivalence of the two criteria should be apparent from the relation between $\Delta\mathbf{k}^i$ and the product of the last three factors of the right-hand side of Eq. (3.180).

Caveat: To the uninitiated, it may appear tempting to transfer $\mathbf{D}_i^{-1}$ to the right-hand side of Eq. (3.181), and obtain $\Delta\mathbf{k}^i$ as the least-square approximation of $\mathbf{S}_i\Delta\mathbf{k}^i = -\mathbf{D}_i\mathbf{S}_i$. This is a bad idea, as the least-square solution obtained from the foregoing equation will be different from that obtained from Eq. (3.181). Reason: Eq. (3.181) *is not necessarily verified, but only approximated.* $\mathbf{D}_i$, or its inverse, carries weights that dictate the value of the least-square approximation.

3.7.2 *Branch-Switching Detection*

This Subsubsection is limited to planar linkages, its generalization to spherical and spatial linkages should be doable, as the problem under study is based on the concept of the sign of the transmission index. The latter was studied in Sect. 3.6.

In the foregoing analysis an implicit assumption was adopted: all generated values $\{\phi_i\}_1^m$ lie on the same linkage branch. However, all four-bar linkages studied in this chapter, planar, spherical and spatial, were shown in Sect. 3.4 to be *bimodal*, i.e., they all entail two solution branches of their input-output equation. This means that, within an iteration loop, the occurrence of branch-switching should be monitored. Below we explain a simple means of doing this, as applicable to planar linkages. The two branches of a typical planar four-bar linkage are apparent in Fig. 3.8a. In this figure, the transmission angle is $\mu = \angle BCD$ in one branch, $\mu' = \angle BC'D$ in the second. The qualitative difference between the two branches lies in *the sign of the sine of the transmission angle*, for we have $\sin\mu > 0$ in the first branch and $\sin\mu' < 0$ in the second. Moreover, $\sin\mu$ vanishes at *deadpoints*, when the input angle reaches either a maximum or a minimum—linkages of this kind have an input rocker. Hence, a simple way of deciding whether all values $\{\phi_i\}_1^m$ lie in the same branch relies on the computation of the sign of $\sin\mu$. This is most simply done by means of the 2D version of the cross product[14] of vectors $\overrightarrow{CB} = \mathbf{b} - \mathbf{c}$ and $\overrightarrow{CD} = \mathbf{d} - \mathbf{c}$, in this order, where $\mathbf{b}$, $\mathbf{c}$ and $\mathbf{d}$ are the position vectors of points B, C and D, respectively, in the given coordinate frame. The product at stake is given by

$$p \equiv (\mathbf{b} - \mathbf{c})^T\mathbf{E}(\mathbf{d} - \mathbf{c}) = \|\mathbf{b} - \mathbf{c}\|\,\|\mathbf{d} - \mathbf{c}\|\sin\mu = a_3a_4\sin\mu \tag{3.182}$$

[14] See Sect. 1.4.1.

with $\mathbf{E}$ introduced in Eq. (1.1a). Given that the link lengths are positive, we have the relation

$$\mathrm{sgn}(\sin\mu) = \mathrm{sgn}(p) \tag{3.183}$$

which now can be used to monitor branch-switching.

3.7.3 *Introducing a Massive Number of Data Points*

As illustrated by Hayes et al. (1999), one simple way of minimizing the structural error is via design-error minimization *in the presence of a large number of prescribed data-pairs*. We show with one example below that, as the *cardinality m* of the data set increases, the design and structural errors converge. The results are taken from the foregoing reference.

In the example below, the *weighted Euclidean norm* of the design and the structural error, $\|\mathbf{e}\|_{2W}$ and $\|\mathbf{s}\|_{2W}$, respectively, are minimized. For any m-dimensional vector $\mathbf{v}$, this norm is defined as the rms value of its components, namely,

$$\|\mathbf{v}\|_{2W} \equiv \sqrt{\frac{1}{m}\mathbf{v}^T\mathbf{v}} \tag{3.184}$$

Example 3.7.1 We synthesize here a planar and a spherical RRRR four-bar linkage to generate a quadratic IO function for the values given below:

$$\psi_i = \alpha + \Delta\psi_i, \ \ \phi_i = \beta + \Delta\phi_i, \quad \Delta\phi_i = \frac{9\Delta\psi_i^2}{8\pi}, \ \ i = 1, \ldots, m$$

For each linkage the IO dial zeros (α and β) are selected to minimize the condition number κ of $\mathbf{S}$ for each data-set, in following the procedure proposed by Liu and Angeles (1992a). Then, both the design and structural errors are determined for the linkages that minimize the respective Euclidean norms for data-sets with cardinalities of $m = \{10, 40, 70, \text{ and } 100\}$. These results are listed in Tables 3.5, 3.6, 3.7 and 3.8. Finally the structural errors, corresponding to $m = 40$, of the linkages that minimize the Euclidean norms of the design and structural errors are graphically displayed in Fig. 3.20.

3.8 Synthesis Under Mobility Constraints

In approximate synthesis of four-bar linkages, additional constraints can be included, such as design error, structural error, or mobility constraints. Of them, the mobility constraints are highly practical for a linkage, which means that either (*i*) the range

Table 3.5 Results for $m = 10$

	Planar RRRR	Spherical RRRR
α_{opt} (°)	123.8668	43.3182
β_{opt} (°)	91.7157	89.5221
κ_{opt}	33.2974	200.5262
$\|\mathbf{e}\|_{2W}$	2.2999×10^{-3}	2.4033×10^{-4}
$\|\mathbf{s}\|_{2W}$	1.8863×10^{-3}	1.3187×10^{-4}

Table 3.6 Results for $m = 40$

	Planar RRRR	Spherical RRRR
α_{opt} (°)	117.4593	42.7696
β_{opt} (°)	89.4020	88.8964
κ_{opt}	32.5549	203.0317
$\|\mathbf{e}\|_{2W}$	2.484×10^{-3}	2.984×10^{-4}
$\|\mathbf{s}\|_{2W}$	2.375×10^{-3}	1.671×10^{-4}

Table 3.7 Results for $m = 70$

	Planar RRRR	Spherical RRRR
α_{opt} (°)	116.4699	42.7014
β_{opt} (°)	89.0488	88.8045
κ_{opt}	32.5242	204.7696
$\|\mathbf{e}\|_{2W}$	2.496×10^{-3}	3.031×10^{-4}
$\|\mathbf{s}\|_{2W}$	2.438×10^{-3}	1.701×10^{-4}

Table 3.8 Results for $m = 100$

	Planar RRRR	Spherical RRRR
α_{opt} (°)	116.0679	42.6740
β_{opt} (°)	88.9057	88.7674
κ_{opt}	32.5170	205.5603
$\|\mathbf{e}\|_{2W}$	2.499×10^{-3}	3.047×10^{-4}
$\|\mathbf{s}\|_{2W}$	2.464×10^{-3}	1.712×10^{-4}

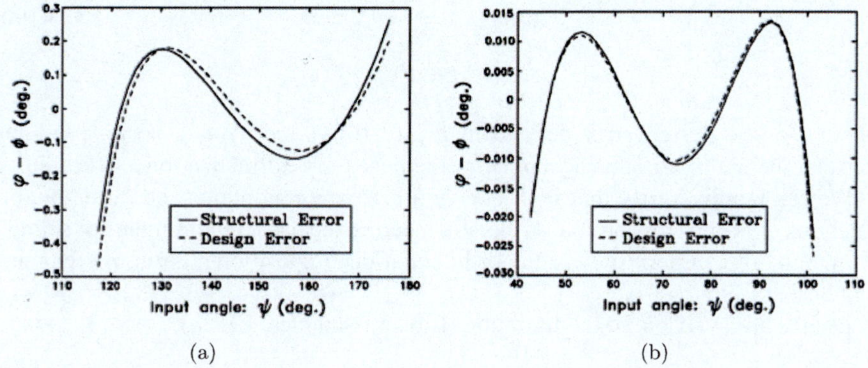

Fig. 3.20 Structural error comparison for: **a** planar and **b** spherical RRRR linkages upon minimizing $\|\mathbf{s}\|_{2W}$ and $\|\mathbf{e}\|_{2W}$

of rotation of either the input or the output link is larger than a specified value, or (ii) the full mobility criteria have to be satisfied.

Many criteria for the full mobility of different types of linkages have been developed. They are generally expressed in the form of inequalities. A simple way to include the mobility constraints in the optimization procedure is to use these constraints directly in the optimization problem formulation. However, such a formulation is less efficient and brings about possible problems of convergence. A method to overcome this limitation was developed by Liu and Angeles (1992b), with which inequality constraints are converted into equality constraints.

We take a planar four-bar linkage as an example to explain this method. It is known that the full mobility of the input link in a four-bar linkage is imposed by the inequalities

$$(k_1 + k_3)^2 - (1 + k_2)^2 \leq 0 \tag{3.185}$$

$$(k_1 - k_3)^2 - (1 - k_2)^2 \leq 0 \tag{3.186}$$

where k_i, $i = 1, \ldots, 3$, are Freudenstein parameters of the linkage, as defined in Eq. (3.8). Two equality constraints can be formulated,

$$g_1 = (k_1 + k_3)^2 - (1 + k_2)^2 - k_5^2 = 0 \tag{3.187}$$

$$g_2 = (k_1 - k_3)^2 - (1 - k_2)^2 - k_6^2 = 0 \tag{3.188}$$

where k_5 and k_6 are two *slack variables* that are included as additional design parameters. The mobility-constraint equations are grouped in vector form as

$$\mathbf{g}(\mathbf{x}) = 0 \tag{3.189}$$

An approximate design synthesis problem can thus be formulated in the form

$$\min z(\mathbf{x}) = \frac{1}{2}\mathbf{e}^T\mathbf{W}\mathbf{e} \quad (3.190)$$
$$\text{s.t. } \mathbf{g}(\mathbf{x}) = \mathbf{0}$$

where $\mathbf{e}$ is the design error defined in Eq. (3.100a) and $\mathbf{W}$ is a positive-definite constant weight matrix that accounts for the necessary normalization and scaling.

We can readily verify that both $\mathbf{e}(\mathbf{x})$ and $\mathbf{g}(\mathbf{x})$ are continuous and differentiable functions of the design variable $\mathbf{x}$, thereby meeting the basic requirements of a gradient method of solution, such as the orthogonal-decomposition algorithm (Teng and Angeles, 2001).

Details are available in the literature (Liu and Angeles, 1992b).

References

Alizadeh, D., Angeles, J., & Nokleby, S. (2013). Optimum design of a spherical quasi-homokinetic linkage for motion transmission between orthogonal axes. *Mechanism and Machine Theory,59*, 107–118.

Angeles, J. (2012). The dual generalized inverses and their applications in kinematic synthesis. In J. Lenarčič & M. Husty (Eds.), *Latest advances in robot kinematics* (pp. 1–10). Springer.

Angeles, J. (2014). *Fundamentals of robotic mechanical systems: Theory, methods, and algorithms* (4th ed.). Springer.

Bai, S. (2017). Geometric analysis of coupler-link mobility and circuits for planar four-bar linkages. *Mechanism and Machine Theory,118*, 53–64.

Dimentberg, F. M. (1965). *The screw calculus and its applications in mechanics*. Izdat.

Duditçă, F. D., Diaconescu, D., & Gogu, Gr. (1989). *Mecanisme Articulate*. Inventica şi Cinematica în Abordare Filogenetica: Tehnicá Publishers.

Forsythe, G. E. (1970). Pitfalls in computation, or why a math book isn't enough. *American Mathematical Monthly,27*, 931–956.

Freudenstein, F. (1955). Approximate synthesis of four-bar linkages. *Transactions of the ASME,77*, 853–861.

Hartenberg, R., & Denavit, J. (1964). *Kinematic synthesis of linkages*. McGraw-Hill Book Company.

Hayes, M., Parsa, K., and Angeles, J. (1999). The effect of data-set cardinality on the design and structural errors of four-bar function-generators. In *Proceedings of the Tenth World Congress on the Theory of Machines and Mechanisms*, (vol. 1, pp. 437–442). Oulu, Finland.

Liu, Z., & Angeles, J. (1992). Least-Square Optimization of Planar and Spherical Four-Bar Function Generators Under Mobility Constraints. *ASME Journal of Mechanical Design,114*(4), 569–573.

Liu, Z., & Angeles, J. (1992). The constrained least-square optimization of spherical four-bar linkages for rigid-body guidance. *Transuactions of the Canadian Society for Mechanical Engineering,16*(1), 47–60.

McCarthy, J. M., & Soh, G. S. (2011). *Geometric design of linkages*. Springer.

Nash, S. G., & Sofer, A. (1996). *Linear and nonlinear programming*. The McGraw-Hill Companies Inc.

Rico Martínez, J. M., & Duffy, J. (1993). The principle of transference: History, statement and proof. *Mechanism and Machine Theory,28*(1), 165–177.

Teng, C. P., & Angeles, J. (2001). A sequential-quadratic-programming algorithm using orthogonal decomposition with Gerschgorin stabilization. *ASME Journal of Mechanical Design,123*(4), 501–509.

Tinubu, S. O., & Gupta, K. C. (1984). Optimal synthesis of function generators without the branch defect. *ASME Journal of Mechanical Design,106*(3), 348–354.

Waldron, K. J., & Kinzel, G. L. (1999). *Kinematics, dynamics, and design of machinery*. Wiley.

Wright, D. G., Desai, S. M., & Henderson, W. H. (1964). Action of the subtalar and ankle-joint complex during the stance phase of walking. *The Journal of Bone and Joint Surgery,46*(2), 361–464.

Yang, A. T., & Freudenstein, F. (1964). Application of dual-number quaternion algebra to the analysis of spatial mechanisms. *Journal of Applied Mechanics,31*(2), 300–308.

Chapter 4
Motion Generation

4.1 Introduction

In the classical problem of *motion generation*, a.k.a. *rigid-body guidance*, a *discrete* sequence of poses of a rigid-body is given. These poses are to be visited, in the prescribed order, by the coupler link of a four-bar linkage of *a given type*, whose geometric parameters are to be determined. In this chapter we study the four-bar linkages of the three types studied in Chap. 3, namely, *planar, spherical* and *spatial*.

The problem was originally formulated by Ludwig Burmester (1840–1927), for the synthesis of a planar four-bar linkage (Koetsier, 2010). Moreover, the problem discussed here is limited to *finitely separated poses*, which means that velocity and acceleration conditions of the body in question are not considered. Furthermore, the problem reduces to one of *dyad synthesis*, a *dyad*, as the name indicates, being a link that carries *two kinematic pairs*, fixed at distinct points of the link. In the case of planar linkages, only R and P pairs are considered; in the spherical case, only R joints; in the case of spatial linkages, a dyad can carry any two of the six lower kinematic pairs.

4.2 Planar Four-Bar Linkages

A dyad is defined as a link coupled to two other links, one fixed, one mobile. In fact, this definition is not limited to planar linkages; it applies to spherical and spatial linkages as well. Regarding planar linkages, the coupling can take place via a R or a P joint, a dyad can thus be of any of four types: RR; RP; PR or PP. All four cases are considered in this chapter. Furthermore, a planar R pair is defined by one point, its center, a P joint by a unit vector indicating its direction. The R joint is thus known when the position vector of its center is known, the P-joint when a unit vector parallel to its direction is known.

J. Angeles and S. Bai, *Kinematics of Mechanical Systems*, Mathematical Engineering,
https://doi.org/10.1007/978-3-031-09544-3_4

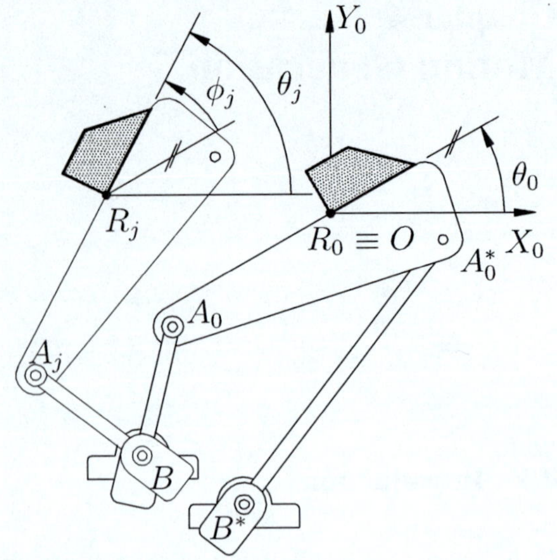

Fig. 4.1 Two finitely separated poses of a rigid body carried by the coupler link of a four-bar linkage

Shown in Fig. 4.1 is a four-bar linkage resulting from the arrangement of two dyads, A_0B and $A_0^*B^*$, sharing two common links, the *fixed* BB^* and the *floating* or *coupler* link $A_0A_0^*$. In the problem of linkage synthesis for *motion generation*, a set of poses to be attained by the coupler link is given via the successive locations of a landmark point R of the body, $\{ R_j \}_0^m$, and the corresponding set of position vectors $\{ \mathbf{r}_j \}_0^m$, together with the attitude attained by the body at the given poses, denoted by the set $\{ \theta_j \}_0^m$.

Henceforth we use X_0-Y_0 of Fig. 4.1 as the reference coordinate frame. Under the usual rigid-body assumption, the *dyad-synthesis equation* is readily derived: the distance between points A_0 and B equals that between A_j and B, for $j = 1, \ldots, m$, and hence,

$$\| \underbrace{(\mathbf{r}_j - \mathbf{b}) + \mathbf{Q}_j\mathbf{a}_0}_{\mathbf{a}_j - \mathbf{b}} \|^2 = \|\mathbf{a}_0 - \mathbf{b}\|^2 , \qquad j = 1, \ldots, m \tag{4.1}$$

where $\mathbf{a}_0$ and $\mathbf{b}$ are the position vectors of points A_0 and B, respectively, which are the *design parameters* of the RR dyad, while $\mathbf{Q}_j$ denotes the rotation matrix carrying the guided body, $A_0A_0^*$, from pose 0 to pose j, i.e.,

$$\mathbf{Q}_j = \begin{bmatrix} \cos\phi_j & -\sin\phi_j \\ \sin\phi_j & \cos\phi_j \end{bmatrix}, \quad \text{with} \quad \phi_j \equiv \theta_j - \theta_0 \tag{4.2}$$

Upon expansion of Eq. (4.1), and simplification of the expression thus resulting, we obtain

$$\mathbf{b}^T(\mathbf{1}-\mathbf{Q}_j)\mathbf{a}_0+\mathbf{r}_j^T\mathbf{Q}_j\mathbf{a}_0-\mathbf{r}_j^T\mathbf{b}+\frac{\mathbf{r}_j^T\mathbf{r}_j}{2}=0\,,\quad j=1,\ldots,m \tag{4.3}$$

where $\mathbf{1}$ is the 2×2 identity matrix, thereby obtaining the *dyad-synthesis equations* allowing the computation of the design parameters.

It will be shown that a planar dyad can be synthesized *exactly* for up to five poses, i.e., for $m=2,\ 3,\ 4$. These cases are studied below, following which *approximate synthesis* is discussed.

4.2.1 Dyad Synthesis for Three Poses

For the case at hand, $m=2$, and hence, two equations of the form of Eq. (4.3) are available. However, these equations involve four real unknowns, the two components of $\mathbf{a}_0$ and the two of $\mathbf{b}$. The designer thus has the freedom to specify two of these unknowns. Although any of the four unknowns can be prescribed, it makes sense to specify the two coordinates of the same point, A_0 or B. Moreover, since the space available to locate the anchor joint centers, B and B^*, is known, it is common practice to specify these. In this light, the two equations at hand are now arranged in a convenient form, with $\mathbf{a}_0$ as unknown:

$$\underbrace{\begin{bmatrix}\mathbf{b}^T(\mathbf{1}-\mathbf{Q}_1)+\mathbf{r}_1^T\mathbf{Q}_1\\ \mathbf{b}^T(\mathbf{1}-\mathbf{Q}_2)+\mathbf{r}_2^T\mathbf{Q}_2\end{bmatrix}}_{\mathbf{B}\in\mathbb{R}^{2\times 2}}\mathbf{a}_0=\underbrace{\begin{bmatrix}\mathbf{r}_1^T(\mathbf{b}-(1/2)\mathbf{r}_1)\\ \mathbf{r}_2^T(\mathbf{b}-(1/2)\mathbf{r}_2)\end{bmatrix}}_{\mathbf{r}\in\mathbb{R}^2} \tag{4.4}$$

whose solution can be obtained in closed form:

$$\mathbf{a}_0=\mathbf{B}^{-1}\mathbf{r} \tag{4.5}$$

with $\mathbf{B}^{-1}$ given from Fact 1.4.2 as

$$\mathbf{B}^{-1}=\frac{1}{\det(\mathbf{B})}\mathbf{E}\left[-(\mathbf{1}-\mathbf{Q}_2^T)\mathbf{b}-\mathbf{Q}_2^T\mathbf{r}_2\ \ (\mathbf{1}-\mathbf{Q}_1^T)\mathbf{b}+\mathbf{Q}_1^T\mathbf{r}_1\right] \tag{4.6}$$

and $\mathbf{E}$ representing the 2×2 rotation matrix through 90°, as introduced in Eq. (1.1a). In turn, $\det(\mathbf{B})$ is obtained by resorting to Fact 1.4.1:

$$\det(\mathbf{B})=[\mathbf{b}^T(\mathbf{1}-\mathbf{Q}_2)+\mathbf{r}_2^T\mathbf{Q}_2]\mathbf{E}[(\mathbf{1}-\mathbf{Q}_1^T)\mathbf{b}+\mathbf{Q}_1^T\mathbf{r}_1] \tag{4.7}$$

thereby completing the computation of $\boldsymbol{a}_0$ corresponding to the given position vector $\boldsymbol{b}$ of B. This completes, in turn, the synthesis problem at hand.

4.2.2 Dyad Synthesis for Four Poses

We start with some definitions and concepts of *algebraic geometry*, that will become useful in the balance of the chapter. For starters, according to Todd[1] "Algebraic geometry is the study of geometries that come from algebra, in particular, from rings." A ring, in turn, is a set S together with two binary operators, denoted by "+" and "$*$," (commonly interpreted as addition and multiplication, respectively) that satisfy certain conditions that allow for operations similar to those of real numbers, except for division. The instance of rings that are of interest in our study is polynomials, and multivariate polynomials for that matter.

Of particular interest to our study is multivariate polynomials involving the two coordinates of a point in a plane, $(x,\ y)$, or the three coordinates of a point in three-dimensional space, $(x,\ y,\ z)$. More facts from algebraic geometry, as pertaining specifically to cubic curves in the plane and in the unit sphere, are provided at the end of this subsection.

In the four-pose problem at hand, $m = 3$, and hence, this case leads to an underdetermined system of three equations in four unknowns. The approach to the solution of this case is different from the previous case. Rather than finding one specific solution, we will find a *locus* of solutions, or rather *two loci*, one for point A_0, henceforth termed the *circlepoint*, one for B, henceforth termed the *centerpoint*. The names arise from the geometry of the linkage, as A_0 lies in a *circle* with *center* at B.

In order to obtain the locus of points A_0, we eliminate $\mathbf{b}$ from the three synthesis equations at hand. To this end, we rewrite Eq. (4.3) in a more appropriate form, namely,

$$[\mathbf{a}_0^T(\mathbf{1}-\mathbf{Q}_j^T)-\mathbf{r}_j^T]\mathbf{b}+\frac{\mathbf{r}_j^T\mathbf{r}_j}{2}+\mathbf{r}_j^T\mathbf{Q}_j\mathbf{a}_0=0\,,\quad j=1,2,3 \tag{4.8}$$

To eliminate $\mathbf{b}$ from the above equations, let $\mathbf{z}$ denote the vector of homogeneous coordinates, introduced in Eq. (2.18), of B, the three equations (4.8) then leading to a *homogeneous system*, namely,

$$\mathbf{G}\mathbf{z}=\mathbf{0}_3 \tag{4.9}$$

where $\mathbf{0}_3$ is the three-dimensional zero vector, while $\mathbf{G}$ is a 3×3 matrix linear function of $\mathbf{a}_0$, whose jth row is the three-dimensional vector $\mathbf{g}_j^T$, for $j = 1,\ 2,\ 3$, all defined below:

$$\mathbf{z}\equiv\begin{bmatrix}\mathbf{b}\\1\end{bmatrix}=\begin{bmatrix}x\\y\\1\end{bmatrix},\quad \mathbf{G}\equiv\begin{bmatrix}\mathbf{g}_1^T\\\mathbf{g}_2^T\\\mathbf{g}_3^T\end{bmatrix}\in\mathbb{R}^{3\times 3} \tag{4.10}$$

[1] Todd, R., "Algebraic Geometry." From MathWorld–A Wolfram Web Resource, created by Eric W. Weisstein: http://mathworld.wolfram.com/AlgebraicGeometry.html.

with

$$\mathbf{g}_j = \begin{bmatrix} (\mathbf{1} - \mathbf{Q}_j)\mathbf{a}_0 - \mathbf{r}_j \\ \mathbf{r}_j^T(\mathbf{Q}_j\mathbf{a}_0 + \mathbf{r}_j/2) \end{bmatrix} \in \mathbb{R}^3, \quad j = 1,\ 2,\ 3 \tag{4.11}$$

Now, for Eq. (4.9) to admit a non-trivial solution, as required by the third component of $\mathbf{z}$, $\mathbf{G}$ must be singular, and hence,

$$G(x, y) \equiv \det(\mathbf{G}) \equiv \mathbf{g}_1 \times \mathbf{g}_2 \cdot \mathbf{g}_3 = 0 \tag{4.12}$$

Since each row of $\mathbf{G}$ is a *linear function* of $\mathbf{a}_0$, and the matrix is of 3×3, it follows that Eq. (4.12) defines an (implicit) *cubic* function[2] $G(x, y)$ in the components of $\mathbf{a}_0$, and hence, a cubic curve $\mathcal{K}$ in the X_0-Y_0 plane, termed the *circlepoint curve*, as it is the locus of the circlepoint A_0. Therefore, any point on $\mathcal{K}$ can be used to locate the centerpoint A_0.

Once a given point A_0 of $\mathcal{K}$ has been chosen, Eq. (4.8) represents an *overdetermined* linear system of three equations in only two unknowns, the components of $\mathbf{b}$. However, because $\mathbf{a}_0$ is the position vector of a point in $\mathcal{K}$, the three equations in question are bound to be linearly dependent—remember that $\mathbf{G}$ has been forced to be singular! Hence, any two of these equations yields the corresponding $\mathbf{b}$. The only difference in the three possible solutions obtained upon choosing two out of the three equations (4.8) is due to *roundoff error*. The difference can be large in the presence of an ill-conditioned 2×2 submatrix of $\mathbf{G}$. To alleviate the roundoff-error effect, it is highly advisable to regard the three equations *as if they were independent*, and compute $\mathbf{b}$ as the *linear* least-square approximation of the three equations using one of the methods of Sect. 1.4.5. This is one more instance of our *robustness philosophy*, which we stress throughout the book when numerical computations are at stake.

By the same token, $\mathbf{a}_0$ can be eliminated from Eqs. (4.3). To this end, these are rewritten in the form

$$\mathbf{H}\mathbf{w} = \mathbf{0}_3 \tag{4.13}$$

with $\mathbf{H}$ defined as a 3×3 matrix linear function of $\mathbf{b}$, while $\mathbf{w}$ is the *three-dimensional array of homogeneous coordinates* of A_0, i.e.,

$$\mathbf{w} \equiv \begin{bmatrix} \mathbf{a}_0 \\ 1 \end{bmatrix} = \begin{bmatrix} u \\ v \\ 1 \end{bmatrix}, \quad \mathbf{H} \equiv \begin{bmatrix} \mathbf{h}_1^T \\ \mathbf{h}_2^T \\ \mathbf{h}_3^T \end{bmatrix} \in \mathbb{R}^{3\times 3} \tag{4.14}$$

with

$$\mathbf{h}_j = \begin{bmatrix} (\mathbf{1} - \mathbf{Q}_j^T)\mathbf{b} + \mathbf{Q}_j^T\mathbf{r}_j \\ -\mathbf{r}_j^T\mathbf{b} + \mathbf{r}_j^T\mathbf{r}_j/2 \end{bmatrix} \in \mathbb{R}^3, \quad j = 1,\ 2,\ 3 \tag{4.15}$$

Following the same reasoning as that leading to Eq. (4.12), matrix $\mathbf{H}$ is bound to be singular, and hence,

[2] The cubic nature of $\det(\mathbf{G})$ is made apparent in Eq. (4.12), which shows three factors *linear* in $\mathbf{a}_0$.

$$\det(\mathbf{H}) = 0 \tag{4.16}$$

which defines a cubic equation in **b**, and hence, a cubic curve $\mathcal{M}$ in the X_0-Y_0 plane, termed the *centerpoint curve*. This curve is the locus of all centerpoints B that satisfy the three given synthesis equations. Again, once a point B in $\mathcal{M}$ has been chosen, and its position vector substituted into the synthesis equations (4.8), an "overdetermined" system of three redundant equations in the two components of $\mathbf{a}_0$ is obtained. The quotation marks indicate that the system is "redundant" only in form, as its three equations are bound to be consistent. The *putative least-square approximation* of Eq. (4.9) determines point A_0 in a form *robust to roundoff-error effects*.[3]

A Note on Cubic Equations

Curves $\mathcal{K}$ and $\mathcal{M}$ are cubic, meaning that they are endowed with special features worth highlighting. For starters, these curves are what is known as *algebraic*. A generic *planar* curve of this kind is represented by an *implicit function* $f(x, y) = 0$, whose left-hand side is a *bivariate polynomial*, namely, a linear combination of expressions $x^i y^j$, for integers i and j. The degree N of an algebraic curve is defined as $N = \max(i + j)$. As a result, an algebraic curve of degree N intersects a line at N points *at most*. In specific cases, the number of visible intersections of the plots of the curve and the line may be smaller than N, the reason being that some of the intersections are either points with complex coordinates, which are, hence, invisible in a plot, or points at infinity. Intersection points that represent a *multiple root* also fall into this category.

Furthermore, cubic curves bear the property of having one *asymptote*. This is *a line that is tangent to the curve, with the points of tangency lying at infinity*.

4.2.3 Dyad Synthesis for Five Poses

When $m = 4$, Eq. (4.8) can be cast in a homogeneous form, similar to Eq. (4.9), but now **G** becomes a 4×3 matrix, and hence,

$$\mathbf{G}\mathbf{z} = \mathbf{0}_4 \tag{4.17}$$

where $\mathbf{0}_4$ is the four-dimensional zero vector, while **G**, still a matrix linear function of $\mathbf{a}_0$, has won a fourth row, $\mathbf{g}_4^T$, vector **z** remaining as in Eq. (4.10), and all four vectors $\mathbf{g}_j$ keeping the same form as in the aforementioned equation.

The condition under which Eq. (4.17) admits a non-trivial solution **z** leads to the rank-deficiency of **G**, which means that $\text{rank}(\mathbf{G}) < 3$. This condition can be

[3] Robustness has been obtained by virtue of *redundancy*, a common engineering practice, as any civil-engineering structure, e.g., a *statically indeterminate structure*, like a bridge, can attest.

imposed by noticing that it implies that every triplet of rows of $\mathbf{G}$ is bound to be linearly dependent, thereby rendering every 3×3 submatrix of $\mathbf{G}$ singular.

If $\mathbf{G}_j$ is the 3×3 matrix formed by deleting the jth row from $\mathbf{G}$, then the singularity of every such square matrix defines one circlepoint curve $\mathcal{K}_j$, i.e.,

$$\mathcal{K}_j : \Delta_j(\mathbf{a}_0) \equiv \Delta_j(x, y) \equiv \det(\mathbf{G}_j) = 0, \quad j = 1, \ldots, 4 \tag{4.18}$$

Each of the four equations (4.18) defining one circlepoint curve in the X_0-Y_0 plane, one common intersection of the four curves $\mathcal{K}_j$ yields one real circlepoint of the dyad sought. If no such intersection occurs, the problem admits *no real solution*. Of course, *any* pair of equations (4.18) suffices to determine a set of circlepoints $\mathbf{a}_0$. Moreover, a pair of cubic equations bearing a Bezout number[4] of 9, one can expect up to nine real intersections of two cubics. However, upon (1) expressing the rotation matrix as $\mathbf{Q}_j = \cos\phi_j \mathbf{1} + \sin\phi_j \mathbf{E}$, with $\mathbf{E}$ introduced in Eq. (1.1a), and (2) introducing a change of variable, as suggested by Bottema and Roth (1978), the four cubic equations can be reduced to two quadratics, with a Bezout number of 4. Hence, the number of real solutions to be expected in this case is 0, 2, or 4, as demonstrated by Chang et al. (1991). Moreover, given that two dyads are needed to produce a planar four-bar linkage, the total number N of possible linkages is

$$N = \binom{4}{2} = 6$$

An alternative solution of the problem at hand is purely numeric: Regard the four cubic equations in $\mathbf{a}_0$ as an overdetermined system of four equations in two unknowns, and solve it by means of the Newton-Gauss method, introduced in Sect. 1.6.1. Notice that, in this case, various initial guesses must be tried, to obtain all possible solutions.

Likewise, the centerpoints of a RR dyad can be found from relations similar to those of Eq. (4.14), but now with a 4×3 matrix $\mathbf{H}$, namely.

$$\mathbf{H}\mathbf{w} = \mathbf{0}_4 \tag{4.19}$$

with $\mathbf{H}$ now gaining one fourth row, $\mathbf{h}_4^T$, $\mathbf{w}$ staying as in Eq. (4.14), while the four vectors $\mathbf{h}_j$ bear the same *gestalt* as in the same equation.

Therefore, for Eq. (4.19) to admit a non-trivial solution, $\mathbf{H}$ must be rank-deficient, and hence, each of its four 3×3 submatrices must be singular, which then defines four corresponding centerpoint curves $\mathcal{M}_j$:

$$\mathcal{M}_j : \tilde{\Delta}_j(\mathbf{b}) \equiv \tilde{\Delta}_j(u, v) \equiv \det(\mathbf{H}_j) = 0, \quad j = 1, \ldots, 4 \tag{4.20}$$

where $\mathbf{H}_j$ is formed by deleting the jth row from $\mathbf{H}$. The four equations thus obtained are functions of $\mathbf{b} = [u,\ v]^T$, and provide four centerpoint curves $\mathcal{M}_j$ in the X_0-Y_0 plane, for $j = 1, \ldots, 4$. The centerpoints are determined as the intersections of all

[4] See Sect. 1.3.1.

four curves $\mathcal{M}_j$. Again, if no common intersection occurs, then the problem admits no real solution.

While *algebraically* any pair of Eqs. (4.18) and (4.20) suffices to determine $\mathbf{a}_0$ or, correspondingly, $\mathbf{b}$, for *numerical robustness we recommend the use of all four equations involved*. Upon regarding this system of equations in the two unknown components of $\mathbf{a}_0$ ($\mathbf{b}$)—the Cartesian coordinates of A_0 (B)—as an overdetermined system of *four* equations in *two* unknowns, we compute the two unknowns as the *least-square approximation* of Eqs. (4.18) or (4.20), as the case may be. This approach is at the core of the robustness of the methodology proposed here, and stressed throughout the book.

4.2.4 Case Study: Synthesis of a Landing Gear Mechanism

Throughout this subsection we illustrate the synthesis procedure with the case study described below, for various numbers of given poses.

In developing a compact landing gear for *small aircraft*, a planar four-bar linkage is being considered. The linkage is to be anchored to the fuselage, indicated as the shaded region in Fig. 4.2. The task now consists in producing a design that will do the job, with the fixed R joints as close as possible to the fuselage boundary, and outside of the *working region*, i.e., the region swept by the wheel when it is being either deployed or retracted, as shown in the same figure. It has not as yet been decided what tire model will be used. For this reason, the design should be proposed in terms of the tire radius r. To this end, the relations below, deemed pertinent, are specified:

$$a = 2r, \quad b = 4r$$

Moreover, the tire can be assumed to fit in a rectangle of width $r/3$ and height $2r$.

In order to make the operations of *deployment* and *retraction* as smooth as possible, a *smooth* trajectory Γ to be plausibly followed by the *midpoint* of the tire, is defined as the intersection of the tire axis with the tire midplane, while the axis of the tire is to remain tangent to Γ. More precisely, Γ is defined as a *fourth-degree Lamé curve*[5] (Angeles and Pasini, 2020), namely,

$$\left(\frac{x}{a}\right)^4 + \left(\frac{y}{b}\right)^4 = 1 \tag{4.21}$$

A set of 11 poses is given in Table 4.1, from which subsets of poses are to be taken to illustrate the synthesis procedure for $m = 2,\ 3,\ 4,$ and 11, in the coordinate frame X-Y of the figure. A note on *data-conditioning* is in order here: The Lamé curve is as smooth as an ellipse—it bears the nickname of "superellipse"—but this smoothness carries certain features that need careful attention, as discussed below.

[5] Named after the French mathematician Gabriel Lamé (1795–1870).

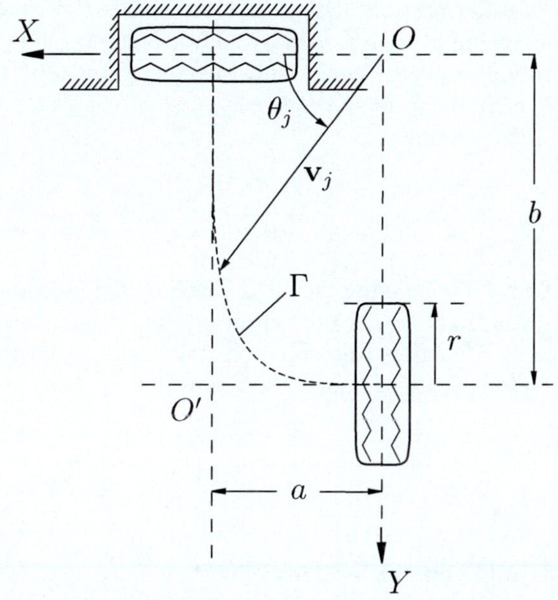

Fig. 4.2 A landing gear for small aircraft

Table 4.1 A set of prescribed poses for rigid-body guidance

j	$\mathbf{r}_j$ (mm)	θ_j (deg)
0	$r[2.0, 0.0]^T$	0.0
1	$r[1.99998, 0.3168]^T$	0.0142
2	$r[1.9997, 0.6497]^T$	0.1228
3	$r[1.9979, 1.0180]^T$	0.4737
4	$r[1.9914, 1.4468]^T$	1.3731
5	$r[1.9699, 1.9699]^T$	3.5763
6	$r[1.9013, 2.6170]^T$	9.2559
7	$r[1.6974, 3.3314]^T$	25.2897
8	$r[1.2474, 3.8392]^T$	61.2401
9	$r[0.6320, 3.98999]^T$	86.3626
10	$r[0.0, 4.0]^T$	90.0

For starters, if the Cartesian coordinates of the sample points are given in a frame with origin at O' of Fig. 4.2, with axis X coinciding with the axis of the wheel at its *deployed* pose and axis Y coinciding with the same axis at its *retracted* pose, then the resolution of the y-coordinates at points close to the value $x = a$ becomes rather poor. The same holds for the x-coordinates at points close to the value $y = b$. For this reason, the coordinate axes have been defined as shown in the same figure, with origin at O.

Moreover, calculating the coordinates from the *implicit equation* (4.21) upon assigning a value to, say x, and then solving for y is a *bad idea*, as this requires solving a quartic equation, which incurs *roundoff error*. A better, robust, approach is to resort to the *parametric equations* of the curve, as they *require no equation solving*, namely,

$$x(\theta) = \frac{a}{(1+\tan^m\theta)^{1/m}}, \quad y(\theta) = \frac{b\tan\theta}{(1+\tan^m\theta)^{1/m}}, \quad 0 \le \theta < \frac{\pi}{2} \tag{4.22}$$

with (a) θ denoting the angle made by the position vector of a point P of Γ in the given frame and (b) the strict inequality intended to avoid the *singularity*[6] of the $\tan(\cdot)$ function. Moreover, the unit vector $\mathbf{e}_t$, tangent to the curve at any value of θ within the above range, is given by

$$\mathbf{e}_t \equiv \frac{1}{\sqrt{x'(\theta)^2 + y'(\theta)^2}}\begin{bmatrix} -x'(\theta) \\ y'(\theta) \end{bmatrix} = \frac{1}{\sqrt{a^2\tan^6\theta + b^2}}\begin{bmatrix} -a\tan^3\theta \\ b \end{bmatrix}$$

Dyad Synthesis for Three Poses

Entries $j = 0, 5$ and 10 of Table 4.1 are chosen[7] here, the two ensuing rotation matrices, along with the two vectors $\mathbf{r}_j$, are now evaluated:

$$\mathbf{Q}_1 = \begin{bmatrix} 0.9981 & -0.0624 \\ 0.0624 & 0.9981 \end{bmatrix}, \quad \mathbf{Q}_2 = \begin{bmatrix} 0.0 & -1.0 \\ 1.0 & 0.0 \end{bmatrix}, \quad \mathbf{r}_1 = r\begin{bmatrix} 1.9699 \\ 1.9699 \end{bmatrix}, \quad \mathbf{r}_2 = r\begin{bmatrix} 0 \\ 4.0000 \end{bmatrix}$$

Further, let us choose B and B^*, of Fig. 4.1, symmetrically located on the X-axis with respect to the wheel center at the retracted pose, and with coordinates $x = 0$ and $x^* = 2a$, respectively. Upon plugging these values into matrix $\mathbf{B}$ and vector $\mathbf{r}$ of Eq. (4.4), the numerical values of $\mathbf{B}$, $\mathbf{B}^*$, $\mathbf{r}$ and $\mathbf{r}^*$ are obtained below, with an *intuitive* notation:

$$\mathbf{B} = r\begin{bmatrix} 2.0890 & 1.8432 \\ 4.0 & 0.0 \end{bmatrix}, \quad \mathbf{r} = r^2\begin{bmatrix} -3.8806 \\ -8.0 \end{bmatrix},$$
$$\mathbf{B}^* = r\begin{bmatrix} 2.0929 & 1.9680 \\ 6.0 & 2.0 \end{bmatrix}, \quad \mathbf{r}^* = r^2\begin{bmatrix} 0.0593 \\ -8 \end{bmatrix}$$

from which $\mathbf{B}^{-1}$ and $(\mathbf{B}^*)^{-1}$ are readily computed:

---

[6] Points at which the value of a function becomes unbounded are termed *singular*.

[7] The choice is made *strategically* in this and the ensuing cases, so as to have the three poses as evenly separated as possible. This, again, is in light of our robustness philosophy.

$$\mathbf{B}^{-1} = \frac{1}{r}\begin{bmatrix} 0.0 & 0.25 \\ 0.5425 & -0.2833 \end{bmatrix}, \quad (\mathbf{B}^{*})^{-1} = \frac{1}{r}\begin{bmatrix} -0.2624 & 0.2582 \\ 0.7872 & -0.2746 \end{bmatrix}$$

and hence,

$$\mathbf{a}_0 = r\begin{bmatrix} -2 \\ 0.1613 \end{bmatrix}, \quad \mathbf{a}_0^* = r\begin{bmatrix} -2.0811 \\ 2.2433 \end{bmatrix}$$

thereby completing the solution.

Dyad Synthesis for Four Poses

For this case we choose entries $j = 0, 3, 6$ and 10 of Table 4.1, which lead to the three rotation matrices and the three position vectors $\mathbf{r}_j$ below:

$$\mathbf{Q}_1 = \begin{bmatrix} 0.99997 & -0.00827 \\ 0.00827 & 0.99997 \end{bmatrix}, \quad \mathbf{Q}_2 = \begin{bmatrix} 0.9870 & -0.1608 \\ 0.1608 & 0.9870 \end{bmatrix}, \quad \mathbf{Q}_3 = \begin{bmatrix} 0.0 & -1.0 \\ 1.0 & 0.0 \end{bmatrix}$$

$$\mathbf{r}_1 = r\begin{bmatrix} 1.9979 \\ 1.0180 \end{bmatrix}, \quad \mathbf{r}_2 = r\begin{bmatrix} 1.9013 \\ 2.6170 \end{bmatrix}, \quad \mathbf{r}_3 = r\begin{bmatrix} 0.0 \\ 4.0 \end{bmatrix}$$

while vectors $\mathbf{g}_j$, for $j = 1, 2, 3$, are readily obtained:

$$\mathbf{g}_1 = r\begin{bmatrix} 0.0083\,\eta - 1.9979 \\ -0.0083\,\xi - 1.0180 \\ r(2.0062\,\xi + 1.0014\,\eta + 2.5139) \end{bmatrix},$$

$$\mathbf{g}_2 = r\begin{bmatrix} 0.0130\,\xi + 0.1608\,\eta - 1.9013 \\ -0.1608\,\xi + 0.0130\,\eta - 2.6170 \\ r(2.2975\,\xi + 2.2771\,\eta + 5.2318) \end{bmatrix},$$

$$\mathbf{g}_3 = r\begin{bmatrix} \xi + \eta \\ -\xi + \eta - 4 \\ 4r(\xi + 2) \end{bmatrix}, \quad \text{with } \xi \equiv \frac{x}{r} \ \& \ \eta \equiv \frac{y}{r}$$

Hence, $\det(\mathbf{G})$ follows:

$$\det(\mathbf{G}) = r^4(0.2781\,\xi^3 + 3.6362\,\xi^2 + 8.0696\,\xi + 0.1293\,\eta^3$$
$$+2.7084\,\eta^2 - 4.1706\,\eta + 0.1293\,\xi^2\eta + 0.2781\,\xi\eta^2 + 0.5424\,\xi\,\eta + 3.6524)$$

whose plot, representing the circlepoint curve $\mathcal{K}$, is displayed in Fig. 4.3, with a value of $r = 1$ unit—whatever the unit may be.

Next, the centerpoint curve $\mathcal{M}$ is obtained for the same set of poses. To this end, the 3×3 matrix $\mathbf{H}$ is produced from the three three-dimensional vectors $\mathbf{h}_j$, defined in Eq. (4.15), namely,

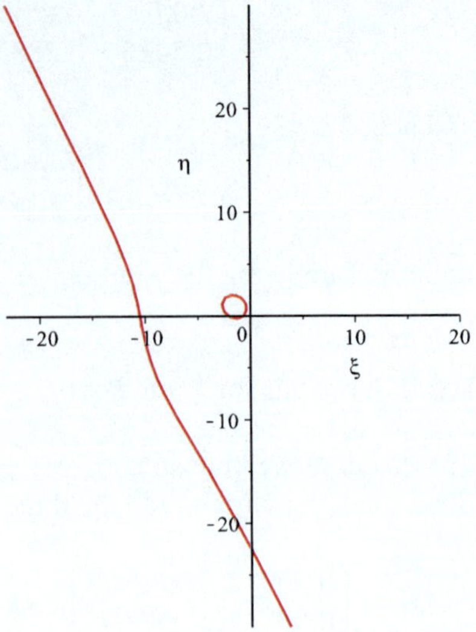

Fig. 4.3 Circlepoint curve $\mathcal{K}$ for $m = 3$

$$\mathbf{h}_1 = r\begin{bmatrix} -0.0083\,\eta^* + 2.0062 \\ 0.0083\,\xi^* + 1.0014 \\ -r(1.9979\,\xi^* + 1.0180\,\eta^* - 2.5139) \end{bmatrix},$$

$$\mathbf{h}_2 = r\begin{bmatrix} 0.0130\,\xi^* - 0.1608\,\eta^* + 2.2975 \\ 0.1608\,\xi^* + 0.0130\,\eta^* + 2.2771 \\ -r(1.9013\,\xi^* + 2.6170\,\eta^* - 5.2318) \end{bmatrix},$$

$$\mathbf{h}_3 = r\begin{bmatrix} \xi^* - \eta^* + 4 \\ \xi^* + \eta^* \\ 4r(2 - \eta^*) \end{bmatrix}, \quad \text{with} \quad \xi^* \equiv \frac{u}{r} \;\&\; \eta^* \equiv \frac{v}{r}$$

With the above vectors, **H** is obtained, its determinant being

$$\det(\mathbf{H}) = r^4(0.2797\,\xi^{*3} + 2.7628\,\xi^{*2} + 6.2571\,\xi^* + 0.1293\,\eta^{*3} + 2.2643\,\eta^{*2} \\ -13.3002\,\eta^* + 0.1293\,\xi^{*2}\eta^* + 0.2797\,\xi^*\eta^{*2} - 1.3029\,\xi^*\,\eta^* + 16.1999)$$

whose plot is also obtained for a value of $r = 1$ unit, as displayed in Fig. 4.4.

The reader is invited to choose points B and B^* on $\mathcal{K}$ at convenient locations, and then find the corresponding points A_0 and A_0^*, while ensuring that these points meet the design specs given at the outset.

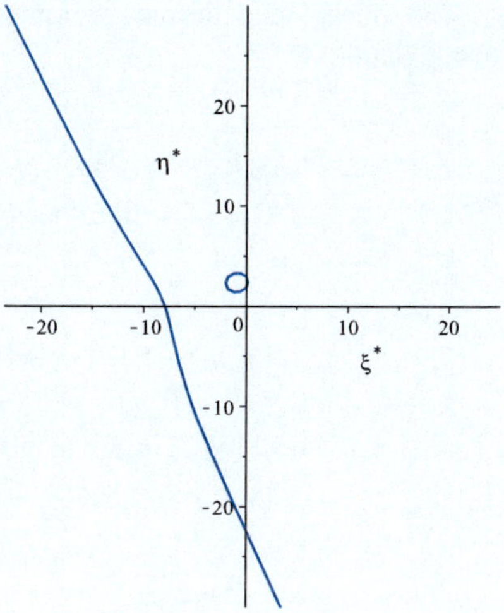

Fig. 4.4 Centerpoint curve $\mathcal{M}$ for $m = 3$

Dyad Synthesis for Five Poses

Now five poses are chosen from Table 4.1, namely, those of entries 0, 2, 5, 8 and 10. In this case, we have four $\mathbf{g}_j$ and four $\mathbf{h}_j$ vectors, namely,

$$\mathbf{g}_1 = r \begin{bmatrix} 0.0021\,\eta - 1.9997 \\ -0.0021\,\xi - 0.6497 \\ r(2.0010\,\xi + 0.6454\,\eta + 2.2104) \end{bmatrix},$$

$$\mathbf{g}_2 = r \begin{bmatrix} 0.0019\,\xi + 0.0624\,\eta - 1.9699\,r \\ -0.0624\,\xi + 0.0019\,\eta - 1.9699\,r \\ r\,(2.0890\,\xi + 1.8432\,\eta + 3.8806) \end{bmatrix}$$

$$\mathbf{g}_3 = r \begin{bmatrix} 0.5189\,\xi + 0.8766\,\eta - 1.2474 \\ -0.8766\,\xi + 0.5189\,\eta - 3.8392 \\ r\,(3.9658\,\xi + 0.7536\,\eta + 8.1479) \end{bmatrix}, \quad \mathbf{g}_4 = r \begin{bmatrix} \xi + \eta \\ -\xi + \eta - 4 \\ 4\,r\,(\xi + 2) \end{bmatrix}$$

$$\mathbf{h}_1 = r \begin{bmatrix} -0.0021\,\eta^* + 2.0010 \\ 0.0021\,\xi^* + 0.6454 \\ r(-1.9997\,\xi^* - 0.6497\,\eta^* + 2.2104) \end{bmatrix},$$

$$\mathbf{h}_2 = r \begin{bmatrix} 0.0019\,\xi^* - 0.0624\,\eta^* + 2.0890 \\ 0.0624\,\xi^* + 0.0019\,\eta^* + 1.8432 \\ r(-1.9699\,\xi^* - 1.9699\,\eta^* + 3.8806) \end{bmatrix}$$

$$\mathbf{h}_3 = r \begin{bmatrix} 0.5189\,\xi^* - 0.8766\,\eta^* + 3.9658 \\ 0.8766\,\xi^* + 0.5189\,\eta^* + 0.7536 \\ r(-1.2474\,\xi^* - 3.8392\,\eta^* + 8.1479) \end{bmatrix}, \quad \mathbf{h}_4 = r \begin{bmatrix} \xi^* - \eta^* + 4 \\ \xi^* + \eta^* \\ 4r(-\eta^* + 2) \end{bmatrix}$$

The four 3×3 determinants out of the corresponding 3×3 submatrices of $\mathbf{G}$ are displayed below[8]:

$$\det(\mathbf{G}_1) = r^4(0.6304\,\xi^3 + 2.4928\,\xi^2 + 0.5538\,\xi + 0.6139\,\eta^3 \\ -0.3568\,\eta^2 + 5.4837\,\eta + 0.6139\,\xi^2\eta + 0.6304\,\xi\eta^2 + 3.5365\,\xi\,\eta - 3.9946)$$

$$\det(\mathbf{G}_2) = r^4(0.7119\,\xi^3 + 4.6506\,\xi^2 + 7.4987\,\xi + 0.2293\,\eta^3 \\ 1.2157\,\eta^2 + 2.4298\,\eta + 0.2293\,\xi^2\eta + 0.7119\,\xi\eta^2 + 1.5992\,\xi\,\eta + 0.7906)$$

$$\det(\mathbf{G}_3) = r^4(0.1165\,\xi^3 + 2.9461\,\xi^2 + 9.0801\,\xi + 0.0351\,\eta^3 \\ +2.4682\,\eta^2 - 4.6776\,\eta + 0.0351\,\xi^2\eta + 0.1165\,\xi\eta^2 + 0.1228\,\xi\,\eta + 7.6518)$$

$$\det(\mathbf{G}_4) = r^4(0.0590\,\xi^3 + 1.5161\,\xi^2 + 5.1982\,\xi + 0.0177\,\eta^3 \\ +1.2675\,\eta^2 - 4.2112\,\eta + 0.0177\,\xi^2\eta + 0.0590\,\xi\eta^2 + 0.0228\,\xi\,\eta + 6.3060)$$

those of $\mathbf{H}$ being

$$\det(\mathbf{H}_1) = r^4(0.6294\,\xi^{*3} - 0.5660\,\xi^{*2} + 0.6615\,\xi^* + 0.5954\,\eta^{*3} - 1.8632\,\eta^{*2} \\ +0.0194\,\eta^* + 0.5954\,\xi^{*2}\eta^* + 0.6294\,\xi^*\eta^{*2} - 2.2636\,\xi^*\eta^* + 2.4905)$$

$$\det(\mathbf{H}_2) = r^4(0.7128\,\xi^{*3} + 1.4868\,\xi^{*2} + 2.4679\,\xi^* + 0.2287\,\eta^{*3} + 1.2504\,\eta^{*2} \\ -6.6459\,\eta^* + 0.2287\,\xi^{*2}\eta^* + 0.7128\,\xi^*\eta^{*2} - 2.2662\,\xi^*\eta^* + 5.9594)$$

$$\det(\mathbf{H}_3) = r^4(0.1166\,\xi^{*3} + 2.5357\,\xi^{*2} + 5.3744\,\xi^* + 0.0351\,\eta^{*3} + 2.3775\,\eta^{*2} \\ -10.9295\,\eta^* + 0.0351\,\xi^{*2}\eta^* + 0.1166\,\xi^*\eta^{*2} - 0.4544\,\xi^*\eta^* + 12.4422)$$

$$\det(\mathbf{H}_4) = r^4(0.0591\,\xi^{*3} + 1.3093\,\xi^{*2} + 3.3642\,\xi^* + 0.0177\,\eta^{*3} + 1.2204\,\eta^{*2} \\ -7.3735\,\eta^* + 0.0177\,\xi^{*2}\eta^* + 0.0591\,\xi^*\eta^{*2} - 0.2692\,\xi^*\eta^* + 10.4695)$$

Therefore, in the case of synthesis for five poses, we end up with four circlepoint curves $\mathcal{K}$ and four centerpoint curves $\mathcal{M}$, while only two of each of these suffice to determine each corresponding pair of points $(A_0,\ B)$ or $(A_0^*,\ B^*)$. Because of unavoidable roundoff error, the four curves in each case will intersect at different pairs of points. For this reason, we recommend the use of all four curves in each case, while regarding each quadruplet as being "independent." Numerically, this means regarding all four equations of interest as an overdetermined system of equations, and look for the *least-square* approximation to the solution in each case. This approach will lead to a *robust solution*.

[8] $\mathbf{G}_i$, for $i = 1, \ldots, 4$, is the 3×3 matrix obtained from $\mathbf{G}$ upon deleting its ith row. Ditto $\mathbf{H}_i$ regarding matrix $\mathbf{H}$.

4.2.5 The Presence of a P Joint in dyad Synthesis

The three cases studied above are considered here. We will do this under the consideration that a PR dyad can occur, i.e., one with a P joint coupling the dyad under synthesis with the fixed link BB^*. A similar procedure follows if a dyad of the RP type is desired. For starters, division of the two sides of Eq. (4.3) by $\|\mathbf{b}\|$ leads to:

$$\underbrace{[(\mathbf{1}-\mathbf{Q}_j)\mathbf{a}_0-\mathbf{r}_j]^T}_{-\mathbf{u}_j}\frac{\mathbf{b}}{\|\mathbf{b}\|}+\left(\mathbf{r}_j^T\mathbf{Q}_j\mathbf{a}_0+\frac{\mathbf{r}_j^T\mathbf{r}_j}{2}\right)\frac{1}{\|\mathbf{b}\|}=0\,,\qquad j=1,\ldots,m \tag{4.23}$$

Furthermore, we define a unit vector $\boldsymbol{\beta}$ as

$$\boldsymbol{\beta}=\frac{\mathbf{b}}{\|\mathbf{b}\|} \tag{4.24}$$

When $\|\mathbf{b}\|\rightarrow\infty$, the second term of the left-hand side of Eq. (4.23) goes to zero, while the centerpoint B goes to infinity, which leads to a PR dyad. The unit vector $\boldsymbol{\beta}$, in turn, gives the *line of sight* of B at infinity, the normal direction to $\boldsymbol{\beta}$ indicating the direction of the translations allowed by the P joint sought.

Moreover, upon substitution of Eq. (4.24) into Eq. (4.23), we obtain

$$\mathbf{u}_j^T\boldsymbol{\beta}=0,\quad j=1,\ldots,m \tag{4.25}$$

where $\mathbf{u}_j\equiv\mathbf{a}_j-\mathbf{a}_0$ is the displacement of the circlepoint A_0 at the jth pose, i.e.,

$$\mathbf{u}_j=\mathbf{r}_j-(\mathbf{1}-\mathbf{Q}_j)\mathbf{a}_0,\quad j=1,\ldots,m \tag{4.26}$$

With reference to Fig. 4.5, $\mathbf{u}_i$ $(i=1,\ldots,m)$ is the ith displacement vector of the circlepoint. For a PR dyad, all m vectors $\mathbf{u}_i$ must be parallel. In other words, the cross product of any two displacement vectors must vanish. However, rather than working with cross products, we simplify the analysis by resorting to the two-dimensional representation of the cross product introduced in Sect. 1.4.1. This is based on matrix $\mathbf{E}$ rotating vectors in the plane through 90° ccw. Hence, the parallelism condition between $\mathbf{u}_i$ and $\mathbf{u}_j$ can be expressed as

$$\Delta_{ij}=\mathbf{u}_j^T\mathbf{E}\mathbf{u}_i=0,\qquad i,j=1,\ldots,m,\qquad i\neq j \tag{4.27}$$

which expands to

$$\begin{aligned}\Delta_{ij}=\;&\mathbf{a}_0^T(-\mathbf{E}\mathbf{Q}_i-\mathbf{Q}_j^T\mathbf{E}+\mathbf{Q}_j^T\mathbf{E}\mathbf{Q}_i)\mathbf{a}_0\\&-(\mathbf{E}\mathbf{r}_i-\mathbf{Q}_j^T\mathbf{E}\mathbf{r}_i-\mathbf{E}\mathbf{r}_j+\mathbf{Q}_i^T\mathbf{E}\mathbf{r}_j)^T\mathbf{a}_0+\mathbf{r}_j^T\mathbf{E}\mathbf{r}_i=0\end{aligned} \tag{4.28}$$

We develop below all quadratic terms of Eq. (4.28), those in the first line of this equation, by writing $\mathbf{Q}_i$ in the form $\mathbf{Q}_i=c_i\mathbf{1}+s_i\mathbf{E}$, in which $c_i\equiv\cos\phi_i$ and

$s_i \equiv \sin\phi_i$. Hence,

$$-\mathbf{a}_0^T\mathbf{E}\mathbf{Q}_i\mathbf{a}_0 \equiv -\mathbf{a}_0^T\mathbf{E}(c_i\mathbf{1}+s_i\mathbf{E})\mathbf{a}_0 = -c_i\mathbf{a}_0^T\mathbf{E}\mathbf{a}_0 - s_i\mathbf{a}_0^T\mathbf{E}^2\mathbf{a}_0 = s_i||\mathbf{a}_0||^2 \quad (4.29a)$$

$$-\mathbf{a}_0^T\mathbf{Q}_j^T\mathbf{E}\mathbf{a}_0 \equiv -\mathbf{a}_0^T\mathbf{E}^T\mathbf{Q}_j\mathbf{a}_0 = \mathbf{a}_0^T\mathbf{E}\mathbf{Q}_j\mathbf{a}_0 = -s_j||\mathbf{a}_0||^2 \quad (4.29b)$$

$$\begin{aligned}\mathbf{a}_0^T\mathbf{Q}_j^T\mathbf{E}\mathbf{Q}_i\mathbf{a}_0 &\equiv \mathbf{a}_0^T(c_j\mathbf{1}-s_j\mathbf{E})(c_i\mathbf{E}+s_i\mathbf{1})\mathbf{a}_0\\ &= \mathbf{a}_0^T[-(c_js_i-s_jc_i)\mathbf{1}+(s_js_i+c_jc_i)\mathbf{E}]\mathbf{a}_0\\ &= (-c_js_i+s_jc_i)||\mathbf{a}_0||^2 = -\sin(\phi_j-\phi_i)||\mathbf{a}_0||^2 \end{aligned} \quad (4.29c)$$

Notice that the foregoing relations could have been obtained by pure geometric reasoning, as $\mathbf{E}\mathbf{Q}_i$ is a rotation through an angle $\phi_i + \pi/2$, given that angles of rotation are additive in the case of planar motion. By the same token, $\mathbf{Q}_j^T\mathbf{E}$ is a rotation through an angle $\pi/2 - \phi_j$, while $\mathbf{Q}_j^T\mathbf{E}\mathbf{Q}_i$ is one through an angle $\phi_i + \pi/2 - \phi_j$. Moreover, terms of the form $\mathbf{a}_0^T\mathbf{E}\mathbf{a}_0$ vanish because $\mathbf{E}$ is skew-symmetric.

Further, let

$$\mathbf{v}_{ij} \equiv -\mathbf{E}\mathbf{r}_i + \mathbf{Q}_j^T\mathbf{E}\mathbf{r}_i + \mathbf{E}\mathbf{r}_j - \mathbf{Q}_i^T\mathbf{E}\mathbf{r}_j \quad (4.30)$$

as appearing in the second line of Eqs. (4.28) the expression thus resulting now taking the form

$$\Delta_{ij} = (s_i - s_j - s_{i\bar{j}})\|\mathbf{a}_0\|^2 + \mathbf{v}_{ij}^T\mathbf{a}_0 + \mathbf{r}_j^T\mathbf{E}\mathbf{r}_i = 0, \quad i, j = 1, \ldots, m, \quad i \neq j \quad (4.31)$$

which represents the loci of A_0, of position vector $\mathbf{a}_0$, namely, a family of circles $\{C_{ij}\}_{i,j=1;i\neq j}^m$, where $s_{i\bar{j}} \equiv \sin(\phi_i - \phi_j)$. A line $\mathcal{L}$ passing through A_0 and parallel to the direction of sliding of the P joint is shown normal to vector $\boldsymbol{\beta}$ in Fig. 4.5. Given

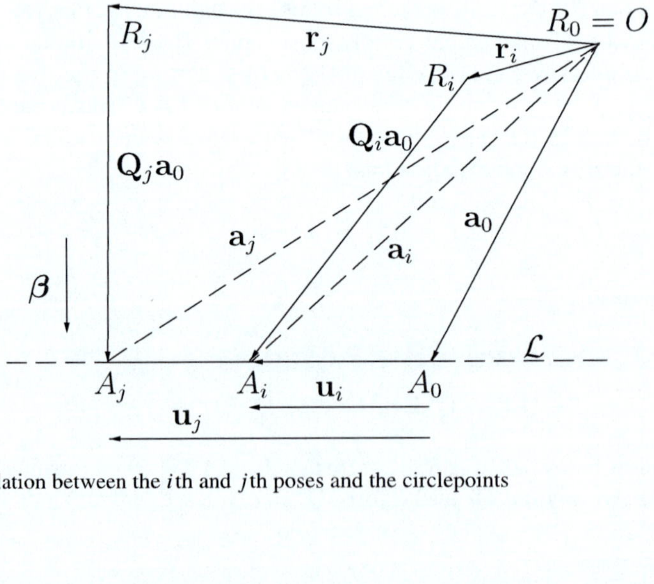

Fig. 4.5 Relation between the ith and jth poses and the circlepoints

that P joints have a direction, *but no position*, the slider implementing this joint can be placed anywhere on the fixed frame, as long as its sliding direction is parallel to $\mathcal{L}$. It is customary to represent the P joint of a PR dyad as a line passing through the center of the R joint, but this (i) *can be misleading* and (ii) *risks limiting the actual implementation of the joint in question.*

Remarks

1. The foregoing relations have been derived from the condition of the vanishing of the product $\mathbf{u}_j^T \mathbf{E} \mathbf{u}_i$ in Eq. (4.27). This product can be shown to be identical to the determinant of a 2×2 matrix $\mathbf{D}$, namely,

$$\mathbf{D} = \begin{bmatrix} \mathbf{u}_j & \mathbf{u}_i \end{bmatrix} \tag{4.32}$$

   Obviously, the vanishing of $\det(\mathbf{D})$ is equivalent to the linear dependence, and hence, the parallelism of vectors $\mathbf{u}_i$ and $\mathbf{u}_j$. Fact 1.4.1 should shed light on this relation.
2. The locus of the centerpoints of a PR dyad, for every pair of parallel unit vectors $(\mathbf{u}_i,\ \mathbf{u}_j)$, is a circle $\mathcal{C}_{ij}$, which is sometimes referred to as the "circle of sliders." Indeed, this circle was derived from geometric arguments by Hall (1961), although without giving a specific name to the circle. In the same reference, Hall points out that, in the limit, as the three poses defining that circle become infinitesimally separated, the circle becomes the *inflection circle* of curvature theory (Hartenberg and Denavit, 1964a).

Three-Pose Synthesis

In this case $m = 2$, Eqs. (4.31) thus reducing to a single one, with $i = 1$ and $j = 2$, i.e.,

$$(s_1 - s_2 - s_{1\bar{2}})\|\mathbf{a}_0\|^2 + \mathbf{v}_{12}^T \mathbf{a}_0 + \mathbf{r}_2^T \mathbf{E} \mathbf{r}_1 = 0$$

which can be cast in the form

$$\|\mathbf{a}_0\|^2 + \frac{\mathbf{v}_{12}^T}{s_1 - s_2 - s_{1\bar{2}}} \mathbf{a}_0 + \frac{\mathbf{r}_2^T \mathbf{E} \mathbf{r}_1}{s_1 - s_2 - s_{1\bar{2}}} = 0 \tag{4.33}$$

The radius and the position vector $\mathbf{c}$ of the center of the circle are readily found when comparing this equation with that of the circle with the foregoing characteristics, and applied to one point A_0 of the circle, of position vector $\mathbf{a}_0$:

$$\|\mathbf{a}_0\|^2 - 2\mathbf{c}^T \mathbf{a}_0 + \|\mathbf{c}\|^2 - r^2 = 0 \tag{4.34}$$

In summary, for three poses, the problem admits infinitely many solutions, all with one circlepoint A_0 lying in a circle of center C, of position vector $\mathbf{c}$, and radius r, as given above.

Four-Pose Synthesis

It is noteworthy that a PR dyad always exists in this case, as we have one single centerpoint curve, the corresponding P joint being equivalent to a R joint with its center at infinity. The direction of the unit vector $\boldsymbol{\beta}$, identical to the direction of the line of sight of point B lying at infinity and normal to the sliding of the P joint, is given by the asymptote of the centerpoint curve. Hence, one single P joint is to be expected here.

Moreover, we have three displacements $\{\mathbf{u}_i\}_1^3$, and hence, three *independent* parallelism conditions, namely, $\mathbf{u}_1 \| \mathbf{u}_2$, $\mathbf{u}_2 \| \mathbf{u}_3$ and $\mathbf{u}_3 \| \mathbf{u}_1$, as given by Eq. (4.31). In principle, two of these conditions imply the third one. However, if $\mathbf{u}_2$ happens to vanish, then, while the first two conditions still hold, the third does not necessarily do so. To guarantee the parallelism condition in any event, we use the three equations available in this case, which is in line with the *robustness* philosophy adopted at the outset in this book.

Now, each of the three equations at hand represents a circle in the X_0-Y_0 plane. It is apparent that we can always find a suitable linear combination of two distinct pairs out of the three equations (4.31), with $m = 3$, that will yield, correspondingly, two lines, upon subtraction of one from the other two, which will lead to the elimination of the quadratic terms $\|\mathbf{a}_0\|^2$ in the latter. Hence, the parallelism condition leads to one circle $\mathcal{C}$ and two lines $\mathcal{L}_1$ and $\mathcal{L}_2$. The geometric interpretation of the problem of finding the point A_0 is then straightforward: the circlepoint sought (a) is the intersection of the two lines and (b) lies on the circle.

Remarks

1. If the coefficient of $\|\mathbf{a}_0\|^2$ in one of Eqs. (4.31) vanishes, then the resulting equation is already a line. A second line is then obtained by a suitable linear combination of the remaining two circle equations, which will lead to the general case.
2. If the same coefficient vanishes in two of Eqs. (4.31), then we need not look for any linear combination to obtain the two lines of the general case.
3. If the same coefficient vanishes in the three Eqs. (4.31), then we have three lines that must be concurrent at one common point, the circlepoint A_0 sought.
4. The above statement on the existence of one single circlepoint A_0 is best explained by noticing that, from the plot of the centerpoint curve, its asymptote can be estimated by inspection—a precise value can be obtained from the equation of the asymptote, of course. If this estimate is plugged into Eq. (4.8), and the second term of the equation is deleted because, as $\|\mathbf{b}\| \to \infty$, this term tends to zero, we obtain three linear equations in $\mathbf{a}_0$. In the absence of roundoff error, only two are independent,[9] and hence, determine $\mathbf{a}_0$ uniquely.

[9] To account for roundoff error and within the spirit of our *robustness philosophy*, we recommend to regard the three equations as "independent", and compute the unique value of $\mathbf{a}_0$ as their least-square approximation.

Once $\mathbf{a}_0$ is obtained, solving for β is straightforward[10]:

$$\bar{\mathbf{u}} = \frac{1}{m}\sum_{j=1}^{m}\mathbf{u}_j, \quad \beta = \frac{\mathbf{E}\bar{\mathbf{u}}}{\|\mathbf{E}\bar{\mathbf{u}}\|} \tag{4.35}$$

Five-Pose Synthesis

In this case we have four displacements $\{\mathbf{u}_i\}_1^4$, and hence, six possible pairs[11] $(\mathbf{u}_i, \mathbf{u}_j)$, for $j \neq i$. It is noteworthy that the parallelism relation is *transitive*, i.e., if $\mathbf{u}_1 \| \mathbf{u}_2$ and $\mathbf{u}_2 \| \mathbf{u}_3$, then $\mathbf{u}_1 \| \mathbf{u}_3$. Hence, the number of independent relations reduces to three. However, and within the spirit of *numerical robustness*, we recommend the use of the full six parallelism conditions available.

We can always find a suitable linear combination of the first equation with each of the remaining five equations (4.31) that will yield, correspondingly, five lines, upon elimination of the quadratic terms $\|\mathbf{a}_0\|^2$, in the four-pose case. Hence, the parallelism conditions lead to one circle $\mathcal{C}$ and five lines $\mathcal{L}_1, \ldots, \mathcal{L}_5$. A geometric interpretation of the solution then follows: the circlepoint A_0 sought (a) is the common intersection of the five lines and (b) lies on the circle.

A similar discussion for the cases in which the coefficient of $\|\mathbf{a}_0\|^2$ in Eq. (4.31) vanishes in all four equations is straightforward, and follows the same line of reasoning as Remark 3 of the four-pose case.

If there is one point common to the five lines and the circle, this point is then the solution A_0 sought. Otherwise, there is no solution. Once $\mathbf{a}_0$ is obtained, β can be found from Eq. (4.35).

This whole subsection was devoted entirely to the synthesis of a PR joint. To handle the case of a RP joint the procedure is the same: divide both sides of Eq. (4.3) by $\|\mathbf{a}_0\|$ and then let $\|\mathbf{a}_0\| \to \infty$. The ensuing procedure parallels that described above. The reader is invited to develop it on her/his own, but details are available in the literature (Chen and Angeles, 2008; Chen et al., 2008).

Finally, notice that, in this case, a P joint is not to be taken for granted. Indeed, for a centerpoint to lie at infinity, and hence, for a PR dyad to be possible, the asymptotes of all centerpoint curves $\mathcal{K}_j$ must be parallel. Likewise, for a RP dyad to be possible, the asymptotes of all circlepoint curves $\mathcal{M}_j$ must be parallel. The condition for a PP dyad to occur in this case follows suit.

---

[10] One single vector $\mathbf{u}_j$ would suffice. We take the mean value here in order to filter out roundoff error.

[11] The combinatoric number of four objects taking two at a time.

4.2.6 Approximate Synthesis

When the number of prescribed poses is greater than five, i.e., when $m > 4$ in Eq. (4.3), the problem admits no solution. However, it is always possible to find a linkage that *best* meets the given poses with *a minimum error in the least-square sense.*[12] To this end, we regard the left-hand side of Eqs. (4.3) as functions e_j of $\mathbf{a}_0$ and $\mathbf{b}$ that represent the error incurred in the approximation of the jth pose. We thus define this error as

$$e_j \equiv \mathbf{b}^T(\mathbf{1} - \mathbf{Q}_j)\mathbf{a}_0 + \mathbf{r}_j^T\mathbf{Q}_j\mathbf{a}_0 - \mathbf{r}_j^T\mathbf{b} + \frac{\mathbf{r}_j^T\mathbf{r}_j}{2} \quad j = 1, \ldots, m \geq 5 \qquad (4.36)$$

Caveat: A *scalar pose error*, in fact, is impossible to define (Angeles, 2006), the reason being that the concept includes items with different dimensions, length for the position and dimensionless for the orientation. Nevertheless, the pose error defined in Eq. (4.36) is dimensionally homogeneous, as it is measured in units of surface, i.e., in m^2.

The problem is now handled as one of *optimum design*. In this vein, an *objective function* $z(\mathbf{x})$ is defined, that is positive-definite—i.e., positive, unless the error vanishes, in which case the function vanishes as well—with $\mathbf{x}$ denoting the vector of *design variables*, namely, the position vectors of the circlepoint and the centerpoint, $\mathbf{a}_0$ and $\mathbf{b}$, respectively. To guarantee positive-definiteness, moreover, $z(\mathbf{x})$ is defined as a norm of the error vector $\mathbf{e}$, as introduced below. Moreover, to ease the ensuing calculations, the Euclidian norm is adopted, as the gradient of its square with respect to the design-variable vector is a vector linear function of the error, as we know from the experience gained in Chap. 1.

The optimum-design problem at hand is thus formulated as

$$z \equiv \frac{1}{2}\|\mathbf{e}\|^2 = \frac{1}{2}\sum_1^m e_j^2(\mathbf{x}) \quad \rightarrow \quad \min_{\mathbf{x}} \quad \text{with} \quad \mathbf{e} \equiv \begin{bmatrix} e_1 \\ \vdots \\ e_m \end{bmatrix} \quad \& \ \mathbf{x} \equiv \begin{bmatrix} \mathbf{a}_0 \\ \mathbf{b} \end{bmatrix} \in \mathbb{R}^4 \qquad (4.37)$$

Since the optimization problem requires no constraints, its *first-order normality conditions* (FONC) (Hillier and Liebermann, 1995) reduce to the vanishing of the gradient of z with respect to $\mathbf{x}$:

[12] By no means the only criterion; other are available, but to a greater computational cost.

$$\nabla z \equiv \begin{bmatrix} \sum_1^m e_j \dfrac{\partial e_j}{\partial \mathbf{a}_0} \\ \sum_1^m e_j \dfrac{\partial e_j}{\partial \mathbf{b}} \end{bmatrix} = \mathbf{0}_4 \tag{4.38}$$

where

$$\frac{\partial e_j}{\partial \mathbf{a}_0} = (\mathbf{1} - \mathbf{Q}_j^T)\mathbf{b} + \mathbf{Q}_j^T \mathbf{r}_j \quad \& \quad \frac{\partial e_j}{\partial \mathbf{b}} = (\mathbf{1} - \mathbf{Q}_j)\mathbf{a}_0 - \mathbf{r}_j \tag{4.39}$$

Upon substitution of the two foregoing expressions into the FONC of Eq. (4.38), while recalling the expression for e_j displayed in Eq. (4.36), the FONC decompose into two two-dimensional subsystems, namely,

$$\frac{\partial z}{\partial \mathbf{a}_0} \equiv \sum_1^m [\mathbf{b}^T(\mathbf{1} - \mathbf{Q}_j)\mathbf{a}_0 + \mathbf{r}_j^T \mathbf{Q}_j \mathbf{a}_0 - \mathbf{r}_j^T \mathbf{b} + \frac{\mathbf{r}_j^T \mathbf{r}_j}{2}][(\mathbf{1} - \mathbf{Q}_j^T)\mathbf{b} + \mathbf{Q}_j^T \mathbf{r}_j] = \mathbf{0}_2 \tag{4.40a}$$

and

$$\frac{\partial z}{\partial \mathbf{b}} \equiv \sum_1^m [\mathbf{b}^T(\mathbf{1} - \mathbf{Q}_j)\mathbf{a}_0 + \mathbf{r}_j^T \mathbf{Q}_j \mathbf{a}_0 - \mathbf{r}_j^T \mathbf{b} + \frac{\mathbf{r}_j^T \mathbf{r}_j}{2}][(\mathbf{1} - \mathbf{Q}_j)\mathbf{a}_0 - \mathbf{r}_j] = \mathbf{0}_2 \tag{4.40b}$$

In light of the two foregoing expressions, we now have two useful results:

Lemma 4.2.1 *The partial derivative of z with respect to $\mathbf{a}_0$ is linear in $\mathbf{a}_0$ and quadratic in $\mathbf{b}$, this derivative thus being cubic in $\mathbf{x}$.*

Lemma 4.2.2 *The partial derivative of z with respect to $\mathbf{b}$ is quadratic in $\mathbf{a}_0$ and linear in $\mathbf{b}$, this derivative thus being cubic in $\mathbf{x}$ as well.*

As a result of the two above lemmas, the normality conditions (4.38) are both cubic in $\mathbf{x}$, and hence, lead to a total of four cubic scalar equations in $\mathbf{x}$, the Bezout number of the algebraic system (4.40a) and (4.40b) then being $N_B = 3^4 = 81$. This means that up to 81 solutions are to be expected, some real, some complex. Of course, we are interested only in the real solutions.

Furthermore, note that the first two-dimensional subsystem of Eq. (4.38) is linear in $\mathbf{a}_0$, the second linear in $\mathbf{b}$, although both are quadratic in the other vector block of $\mathbf{x}$, the latter defined in Eq. (4.71). The above feature eases the reduction of the polynomial system to two scalar equations in only two unknowns, either the coordinates of A_0 or those of B. For concreteness, the second two-dimensional subsystem of the system in question is used to solve for $\mathbf{b}$ in terms of $\mathbf{a}_0$, the resulting expression then substituted into the first two-dimensional subsystem of the same system, thereby obtaining a new two-dimensional vector equation in $\mathbf{a}_0$. First, each term of the second summation of Eq. (4.38), i.e., the intermediate side of Eq. (4.40b), is expanded:

$$e_j \frac{\partial e_j}{\partial \mathbf{b}} = [\underbrace{\mathbf{b}^T(\mathbf{1}-\mathbf{Q}_j)\mathbf{a}_0}_{\mathbf{a}_0^T(\mathbf{1}-\mathbf{Q}_j^T)\mathbf{b}} + \mathbf{r}_j^T\mathbf{Q}_j^T\mathbf{a}_0 - \mathbf{r}_j^T\mathbf{b} + \frac{\mathbf{r}_j^T\mathbf{r}_j}{2}][(\mathbf{1}-\mathbf{Q}_j)\mathbf{a}_0 - \mathbf{r}_j]$$

Then, after rearranging the right-hand side of the above equation, to assemble all terms including $\mathbf{b}$ in one single term, one obtains

$$e_j \frac{\partial e_j}{\partial \mathbf{b}} = [(\mathbf{1}-\mathbf{Q}_j)\mathbf{a}_0 - \mathbf{r}_j]\underbrace{[\mathbf{a}_0^T(\mathbf{1}-\mathbf{Q}_j^T) - \mathbf{r}_j^T]}_{[(\mathbf{1}-\mathbf{Q}_j)\mathbf{a}_0-\mathbf{r}_j]^T}\mathbf{b} + \mathbf{r}_j^T(\mathbf{Q}_j^T\mathbf{a}_0 + \frac{\mathbf{r}_j}{2})[(\mathbf{1}-\mathbf{Q}_j)\mathbf{a}_0 - \mathbf{r}_j]$$

That is,

$$e_j \frac{\partial e_j}{\partial \mathbf{b}} = [(\mathbf{1}-\mathbf{Q}_j)\mathbf{a}_0 - \mathbf{r}_j][(\mathbf{1}-\mathbf{Q}_j)\mathbf{a}_0 - \mathbf{r}_j]^T\mathbf{b} + \underbrace{\mathbf{r}_j^T(\mathbf{Q}_j^T\mathbf{a}_0 + \frac{\mathbf{r}_j}{2})}_{(\mathbf{Q}_j^T\mathbf{a}_0+\mathbf{r}_j/2)^T\mathbf{r}_j}[(\mathbf{1}-\mathbf{Q}_j)\mathbf{a}_0 - \mathbf{r}_j]$$

which, when substituted into Eq. (4.40b), leads to a linear equation in $\mathbf{b}$, in terms of $\mathbf{a}_0$ and the data:

$$\mathbf{B}\mathbf{b} + \mathbf{r} = \mathbf{0}_2 \tag{4.41}$$

with $\mathbf{B}$ and $\mathbf{r}$ defined as[13]

$$\mathbf{B} \equiv \sum_1^m [(\mathbf{1}-\mathbf{Q}_j)\mathbf{a}_0 - \mathbf{r}_j][(\mathbf{1}-\mathbf{Q}_j)\mathbf{a}_0 - \mathbf{r}_j]^T, \quad \mathbf{r} \equiv \sum_1^m [(\mathbf{Q}_j^T\mathbf{a}_0 + \frac{\mathbf{r}_j}{2})^T\mathbf{r}_j][(\mathbf{1}-\mathbf{Q}_j)\mathbf{a}_0 - \mathbf{r}_j]$$

and hence,

$$\mathbf{b} = -\mathbf{B}^{-1}\mathbf{r} \quad \text{with} \quad \mathbf{B}^{-1} \equiv \frac{\mathrm{Adj}(\mathbf{B})}{\det(\mathbf{B})} \tag{4.42}$$

Apparently, $\mathbf{B}$ is a sum of $m > 4$ rank-one[14] 2×2 matrices, all, in general, linearly independent. As it takes only two such matrices to make up a 2×2 nonsingular matrix, and the sum has at least five terms, matrix $\mathbf{B}$ is, except for pathological cases, invertible. Furthermore, upon invoking Facts 1.4.1 and 1.4.2, it becomes apparent that $\mathbf{B}^{-1}$ is a rational expression, whose matrix numerator, $\mathrm{Adj}(\mathbf{B})$, is *quadratic* in $\mathbf{a}_0$, its denominator, $\det(\mathbf{B})$ *quartic* in the same. Therefore, the product $\det(\mathbf{B})e_j(\partial e_j/\partial \mathbf{b})$ is *nonic* in $\mathbf{a}_0$, and hence, we have

Lemma 4.2.3 *The vanishing of the partial derivative of* z *with respect to* $\mathbf{a}_0$ *leads to two* nonic *equations in* $\mathbf{a}_0$, *free of* $\mathbf{b}$.

---

[13] $\mathbf{B}$ in Eq. (4.41) is not to be confused with $\mathbf{B}$ in Eq. (4.4).

[14] See Chap. 2 for a definition of *rank-one matrix*.

Since we have two scalar nonic equations to solve for $\mathbf{a}_0$, the Bezout number of the system in question is $N_B = 9^2 = 81$, i.e., the problem admits up to 81 solutions, some real, some complex.

By symmetry, we have one complementary result:

Lemma 4.2.4 *The vanishing of the partial derivative of z with respect to $\mathbf{b}$ leads to two nonic equations in $\mathbf{b}$, free of $\mathbf{a}_0$.*

Therefore, the system of two nonic equations arising from the vanishing of the partial derivative of z with respect to $\mathbf{b}$ also has a Bezout number of 81. However, once one particular circlepoint A_0 has been found, of position vector $\mathbf{a}_0$, the corresponding centerpoint B, of position vector $\mathbf{b}$ can be computed in closed form, namely, via Eq. (4.42). This holds, of course, as long as $\mathbf{B}$ is non-singular.

Each of the two scalar equations in $\mathbf{a}_0 = [x,\ y]^T$ defines a contour $\mathcal{K}_j$, for $j = 1,\ 2$, in the X_0-Y_0 plane. This means that the real values of $\mathbf{a}_0$ can be found *by inspection*, at the intersection of the two contours, which can be fairly regarded as circlepoint curves, as each represents a *locus* of circlepoint candidates. The intersections of the two loci then determine which of the infinitely many candidates is a valid circlepoint. The same holds for the centerpoints. All valid circlepoints and centerpoints, however, are *stationary points* (SPs) of the optimization problem at hand. Each can be a *local maximum*, a *local minimum* or a *saddle point*. The nature of each SP is elucidated by means of the Hessian matrix $\mathbf{H}$ of z with respect to $\mathbf{x}$, namely,

$$\mathbf{H} = \begin{bmatrix} \partial^2 z/\partial \mathbf{a}_0^2 & \partial^2 z/(\partial \mathbf{a}_0 \partial \mathbf{b}) \\ \partial^2 z/(\partial \mathbf{b} \partial \mathbf{a}_0) & \partial^2 z/\partial \mathbf{b}^2 \end{bmatrix} \tag{4.43}$$

Now, since $z(\mathbf{a}_0,\ \mathbf{b})$ is a biquadratic equation in $\mathbf{a}_0$ and $\mathbf{b}$, i.e., *quartic*—and hence, *algebraic*—in $\mathbf{x}$, it satisfies the conditions of Schwarz's Theorem (Brand, 1955); therefore, $\mathbf{H}$ is symmetric, its three distinct blocks[15] being,

$$\frac{\partial^2 z}{\partial \mathbf{a}_0^2} = \left(\frac{\partial^2 z}{\partial \mathbf{a}_0^2}\right)^T, \quad \frac{\partial^2 z}{\partial \mathbf{b}^2} = \left(\frac{\partial^2 z}{\partial \mathbf{b}^2}\right)^T, \quad \frac{\partial^2 z}{\partial \mathbf{b} \partial \mathbf{a}_0} = \left(\frac{\partial^2 z}{\partial \mathbf{a}_0 \partial \mathbf{b}}\right)^T \tag{4.44}$$

Hence, the four eigenvalues of $\mathbf{H}$ are all real, the corresponding eigenvectors mutually orthogonal. Now the identification of the nature of each SP is summarized below:

1. A SP is a local maximum if $\mathbf{H}$ is negative-definite, i.e., if all its eigenvalues are negative at this point;
2. A SP is a local minimum if $\mathbf{H}$ is positive-definite, i.e., if all its eigenvalues are positive at this point; and
3. A SP is a saddle point if $\mathbf{H}$ is sign-indefinite at this point.

[15] Its two diagonal blocks are identical by virtue of Schwarz's Theorem.

Our interest being the minimization of the Euclidian norm of the error, we look only at the local minima. Of these, the smallest one, i.e., the *minimum minimorum*, is the *global minimum*.

Example of Synthesis with 11 Poses

In this case we aim at a planar four-bar linkage that can *best meet* all the 11 poses displayed in Table 4.1. To this end, $\mathbf{b}$ is expressed as a rational vector function of $\mathbf{a}_0$, as per Eqs. (4.42). The expression thus resulting is then substituted into Eq. (4.40a), which leads to two nonic equations in $\mathbf{a}_0$, $F_i(x, y) = 0$, $i = 1, 2$, free of u and v, the components of $\mathbf{b}$. Notice that $F_1(x, y) = 0$ represents the normality condition $\partial z/\partial x = 0$ as one equation in x and y, $F_2(x, y) = 0$ representing $\partial z/\partial y = 0$—or their corresponding nondimensional versions, with ξ replacing x and η replacing y—likewise.

The two foregoing *implicit functions* of x and y—ξ and η, correspondingly, when normalized—which can be regarded as circlepoint curves $\mathcal{K}$ for the case at hand, are plotted in Fig. 4.6. In this figure, $\partial z/\partial \xi$ appears as a continuous curve, $\partial z/\partial \eta$ as a dashed curve, for visualization purpose.

While the system of two nonic equations is expected to admit up to $9^2 = 81$ solutions, only two real solutions were found within a *reasonable* distance to the origin, i.e., within a region commensurate with the dimensions of the landing gear operation of deployment and retraction. Alas, these two solutions are a) too close to each other, which would make a rather too short coupler link, and b) close to the fuselage boundary.

The moral of the story here is that, to enable *geometrically sound* solutions, *geometric constraints* must be introduced. While we will not pursue the solution of constrained-optimization problems in this case, the reader is invited to look at the pertinent techniques in the literature on mathematical programming, e.g., Hillier and Liebermann (1995).

4.3 Spherical Four-Bar Linkages

The spherical four-bar linkage under synthesis is depicted in Fig. 4.7, its four linkage dimensions $\{\alpha_j\}_1^4$ indicated therein. Just as in the planar case of Sect. 4.2, the linkage is composed of four links coupled by four R joints, this time around with their axes intersecting at one common point, O, as depicted in Fig. 4.7. The two fixed axes, OB and OB^*, intersect the *unit sphere* at points B and B^*, their two moving axes, OA and OA^*, at points A and A^*, respectively. For brevity, the axes will be referred to only by their intersection points with the sphere. More accurately, one should refer here not to centerpoints and circlepoints, but, rather, to *cone axis* and *cone element*, respectively, as axis A sweeps a circular conical surface of apex O, one of its elements being denoted, henceforth, A_0.

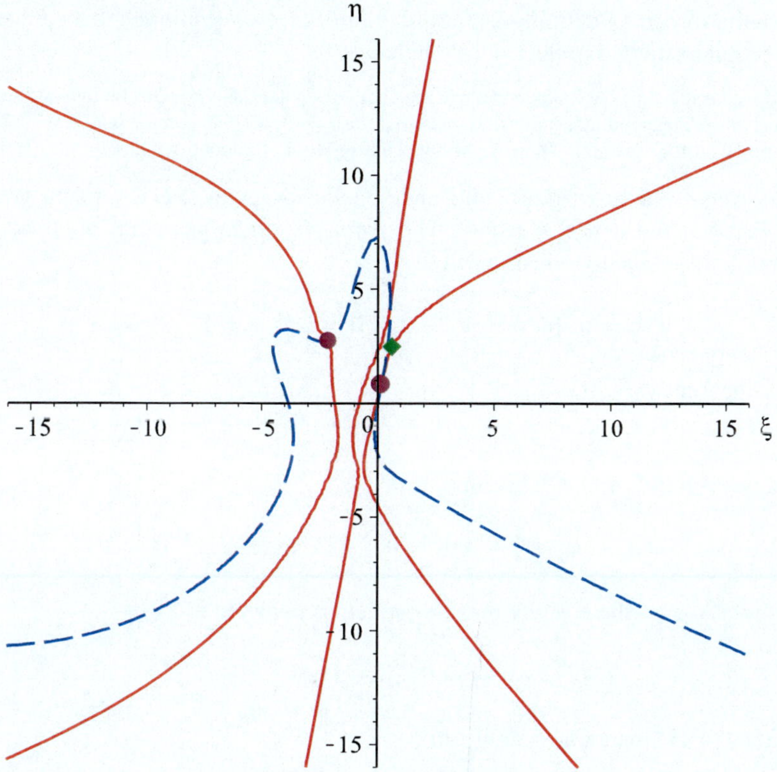

Fig. 4.6 Circlepoint curves $\mathcal{K}$ for $m = 10$

The reference locations of the moving axes are thus denoted A_0 and A_0^*, the linkage then being fully defined by the two dyads BA_0 and $B^*A_0^*$. By analogy with the planar case, points B and B^* are called the *centerpoints*, while A_0 and A_0^* the *circlepoints*.

As the coupler link moves, while visiting the m prescribed *attitudes*, the circlepoint attains positions $A_1, \ldots, A_m$, the moving revolute axis thus attaining locations $\overline{OA_j}$, for $j = 1, \ldots, m$, as shown in Fig. 4.8.

The two dyads of this linkage are thus arc A_0B, defined by axes $\overline{OA_0}$ and $\overline{OB}$, arc A_0^*B, defined by axes $\overline{OA_0^*}$ and $\overline{OB^*}$, respectively; the position vectors of A_0 and B are $\mathbf{a}_0$ and $\mathbf{b}$, respectively, both of unit magnitude:

$$\|\mathbf{a}_0\| = 1, \quad \|\mathbf{b}\| = 1 \tag{4.45}$$

which will be termed henceforth the *unit-magnitude conditions*.

Likewise, the position vectors of points A_0^* and B^* are the unit vectors $\mathbf{a}_0^*$ and $\mathbf{b}^*$, respectively.

With the foregoing definitions, the *spherical Burmester problem*, the counterpart of the problem studied in Sect. 4.2, is stated as:

> Find a spherical four-bar linkage that will conduct its coupler link through a set of m attitudes given by a corresponding set of orthogonal matrices, $\mathcal{Q} = \{\mathbf{Q}_j\}_1^m$, defined with respect to a reference attitude given by $\mathbf{Q}_0 = \mathbf{1}$, $\mathbf{1}$ now denoting the 3×3 identity matrix.

By virtue of the link rigidity, the angle between $\overline{OA_j}$ and $\overline{OB}$ remains equal to α_2 of Fig. 4.7, and hence, constant. The *synthesis equation* is thus obtained upon imposing this geometric constraint:

$$\mathbf{a}_j^T\mathbf{b} = \mathbf{a}_0^T\mathbf{b} \quad \text{or} \quad (\mathbf{a}_j - \mathbf{a}_0)^T\mathbf{b} = 0, \quad j = 1, \ldots, m \tag{4.46}$$

where, apparently,

$$\mathbf{a}_j = \mathbf{Q}_j\mathbf{a}_0 \tag{4.47}$$

Therefore, conditions (4.46) become

$$\mathbf{a}_0^T(\mathbf{Q}_j^T - \mathbf{1})\mathbf{b} = 0, \quad j = 1, \ldots, m \tag{4.48}$$

In order to ease the ensuing discussion, let, as depicted in Fig. 4.8,

$$\mathbf{c}_j \equiv (\mathbf{Q}_j - \mathbf{1})\mathbf{a}_0 \tag{4.49}$$

Equations (4.48) then taking the form

$$\mathbf{c}_j^T\mathbf{b} = 0, \quad j = 1, \ldots, m \tag{4.50}$$

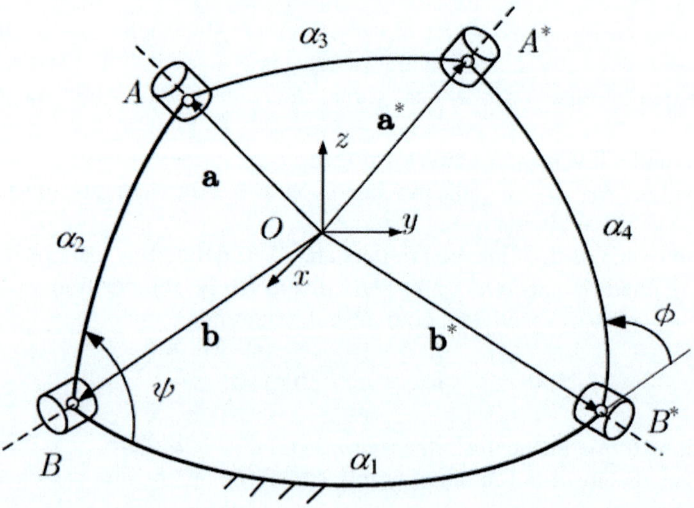

Fig. 4.7 The spherical 4R linkage

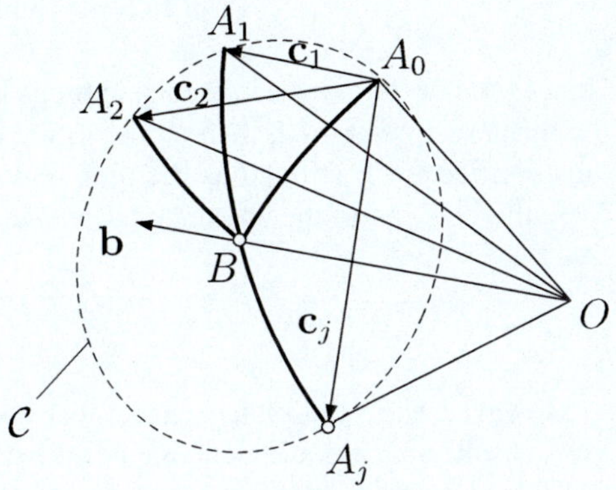

Fig. 4.8 A spherical dyad and the cone formed by the moving axis OA

Geometrically, Eqs. (4.50) state that vector $\mathbf{b}$ is perpendicular to any vector $\mathbf{c}_j$. With reference to Fig. 4.8, this means that segment $\overline{OB}$ is perpendicular to segments $\overline{A_0A_j}$, for $j = 1, \ldots, m$. That is, as link AB attains the set of attitudes $\mathcal{A} = \{A_jB\}_0^m$, all points $\{A_j\}_0^m$ lie in a circle $\mathcal{C}$ whose plane is normal to vector $\mathbf{b}$.

Notice that the rotation matrices $\mathbf{Q}_j$ admit various parametrizations, the one adopted here is in terms of the *natural invariants* of the rotation, namely the unit vector $\mathbf{e}_j$ parallel to the axis of the rotation represented by $\mathbf{Q}_j$, and the corresponding angle of rotation, ϕ_j, as described in Sect. 2.2. The rotation matrix $\mathbf{Q}_j$ thus takes the form of Eq. (2.1b), reproduced below for quick reference.

$$\mathbf{Q}_j = \mathbf{1} + s_j\mathbf{E}_j + (1 - c_j)\mathbf{E}_j^2, \quad c_j \equiv \cos\phi_j, \quad s_j \equiv \sin\phi_j \tag{4.51}$$

with $\mathbf{E}_j \equiv \text{CPM}(\mathbf{e}_j)$ and the operator $\text{CPM}(\cdot)$ having been defined as the cross-product matrix of the three-dimensional vector $(\cdot)$ in Eq. (2.1d). Hence,

$$\mathbf{Q}_j - \mathbf{1} = [s_j\mathbf{1} + (1 - c_j)\mathbf{E}_j]\mathbf{E}_j \tag{4.52}$$

Therefore, Eq. (4.48) becomes

$$\mathbf{a}_0^T\mathbf{E}_j[s_j\mathbf{1} - (1 - c_j)\mathbf{E}_j]\mathbf{b} = 0, \quad j = 1, \ldots, m \tag{4.53}$$

which is the system of *synthesis equations* of the problem at hand. Its solution for various values of m is discussed below.

4.3.1 Dyad Synthesis for Three Attitudes

In this case, $m = 2$, two constraint equations thus occurring in Eq. (4.50), namely,

$$\mathbf{c}_1^T\mathbf{b} = 0, \quad \text{and} \quad \mathbf{c}_2^T\mathbf{b} = 0 \tag{4.54}$$

Hence, one of the two vectors $\mathbf{a}_0$ and $\mathbf{b}$ can be prescribed arbitrarily. If, for example, the former is prescribed, then $\mathbf{c}_1$ and $\mathbf{c}_2$ are known. The conditions of Eq. (4.54) are thus verified for $\mathbf{b}$ defined as the unit vector derived from the cross product of the two other vectors in the above equations, i.e.,

$$\mathbf{b} = \frac{\mathbf{c}_1 \times \mathbf{c}_2}{\|\mathbf{c}_1 \times \mathbf{c}_2\|} \tag{4.55}$$

However, as in the planar case, most commonly the centerpoints B and B^* are given, as the space available to anchor the mechanism is known in a practical design project. In this case, two equations are given explicitly, with $\mathbf{a}_0$ as unknown:

$$\mathbf{d}_j^T\mathbf{a}_0 = 0, \quad j = 1,\ 2 \tag{4.56}$$

and $\mathbf{d}_j$ defined as

$$\mathbf{d}_j \equiv (\mathbf{Q}_j^T - \mathbf{1})\mathbf{b}, \quad j = 1,\ 2 \tag{4.57}$$

Hence, the solution sought takes the form

$$\mathbf{a}_0 = \frac{\mathbf{d}_1 \times \mathbf{d}_2}{\|\mathbf{d}_1 \times \mathbf{d}_2\|} \tag{4.58}$$

thereby showing that, in this case, the problem has a unique solution—obtained in closed form—modulo the *antipodal* of B or of A_0, as the case may be. Indeed, in formulas (4.55) and (4.58), nothing prevents the designer from using the alternative cross product $\mathbf{c}_2 \times \mathbf{c}_1$ or, correspondingly, $\mathbf{d}_2 \times \mathbf{d}_1$. The result would be the same, the only difference being that the new vector $\mathbf{b}$, or $\mathbf{a}_0$, for that matter, would be pointing in the direction opposite to that of its previous counterpart. That is, the designer would be looking at a different branch of the same circular-conical surface, both sharing one common point, the cone apex.

4.3.2 Dyad Synthesis for Four Attitudes

Now we have $m = 3$, the constraints being

$$\mathbf{c}_1^T\mathbf{b} = 0, \quad \mathbf{c}_2^T\mathbf{b} = 0 \quad \text{and} \quad \mathbf{c}_3^T\mathbf{b} = 0 \tag{4.59}$$

or, correspondingly,

$$\mathbf{d}_1^T\mathbf{a}_0 = 0, \quad \mathbf{d}_2^T\mathbf{a}_0 = 0 \quad \text{and} \quad \mathbf{d}_3^T\mathbf{a}_0 = 0 \tag{4.60}$$

In order to be able to find a vector $\mathbf{b}$ simultaneously perpendicular to all three vectors $\mathbf{c}_j$ in Eq. (4.59), these vectors must be coplanar; hence, this triplet must obey the condition

$$\mathcal{K}: \ F(x, y, z) \equiv \mathbf{c}_1 \times \mathbf{c}_2 \cdot \mathbf{c}_3 = 0 \tag{4.61}$$

which is a product of three factors, each linearly homogeneous in $\mathbf{a}_0$, as per Eq. (4.49), the product then being *cubic* and *homogeneous* in $\mathbf{a}_0$. As well, any vector $\mathbf{a}_0$ verifying Eq. (4.61) coincides with an element of a conic surface with apex at the origin O. Its intersection with the unit sphere, described by the first of Eqs. (4.45), is thus a spherical surface, labeled $\mathcal{K}$ to mimic the planar case of Sect. 4.2.2. Notice that every conical surface has two sheets sharing one common point, the surface apex. Therefore, $\mathcal{K}$ consists of two parts, one the *antipodal* of the other. Any point on the curve represents a solution $\mathbf{a}_0$. In this light, we call $\mathcal{K}$ the *circlepoint curve*.

As in the case of a planar cubic curve, its spherical counterpart intersects the spherical-curve equivalent of the planar line, namely, a major circle, at a maximum of three points. Likewise, a *spherical cubic curve* has one asymptote, which in this case is, again, a major circle.

Furthermore, once a point on $\mathcal{K}$ has been chosen as A_0, its corresponding centerpoint B, of position vector $\mathbf{b}$, is chosen as any of the three possible cross products $\mathbf{c}_j \times \mathbf{c}_{j+1}$, for $j = 1, 2, 3$ and $j + 1 \equiv j + 1 \bmod 3$, upon normalizing the product to a unit vector. Algebraically, these three products, although not necessarily identical—the angles between any pair of their factors are, in general, different—yield identical unit vectors upon normalization. However, due to roundoff errors, the three unit vectors will be numerically different. In order to filter the roundoff error, the obvious strategy is to compute all three products, then take their mean value. Normalizing this value to render it of unit norm yields vector $\mathbf{b}$.

By a similar reasoning, the *centerpoint conical surface* $\mathcal{M}$ and its corresponding *centerpoint curve*[16] are obtained as centerpoint curve

$$\mathcal{M}: \ G(u, v, w) \equiv \mathbf{d}_1 \times \mathbf{d}_2 \cdot \mathbf{d}_3 = 0 \tag{4.62}$$

with $\mathbf{d}_j$ defined in Eq. (4.57). Any point on $\mathcal{M}$ gives one corresponding *circlepoint* A_0, while any point of $\mathcal{K}$ gives one *centerpoint* B on the unit sphere.

4.3.3 Dyad Synthesis for Five Attitudes

For $m = 4$, the synthesis equations lead to a system of four homogeneous bilinear equations in the unknown vectors $\mathbf{a}_0$ and $\mathbf{b}$. As these are three-dimensional vectors, the total number of unknowns at hand is six but, then again, two additional equations

[16] McCarthy and Soh (2011) call the conical surfaces generating these curves the *center-axis cone* and the *circling-axis cone*, respectively.

are available, namely, Eqs. (4.45), and the problem is fully determined. The four homogeneous equations can then be cast in the form

$$\underbrace{\begin{bmatrix} \mathbf{a}_0^T\mathbf{E}_1[s_1\mathbf{1}-(1-c_1)\mathbf{E}_1] \\ \mathbf{a}_0^T\mathbf{E}_2[s_2\mathbf{1}-(1-c_2)\mathbf{E}_2] \\ \mathbf{a}_0^T\mathbf{E}_3[s_3\mathbf{1}-(1-c_3)\mathbf{E}_3] \\ \mathbf{a}_0^T\mathbf{E}_4[s_4\mathbf{1}-(1-c_4)\mathbf{E}_4] \end{bmatrix}}_{\mathbf{C}} \mathbf{b} = \mathbf{0}_4 \tag{4.63}$$

whose matrix coefficient $\mathbf{C}$ is a 4×3 array. In light of the second of equations (4.45), moreover, the trivial solution of Eqs. (4.125) is not acceptable, and hence, $\mathbf{C}$ must be *rank-deficient*, i.e., its three columns must be linearly dependent. This happens if and only if the four distinct 3×3 determinants, obtained by taking three rows of $\mathbf{C}$ at a time, vanish. Let

$$\Delta_j(\mathbf{a}_0) \equiv \det(\mathbf{C}_j) = 0, \quad j = 1, \ldots, 4 \tag{4.64}$$

with $\mathbf{C}_j$ denoting the 3×3 matrix obtained from $\mathbf{C}$ upon deleting its jth row. From Sect. 4.3.2 it is known that each of the four determinants defines a conical cubic surface whose apex is the origin. The common intersections of all four surfaces are common elements of these surfaces; they are candidate axes capable of guiding the coupler link through the five prescribed poses.

Likewise, surface equations for $\mathbf{b}$ can be formulated with

$$\mathbf{D}\mathbf{a}_0 = \mathbf{0}_4 \tag{4.65}$$

in which the 4×3 matrix $\mathbf{D} = [\mathbf{d}_1^T, \ldots, \mathbf{d}_4^T]^T$ with $\mathbf{d}_j$ defined in Eq. (4.57) for $m = 2$; the same pattern holds for the case at hand, $m = 4$. The rank-deficiency of matrix $\mathbf{D}$ yields

$$\Delta_j(\mathbf{b}) \equiv \det(\mathbf{D}_j) = 0, \quad j = 1, \ldots, 4 \tag{4.66}$$

with $\mathbf{D}_j$ denoting the 3×3 matrix obtained from $\mathbf{D}$ upon deleting its jth row.

Now, given that one has already one (quadratic) equation in $\mathbf{a}_0$—the first of - Eqs. (4.45)—only two more equations out of the four Eqs. (4.64) are needed to determine $\mathbf{a}_0$, the Bezout number of the equations in question is thus $2 \times 3^2 = 18$. Ditto that of the equations needed to find $\mathbf{b}$—the second of Eqs. (4.45) and two out of the four equations (4.66). This number is an overestimate of the actual number of possible solutions. Indeed, McCarthy and Soh (2011), by invoking the *Burmester-Roth Theorem* (Roth, 1967), showed that the three equations in $\mathbf{a}_0$ can be reduced, by elimination of two of the three coordinates of A_0, to a sextic polynomial. As a consequence, the problem admits six, four, two or zero circlepoint solutions. The same reasoning leads to the conclusion that the problem also admits six, four, two or zero centerpoint solutions. Therefore, the number of possible dyads that solve the problem is six, four, two or zero. Correspondingly, the number of spherical four-bar linkages that can guide their coupler link through the five given attitudes is the number of combinations of six, four, two or zero objects taking two at a time, i.e., 15, six, one or zero.

4.3.4 *Spherical dyads with a P Joint*

While it may appear counterintuitive to associate a P joint with a spherical linkage, such a joint is possible in this linkage type. In fact, the putative P joint is still of the R type, but with a special feature: *it allows the motion of its coupling link via a circular guideway, defined by a great circle on the unit sphere.*

In a spherical dyad, a P joint is a special case of the R dyad, when the arc of the latter subtends an angle of 90°. The foregoing method of synthesis still applies, if with some modifications. Indeed, for the prescribed attitudes to lead to a linkage that admits a P joint, at a circlepoint, for example, the circle traced by A, or A^* for that matter, becomes a major circle. In this case A_j moves on a major circle lying in a plane normal to $\mathbf{b}$, Eq. (6.28) then becoming

$$\mathbf{b}^T\mathbf{a}_j = 0 \quad \Rightarrow \quad \mathbf{b}^T\mathbf{Q}_j\mathbf{a}_0 = 0, \quad j = 0, \ldots, m \tag{4.67}$$

with $\mathbf{Q}_0 = \mathbf{1}$, the 3×3 identity matrix. The foregoing equation means that all vectors $\{\mathbf{a}_j\}_0^m$ are coplanar, $\mathbf{b}$ then being the unit normal to the plane of the set. Furthermore, relations (4.67) can be rewritten as an array of $m + 1$ scalar equations:

$$\mathbf{H}\mathbf{a}_0 = \mathbf{0}_{m+1} \tag{4.68}$$

where $\mathbf{H}$ is a $(m + 1) \times 3$ matrix. Again, matrix $\mathbf{H}$ has to be *rank-deficient*, as only $m \geq 2$ conditions are considered.[17] This implies, for the four-attitude synthesis problem, that

$$\det(\mathbf{H}_i) = 0, \quad i = 1, \ldots, 4 \tag{4.69}$$

the 3×3 matrix $\mathbf{H}_i$ obtained by deleting the ith row from matrix $\mathbf{H}$. In the case of five poses, $\mathbf{H}_i$ is obtained by deleting the ith and $(i + 1)$st rows from matrix $\mathbf{H}$.

4.3.5 *Approximate dyad Synthesis*

As in the case of planar linkages, whenever $m \geq 5$ the synthesis equations, in general, cannot be satisfied exactly. In the present case, as in its planar counterpart, the pertinent approach consists in formulating the synthesis problem as one of *optimum design*. To this end, an objective function z is defined as in the planar case, namely, as one-half the Euclidian norm of an error vector $\mathbf{e}$ defined exactly as in the planar

[17] For $m = 1$ it is always possible to find two linearly independent vectors normal to a given one!

case as well. Contrary to the latter, however, the optimization task at hand is no longer unconstrained, because vectors $\mathbf{a}_0$ and $\mathbf{b}$ are now subjected to the normality conditions (4.45), which play the role of equality constraints in the optimization task. Unfortunately, the literature on the solution of the approximate synthesis problem for spherical four-bar linkages is rather scarce.

The challenge posed by the approximate synthesis of spherical linkages lies in that solving a *constrained optimization problem* is more complex than solving an unconstrained one, as the former must incorporate the constraints, first in the formulation, then in the solution algorithm (Hillier and Liebermann, 1995).

The basic optimization problem is stated below. First, the error incurred at the jth pose is defined:

$$e_j \equiv \mathbf{a}_0^T(\mathbf{Q}_j^T - \mathbf{1})\mathbf{b}, \quad j = 1, \ldots, m \geq 5 \tag{4.70}$$

the optimization problem now being formulated as

$$z \equiv \frac{1}{2}\|\mathbf{e}\|^2 = \frac{1}{2}\sum_1^m e_j^2(\mathbf{x}) \quad \rightarrow \quad \min_{\mathbf{x}}, \quad \text{with} \quad \mathbf{x} \equiv \begin{bmatrix} \mathbf{a}_0 \\ \mathbf{b} \end{bmatrix} \in \mathbb{R}^6 \;\&\; \mathbf{e} \equiv \begin{bmatrix} e_1 \\ \vdots \\ e_m \end{bmatrix} \tag{4.71}$$

subject to[18]

$$\mathbf{h}(\mathbf{x}) \equiv \frac{1}{2}\begin{bmatrix} (\|\mathbf{a}_0\|^2 - 1) \\ (\|\mathbf{b}\|^2 - 1) \end{bmatrix} = \mathbf{0}_2 \tag{4.72}$$

In order to handle the constraints, we can cite two approaches: (i) adjoin the constraints to the objective function $z(\mathbf{x})$, defined above, by means of *Lagrange multipliers*, which become additional unknowns to the optimization problem (Liu and Angeles, 1992a); and (ii) introduce an alternative parametrization of the problem, that allows formulating the synthesis task as an *unconstrained optimization problem* (Léger and Angeles, 2014). There are pros and cons to the two approaches, as described below.

Constrained-Minimization Approach

In this case, a new objective function is defined, that allows the handling of the problem as in the unconstrained case, by means of a vector $\boldsymbol{\lambda}$ of Lagrange multipliers, of dimension identical to the number of scalar constraints, which, in our case, is two. The new objective function then becomes

$$F(\mathbf{x}, \boldsymbol{\lambda}) \equiv z(\mathbf{x}) + \boldsymbol{\lambda}^T\mathbf{h}(\mathbf{x}) \quad \rightarrow \min_{\mathbf{x},\,\boldsymbol{\lambda}}, \quad \boldsymbol{\lambda} \in \mathbb{R}^2 \tag{4.73}$$

[18] The purpose of the 1/2 factor in the definition of $\mathbf{h}(\mathbf{x})$ is to avoid carrying a factor of 2 when calculating $\nabla\mathbf{h}$.

subject to no constraints.

The FONC of the unconstrained problem are now:

$$\nabla F \equiv \frac{\partial F}{\partial \mathbf{x}} = \underbrace{\frac{\partial z}{\partial \mathbf{x}}}_{\nabla z} + \mathbf{J}(\mathbf{x})^T \boldsymbol{\lambda} = \mathbf{0}_6, \quad \mathbf{J}(\mathbf{x}) \equiv \frac{\partial \mathbf{h}}{\partial \mathbf{x}} = \begin{bmatrix} \mathbf{a}_0^T & \mathbf{0}^T \\ \mathbf{0}^T & \mathbf{b}^T \end{bmatrix} \in \mathbb{R}^{2\times 6} \tag{4.74}$$

$$\frac{\partial F}{\partial \boldsymbol{\lambda}} = \mathbf{h}(\mathbf{x}) = \mathbf{0}_2 \tag{4.75}$$

with $\mathbf{J}(\mathbf{x})$ termed the *constraint Jacobian*, and Eq. (4.75) being a restatement of the constraint, Eq. (4.72).

Given the simplicity of $z(\mathbf{x})$, which is a *quartic* scalar function of $\mathbf{x}$, and of $\mathbf{h}(\mathbf{x})$, which is, in turn, a *quadratic* vector function of the same, their gradients $\partial z/\partial \mathbf{x}$ and $\mathbf{J}(\mathbf{x})$, are also algebraic functions, but now *cubic* and, correspondingly, *linear* in $\mathbf{x}$. Their FONC is thus a system of algebraically simple equations, which should lead to simplifications. For one thing, in light of the linearity of Eq. (4.74) in $\boldsymbol{\lambda}$, it is possible to eliminate this vector from the same equation system. However, rather than eliminating $\boldsymbol{\lambda}$ by means of equation-solving, an alternative approach is introduced below.

The concept of orthogonal complement, introduced in Sect. 3.5.3, is key to the elimination of the vector of Lagrange multipliers: let $\mathbf{L}$ be an *orthogonal complement* of the constraint Jacobian $\mathbf{J}$, in the sense that the columns of $\mathbf{L}$ span the null space of $\mathbf{J}$. Since the latter is a 2×6 matrix, $\mathbf{L}$ is bound to be a 6×4 matrix because $\mathbf{J}$ is of full rank, since its two rows are apparently linearly independent. The range of $\mathbf{J}$ is thus 2, and hence, four linearly independent six-dimensional vectors can be found that are all orthogonal to the two columns of $\mathbf{J}$. These four vectors thus span the null space of $\mathbf{J}$, and can thus be chosen as the four columns of $\mathbf{L}$. It should be apparent that the choice of these four vectors is not unique, as long as they are linearly independent.

An obvious choice of the four columns of $\mathbf{L}$ is given below:

$$\mathbf{L} = \begin{bmatrix} \mathbf{A}_0\mathbf{u} & \mathbf{0} & \mathbf{A}_0\mathbf{v} & \mathbf{0} \\ \mathbf{0} & \mathbf{B}\mathbf{u} & \mathbf{0} & \mathbf{B}\mathbf{v} \end{bmatrix} \in \mathbb{R}^{2\times 6} \tag{4.76}$$

where $\mathbf{A}_0 = \text{CPM}(\mathbf{a}_0)$, $\mathbf{B} = \text{CPM}(\mathbf{b})$, while $\mathbf{u}$ and $\mathbf{v}$ are arbitrary linearly independent three-dimensional vectors that, for purposes of numerical stability, might as well be chosen of unit magnitude, i.e., $||\mathbf{u}|| = ||\mathbf{v}|| = 1$. Under these conditions, the reader can readily prove that

$$\mathbf{J}\mathbf{L} = \mathbf{O} \in \mathbb{R}^{2\times 4} \tag{4.77}$$

Upon multiplying both sides of the first of Eqs. (4.74) by $\mathbf{L}^T$ from the left, a new system of first-order normality conditions for the problem at hand is obtained, free of $\boldsymbol{\lambda}$, namely,

$$\mathbf{L}^T \nabla z = \mathbf{0} \in \mathbb{R}^4 \tag{4.78}$$

with ∇z, already introduced in Eq. (4.74), representing the gradient of $z(\mathbf{x})$, i.e., blockwise,

$$\nabla z = \begin{bmatrix} \partial z/\partial \mathbf{a}_0 \\ \partial z/\partial \mathbf{b} \end{bmatrix} \tag{4.79}$$

whose two three-dimensional blocks are computed as shown below:

From the definition of $z(\mathbf{x})$ in Eq. (4.71), $\partial z/\partial \mathbf{a}_0$ readily follows:

$$\frac{\partial z}{\partial \mathbf{a}_0} = \sum_1^m e_j \frac{\partial e_j}{\partial \mathbf{a}_0} = \sum_1^m [\mathbf{a}_0^T(\mathbf{Q}_j^T - \mathbf{1})\mathbf{b}] \frac{\partial e_j}{\partial \mathbf{a}_0} \tag{4.80}$$

with the last partial derivative displayed below:

$$\frac{\partial e_j}{\partial \mathbf{a}_0} = (\mathbf{Q}_j^T - \mathbf{1})\mathbf{b} \tag{4.81}$$

and a similar expression for $\partial e_j/\partial \mathbf{b}$:

$$\frac{\partial e_j}{\partial \mathbf{b}} = (\mathbf{Q}_j - \mathbf{1})\mathbf{a}_0 \tag{4.82}$$

Hence,

$$\frac{\partial z}{\partial \mathbf{a}_0} = \sum_1^m e_j \frac{\partial e_j}{\partial \mathbf{a}_0} = \sum_1^m [\mathbf{a}_0^T(\mathbf{Q}_j^T - \mathbf{1})\mathbf{b}](\mathbf{Q}_j^T - \mathbf{1})\mathbf{b} \tag{4.83a}$$

which can be cast in the form

$$\frac{\partial z}{\partial \mathbf{a}_0} = \mathbf{Z}_a \mathbf{b} \tag{4.83b}$$

with $\mathbf{Z}_a$ defined as

$$\mathbf{Z}_a \equiv \sum_1^m [\mathbf{a}_0^T(\mathbf{Q}_j^T - \mathbf{1})\mathbf{b}](\mathbf{Q}_j^T - \mathbf{1}) \tag{4.83c}$$

Likewise, $\partial z/\partial \mathbf{b}$ is computed as

$$\frac{\partial z}{\partial \mathbf{b}} = \mathbf{Z}_b \mathbf{a} \tag{4.84a}$$

with $\mathbf{Z}_b$ defined, in turn, as

$$\mathbf{Z}_b \equiv \sum_1^m [\mathbf{a}_0^T(\mathbf{Q}_j^T - \mathbf{1})\mathbf{b}](\mathbf{Q}_j - \mathbf{1}) \tag{4.84b}$$

It is noteworthy that the last factor in $\mathbf{Z}_b$ differs from its counterpart in $\mathbf{Z}_a$, in that matrix $\mathbf{Q}_j$ is not transposed, a consequence of the definition of the derivative of a scalar, e_j in this case, with respect to a vector, $\mathbf{a}_0$ or $\mathbf{b}$, as per the definitions introduced in Sect. 1.4.4.

In summary, system (4.78) yields four equations in the six unknowns $\mathbf{x} = [\,\mathbf{a}_0^T\,,\ \mathbf{b}^T\,]^T$, to be complemented with the two scalar constraints, Eqs. (4.72), thereby ending up with a system of six nonlinear equations in six unknowns. The number of possible solutions to the foregoing system can be predicted via its Bezout number. First, notice that $\mathbf{Z}_a$ and $\mathbf{Z}_b$ are bilinear in $\mathbf{a}_0$ and $\mathbf{b}$, i.e., quadratic in $\mathbf{x}$. Since the first block of ∇z is linear in $\mathbf{Z}_a$ and in $\mathbf{b}$, while its second block is linear in $\mathbf{Z}_b$ and $\mathbf{a}_0$, it follows that ∇z is cubic in $\mathbf{x}$. Furthermore, $\mathbf{L}$ is linear in $\mathbf{a}_0$ and $\mathbf{b}$, which means that the *reduced* system of four FONC (4.78) is quartic in $\mathbf{x}$. Moreover, the two scalar constraints (4.72) are quadratic in $\mathbf{x}$, the overall system of six algebraic equations in six unknowns thus having a Bezout number $N_B = 4^4 \times 2 \times 2 = 1024$. Upon removing the antipodal solutions, the total number of distinct solutions can be as high as 512.

This system can be solved numerically via the Newton-Raphson method, whose convergence, as pointed out in Sect. 1.5.1, cannot be predicted. For this reason, a *robust* alternative is investigated.

The alternative solution aims at an overdetermined, although *consistent*, system of nonlinear equations, to be solved numerically via the Newton-Gauss method, whose convergence is guaranteed, as per the results of Sect. 1.6.1. To this end, an alternative, although *expanded* orthogonal complement of $\mathbf{J}(\mathbf{x})$ is explored. What the italicized qualifier means is an orthogonal complement with a *surplus of columns*. This is a 6×6 matrix, namely,

$$\mathbf{L} = \begin{bmatrix} \mathbf{A}_0 & \mathbf{O} \\ \mathbf{O} & \mathbf{B} \end{bmatrix} \in \mathbb{R}^{6\times 6} \tag{4.85}$$

where $\mathbf{O} \in \mathbb{R}^{3\times 3}$ is the 3×3 zero matrix. The reader is invited to prove that $\mathbf{JL} = \mathbf{O} \in \mathbb{R}^{2\times 6}$.

Further, upon premultiplying the first of Eqs. (4.74) by $\mathbf{L}^T$, with $\mathbf{L}$ given as in Eq. (4.85), a new system of first-order normality conditions is obtained, free of Lagrange multipliers:

$$\mathbf{L}^T \nabla F = \begin{bmatrix} -\mathbf{A}_0\mathbf{Z}_a\mathbf{b} \\ -\mathbf{B}\mathbf{Z}_b\mathbf{a}_0 \end{bmatrix} = \mathbf{0} \in \mathbb{R}^6 \tag{4.86}$$

which, together with the constraints (4.72), lead to an *overdetermined* system of eight nonlinear equations in six unknowns, to be solved via the Newton-Gauss method, which is guaranteed to converge to one of the multiple solutions admitted by the problem.

An alternative to the Newton-Gauss method is, of course, a *semigraphical method*, intended to eliminating all but two of the six unknowns of the first six-dimensional system or of the eight-dimensional system in the second approach. All real solutions can then be obtained by *contour intersection*.

An Unconstrained Formulation

If, rather than using Cartesian coordinates, one uses spherical coordinates tailored to points lying on the unit sphere, then not only are the constraints eliminated, but the number of design variables is reduced, from three per point to only two, as the third spherical coordinate, the distance of a point in 3D space to the origin, is constant. Indeed, *all points of interest lie on the surface of the unit sphere*.

However, introduction of spherical coordinates brings about its own complexity, as the point-coordinates entail the harmonic functions, sine and cosine, of the spherical coordinates, thereby destroying the algebraic nature of the problem. While the tan-half identities, introduced in Eq. (3.72), can be invoked, that would bring the system of equations back to algebraic, the degree of the error, Eq. (4.70), would be now higher than quadratic. One alternative to cope with the new algebraic system is to resort to *dialytic elimination* (Salmon, 1964), which offers a systematic means of eliminating unknowns from a system of algebraic equations. This being an open research topic, it is not pursued here. Instead, an alternative formulation is pursued below.

One practical means of coping with the nonlinearity of the problem is by resorting to a purely numerical solution, if with some *massaging* of the relations involved, as outlined presently. Given that the optimization problem is now unconstrained, it can be regarded as one of *unconstrained nonlinear least squares*, as studied in Sect. 1.6, and solved by means of the Newton-Gauss method.

The spherical coordinates to be used here are *longitude* and *latitude*. Let, then,[19] ϑ_A and φ_A be the longitude and the latitude of A_0, ϑ_B and φ_B those of B. Hence, the position vectors of interest, $\mathbf{a}_0$ and $\mathbf{b}$, are now given, each, in terms of only *two design variables*:

$$\mathbf{a}_0 = \begin{bmatrix} \cos\varphi_A \cos\vartheta_A \\ \cos\varphi_A \sin\vartheta_A \\ \sin\varphi_A \end{bmatrix}, \quad \mathbf{b} = \begin{bmatrix} \cos\varphi_B \cos\vartheta_B \\ \cos\varphi_B \sin\vartheta_B \\ \sin\varphi_B \end{bmatrix} \tag{4.87}$$

As the synthesis problem admits two valid solutions, one the antipodal of the other, we will now limit the search of the optimum solution to only one of the two alternatives. This is readily done by specifying the ranges of all spherical coordinates as

$$\{\varphi_A, \vartheta_A, \varphi_B, \vartheta_B\} \in [-\pi/2, \pi/2] \tag{4.88}$$

[19] ϑ and φ are read "vartheta" and "varphi," respectively.

In order to formulate the unconstrained optimization problem we recall the error e_j of Eq. (4.70) for the jth prescribed attitude. The approximate-synthesis problem is now formulated as

$$z \equiv \frac{1}{2}\|\mathbf{e}\|^2 = \frac{1}{2}\sum_{1}^{m} e_j^2(\mathbf{x}) \quad \rightarrow \quad \min_{\mathbf{x}}, \quad \text{with} \;\; \mathbf{e} \equiv \begin{bmatrix} e_1 \\ \vdots \\ e_m \end{bmatrix} \;\&\; \mathbf{x} \equiv \begin{bmatrix} \boldsymbol{\alpha} \\ \boldsymbol{\beta} \end{bmatrix} \in \mathbb{R}^4 \tag{4.89}$$

with vectors $\boldsymbol{\alpha}$ and $\boldsymbol{\beta}$ defined, in turn, as

$$\boldsymbol{\alpha} \equiv \begin{bmatrix} \varphi_A \\ \vartheta_A \end{bmatrix} \quad \text{and} \quad \boldsymbol{\beta} \equiv \begin{bmatrix} \varphi_B \\ \vartheta_B \end{bmatrix} \tag{4.90}$$

The simplest way of solving the foregoing problem is by means of the approach used for overdetermined systems of nonlinear equations, as introduced in Sect. 1.6. That is, we look at an overdetermined system of $m \geq 5$ equations in four unknowns, of the form

$$\mathbf{e}(\mathbf{x}) = \mathbf{0}_m, \quad \text{with} \quad m \geq 5 \text{ and } \mathbf{x} \in \mathbb{R}^4 \tag{4.91}$$

Further, the Newton-Gauss method requires the Jacobian of $\mathbf{e}$ with respect to $\mathbf{x}$, namely,

$$\boldsymbol{\Phi}(\mathbf{x}) \equiv \frac{\partial \mathbf{e}(\mathbf{x})}{\partial \mathbf{x}} \tag{4.92}$$

However, the components e_j of $\mathbf{e}$ are given in Eq. (4.70) in terms of $\mathbf{a}_0$ and $\mathbf{b}$. The gradients of these components w.r.t. $\boldsymbol{\alpha}$ and $\boldsymbol{\beta}$ thus call for application of the chain rule for vector arguments, as introduced in Eq. (1.35). To do this, notice that an increment $\Delta\mathbf{e}$ of $\mathbf{e}$, as a result of an increment $\Delta\mathbf{x}$ in $\mathbf{x}$, is given, to a *first-order approximation*, which is good enough for a *reasonably small* increment $\Delta\mathbf{x}$, by

$$\Delta\mathbf{e} = \boldsymbol{\Phi}\Delta\mathbf{x}, \quad \boldsymbol{\Phi} \equiv \frac{\partial \mathbf{e}}{\partial \mathbf{x}} \in \mathbb{R}^{m\times 4} \tag{4.93}$$

where, in light of the partitioning of $\mathbf{x}$, as per Eq. (4.89), $\boldsymbol{\Phi}$ admits the partitioning

$$\boldsymbol{\Phi} = \begin{bmatrix} \boldsymbol{\Phi}_\alpha & \boldsymbol{\Phi}_\beta \end{bmatrix}, \quad \text{with} \quad \boldsymbol{\Phi}_\alpha \equiv \frac{\partial \mathbf{e}}{\partial \boldsymbol{\alpha}} \;\; \& \;\; \boldsymbol{\Phi}_\beta \equiv \frac{\partial \mathbf{e}}{\partial \boldsymbol{\beta}} \in \mathbb{R}^{m\times 2} \tag{4.94}$$

Further, in order to obtain matrices $\boldsymbol{\Phi}_\alpha$ and $\boldsymbol{\Phi}_\beta$, the Jacobians of $\mathbf{e}$ w.r.t. $\boldsymbol{\alpha}$ and $\boldsymbol{\beta}$, in terms of the corresponding Jacobians of $\mathbf{a}_0$ and $\mathbf{b}$ with respect to $\boldsymbol{\alpha}$ and $\boldsymbol{\beta}$, respectively—notice that $\mathbf{a}_0$ is a function of $\boldsymbol{\alpha}$ alone, while $\mathbf{b}$ is, in turn, a function of $\boldsymbol{\beta}$ alone—*the chain rule of differentiation for vector arrays*, introduced in Eq. (1.35), is now recalled:

$$\frac{\partial e_j}{\partial \boldsymbol{\alpha}} = \left(\frac{\partial \mathbf{a}_0}{\partial \boldsymbol{\alpha}}\right)^T \frac{\partial e_j}{\partial \mathbf{a}_0}, \quad \frac{\partial e_j}{\partial \boldsymbol{\beta}} = \left(\frac{\partial \mathbf{b}}{\partial \boldsymbol{\beta}}\right)^T \frac{\partial e_j}{\partial \mathbf{b}}, \quad j = 1, \ldots, m \tag{4.95}$$

The Jacobians of interest are now computed from Eqs. (4.87):

$$\mathbf{A} \equiv \frac{\partial \mathbf{a}_0}{\partial \boldsymbol{\alpha}} = \begin{bmatrix} -\sin\varphi_A \cos\vartheta_A & -\cos\varphi_A \sin\vartheta_A \\ -\sin\varphi_A \sin\vartheta_A & \cos\varphi_A \cos\vartheta_A \\ \cos\varphi_A & 0 \end{bmatrix} \in \mathbb{R}^{3\times 2} \tag{4.96}$$

and

$$\mathbf{B} \equiv \frac{\partial \mathbf{b}}{\partial \boldsymbol{\beta}} = \begin{bmatrix} -\sin\varphi_B \cos\vartheta_B & -\cos\varphi_B \sin\vartheta_B \\ -\sin\varphi_B \sin\vartheta_B & \cos\varphi_B \cos\vartheta_B \\ \cos\varphi_B & 0 \end{bmatrix} \in \mathbb{R}^{3\times 2} \tag{4.97}$$

while the gradients of e_j w.r.t. $\mathbf{a}_0$ and $\mathbf{b}$ were already obtained, as displayed in Eqs. (4.81) and (4.82), respectively.

Upon substitution of Eqs. (4.96) and (4.97), along with Eqs. (4.81) and (4.82), into Eqs. (4.95), the gradients of e_j with respect to $\boldsymbol{\alpha}$ and $\boldsymbol{\beta}$ are obtained:

$$\frac{\partial e_j}{\partial \boldsymbol{\alpha}} = \mathbf{A}^T(\mathbf{Q}_j^T - \mathbf{1})\mathbf{b}, \quad \frac{\partial e_j}{\partial \boldsymbol{\beta}} = \mathbf{B}^T(\mathbf{Q}_j - \mathbf{1})\mathbf{a}_0 \in \mathbb{R}^2, \quad j = 1, \ldots, m \tag{4.98}$$

Now, upon assembling these arrays into the two arrays Φ_α and Φ_β of Eqs. (4.94), the $m \times 2$ gradients appearing therein, we obtain

$$\Phi_\alpha = \begin{bmatrix} \mathbf{b}^T(\mathbf{Q}_1 - \mathbf{1})\mathbf{A} \\ \mathbf{b}^T(\mathbf{Q}_2 - \mathbf{1})\mathbf{A} \\ \vdots \\ \mathbf{b}^T(\mathbf{Q}_m - \mathbf{1})\mathbf{A} \end{bmatrix}, \quad \Phi_\beta = \begin{bmatrix} \mathbf{a}_0^T(\mathbf{Q}_1^T - \mathbf{1})\mathbf{B} \\ \mathbf{a}_0^T(\mathbf{Q}_2^T - \mathbf{1})\mathbf{B} \\ \vdots \\ \mathbf{a}_0^T(\mathbf{Q}_m^T - \mathbf{1})\mathbf{B} \end{bmatrix} \in \mathbb{R}^{m\times 2} \tag{4.99}$$

the Jacobian Φ of $\mathbf{e}$ w.r.t. $\mathbf{x}$, introduced in Eq. (4.92), thus being completed:

$$\Phi = \begin{bmatrix} \mathbf{b}^T(\mathbf{Q}_1 - \mathbf{1})\mathbf{A} & \mathbf{a}_0^T(\mathbf{Q}_1^T - \mathbf{1})\mathbf{B} \\ \mathbf{b}^T(\mathbf{Q}_2 - \mathbf{1})\mathbf{A} & \mathbf{a}_0^T(\mathbf{Q}_2^T - \mathbf{1})\mathbf{B} \\ \vdots & \\ \mathbf{b}^T(\mathbf{Q}_m - \mathbf{1})\mathbf{A} & \mathbf{a}_0^T(\mathbf{Q}_m^T - \mathbf{1})\mathbf{B} \end{bmatrix} \in \mathbb{R}^{m\times 4} \tag{4.100}$$

Once the Jacobian Φ is available, the Newton-Gauss method can be implemented for a given initial guess $\mathbf{x}^0$, which is all this algorithm needs.

4.3.6 *Examples*

A set of attitudes of a rigid link is given in Table 4.2, from which data will be taken for the synthesis examples below.

Synthesis of a Spherical dyad for Three Attitudes

Attitudes 1 and 2 are taken from Table 4.2 to design a RR dyad for a spherical linkage. First, the rotation matrices $\mathbf{Q}_1$ and $\mathbf{Q}_2$, taking the dyad end-link from the reference attitude, 0 in the table, and represented by **1**, are calculated, based on the angle of rotation ϕ and the unit vector **e** given in rows 1 and 2 of the table. These are

$$\mathbf{Q}_1 = \begin{bmatrix} 0.9794 & 0.1736 & -0.1029 \\ -0.1726 & 0.9848 & 0.0181 \\ 0.1045 & 0 & 0.9945 \end{bmatrix}, \quad \mathbf{Q}_2 = \begin{bmatrix} 0.3876 & 0.6569 & 0.6467 \\ -0.4784 & 0.7430 & -0.4680 \\ -0.7879 & -0.1280 & 0.6023 \end{bmatrix}$$

Furthermore, a *first centerpoint* **b**—of unit magnitude, of course—is chosen a

$$\mathbf{b} = [0.5774,\ 0.5774,\ 0.5774]^T$$

which, as per Eq. (4.57), leads to vectors

$$\mathbf{d}_1 = \begin{bmatrix} -0.1955 \\ 0.0236 \\ -0.0039 \end{bmatrix}, \quad \mathbf{d}_2 = \begin{bmatrix} -0.0072 \\ -0.1492 \\ -0.1907 \end{bmatrix}$$

With the two foregoing vectors, the corresponding circlepoint is found as

$$\mathbf{a}_0 = [0.9199,\ 0.36117,\ 0.1532]^T$$

A *second centerpoint* **b** is specified as

$$\mathbf{b} = [0.0,\ -0.7071,\ 0.7071]^T$$

Table 4.2 Five given attitudes of a rigid link

No.	ϕ_j [rad]	$\mathbf{e}^T$
0	0	Undefined
1	0.2034	$[-0.0449, -0.5133, -0.8569]^T$
2	1.1957	$[0.1827, 0.7709, -0.6101]^T$
3	1.1932	$[0.5212, 0.8414, -0.1422]^T$
4	1.0512	$[0.5384, 0.8114, 0.2271]^T$

which leads, in turn, to vectors

$$\mathbf{d}_1 = \begin{bmatrix} -0.1955 \\ 0.0236 \\ -0.0039 \end{bmatrix}, \quad \mathbf{d}_2 = \begin{bmatrix} -0.0072 \\ -0.1492 \\ -0.1907 \end{bmatrix}$$

the corresponding circlepoint being given by the position vector

$$\mathbf{a}_0 = [-0.1063, \ -0.7812, \ 0.6152]^T$$

thereby completing the synthesis of the two dyads making up the desired spherical four-bar linkage.

Synthesis of a Spherical dyad for Four Attitudes

Attitudes 1, 2 and 3 of Table 4.2 are taken now, which produce one third rotation matrix, besides $\mathbf{Q}_1$ and $\mathbf{Q}_2$ used above for the three-attitude synthesis, namely,

$$\mathbf{Q}_3 = \begin{bmatrix} 0.5403 & 0.4090 & 0.7353 \\ 0.1447 & 0.8157 & -0.5600 \\ -0.8289 & 0.4089 & 0.3816 \end{bmatrix}$$

With $\mathbf{a}_0 = [x, y, z]^T$ and the three foregoing matrices, vectors $\mathbf{c}_j$, for $j = 1, 2, 3$, are obtained, as per Eq. (4.49); when the latter is substituted into Eq. (4.61), a cubic equation in $\mathbf{a}_0 = [x, y, z]^T$ defining the *circlepoint spherical curve* $\mathcal{K}$ is obtained, namely,

$$\begin{aligned}\mathcal{K}: \ F(x, y, z) = & -0.01766\,x^3 + 0.03116\,x^2y + 0.04156\,x^2z + 0.02939\,xy^2 + 0.08747\,xyz \\ & -0.06021\,xz^2 - 0.01747\,y^3 - 0.02155\,y^2z + 0.02673\,yz^2 - 0.00482\,z^3 = 0\end{aligned}$$

Likewise, the *centerpoint spherical curve* $\mathcal{M}$ is obtained for unit vector $\mathbf{b} = [u, v, w]^T$ as

$$\begin{aligned}\mathcal{M}: \ G(u, v, w) = & \ 0.01678\,u^3 + 0.05157\,u^2v + 0.01743\,u^2w - 0.04005\,uv^2 + 0.02803\,uvw \\ & -0.05074\,uw^2 - 0.00259\,v^3 - 0.05317\,v^2w - 0.00933\,vw^2 \\ & +0.02762\,w^3 = 0\end{aligned}$$

The spherical circlepoint and centerpoint curves are shown in Fig. 4.9. Dyad synthesis now proceeds as discussed in Sect. 4.3.2. In the same figure, the reader can appreciate the asymptotes of the two cubic curves, as two major circles.

Synthesis of Spherical dyads for Five Attitudes

In this example, attitudes 1, 2, 3, and 4 of Table 4.2 are to be met by a spherical dyad. In order to enable the visual determination of the solutions from the intersections of the four circlepoint equations arising from the data, we adopt here spherical coordinates, as introduced in Sect. 4.3.5 for the unconstrained solution, to the approximate synthesis of spherical dyads.

The four circlepoint contours in the φ_a-versus-ϑ_a plane are illustrated in Fig. 4.10a, those of the centerpoint in the φ_b-versus-ϑ_b plane in Fig. 4.10b. Four common intersections can be identified *by inspection* on each of the two figures, which provide plausible values for the initial guess of numerical solutions via the Newton-Gauss method. To match each solution for $\mathbf{a}_0$ and its corresponding solution for $\mathbf{b}$, the linear equations derived from a value of $\mathbf{a}_0$ are used: for example, each solution of $\mathbf{a}_0$ is substituted into the synthesis equation (4.48), which yields a system of m linear equations in $\mathbf{b}$. The values of $\mathbf{b}$ that verify these equations correspond to the given value of $\mathbf{a}_0$. The matched solutions are recorded in Table 4.3. Notice that the results are also the solutions of vectors $\mathbf{a}_0^*$ and $\mathbf{b}^*$. Six linkages,[20] shown in Fig. 4.11, are generated from the four dyads. In the foregoing subfigures, notation L#: $i + j$ denotes a linkage generated using solutions i and j of Table 4.3.

Branching-detection on the six mechanisms was conducted by means of the sign of $\sin\mu$, with μ denoting the transmission angle for spherical linkages, defined in Sect. 3.6.2. Sign changes of the sine of the transmission angle were found in linkages L3 and L4, which indicates the presence of branching defect in these two solutions. The remaining linkages were found to be defect-free. Linkage animations confirmed

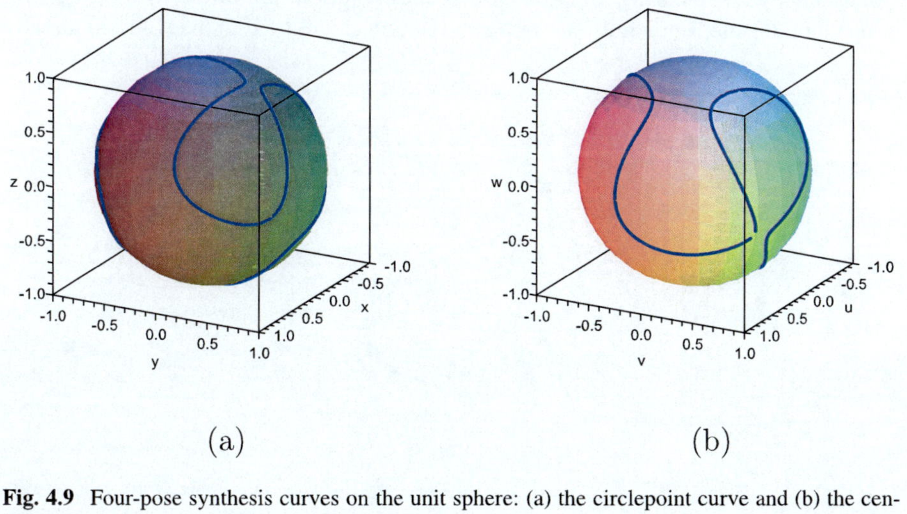

Fig. 4.9 Four-pose synthesis curves on the unit sphere: (a) the circlepoint curve and (b) the centerpoint curve

[20] The number of possible combinations of four objects taken two at a time.

Table 4.3 Solutions for the spherical-dyad 5-attitude synthesis

	$\mathbf{a}_0$(or $\mathbf{a}_0^*$)	$\mathbf{b}$(or $\mathbf{b}^*$)
# 1	$[0.7085, -.6418, -.2932]^T$	$[0.2640, -.6636, -.6998]^T$
# 2	$[0.0385, 0.3163, 0.9478]^T$	$[0.1143, 0.7263, -.6777]^T$
# 3	$[0.1642, 0.6977, 0.6972]^T$	$[0.5218, 0.8413, -.1403]^T$
# 4	$[0.8077, 0.1493, 0.5702]^T$	$[0.9524, -.2535, 0.1686]^T$

this claim. It was also found from animations that linkages L1, L2 and L5 are of the crank-rocker type, while linkage L6 is a double-rocker.

Accurate values of the four circlepoint and centerpoint coordinates, displayed in Table 4.3, were obtained numerically, using nonlinear least-squares.

Synthesis of a Spherical dyad with a P Joint

For concreteness, let us assume that the P joint is located at the centerpoint B, coupling the dyad under synthesis with the fixed link, as per Fig. 4.7. Then, the axis of the R joint at A_0 intersects the foregoing major circle at right angles, i.e., $\alpha_2 = \pi/2$. In the example at hand, five attitudes are given, as displayed in Table 4.4.

By means of Eq. (4.69), and a representation of vector $\mathbf{a}_0$ with spherical coordinates ϑ_a and φ_a, contours of the ensuing four determinant equations (4.69) are plotted in Fig. 4.12. There is apparently only one real solution $\{\vartheta_a, \varphi_a\}$. The corresponding unit vector $\mathbf{b}^*$ is thus obtained, followed by the unit vector $\mathbf{a}_0^*$. For the remaining dyad, consisting of two R joints, four solutions are found with the general procedure. Of the four solutions for $\mathbf{a}_0$ and $\mathbf{b}$, one is identical to the solution already

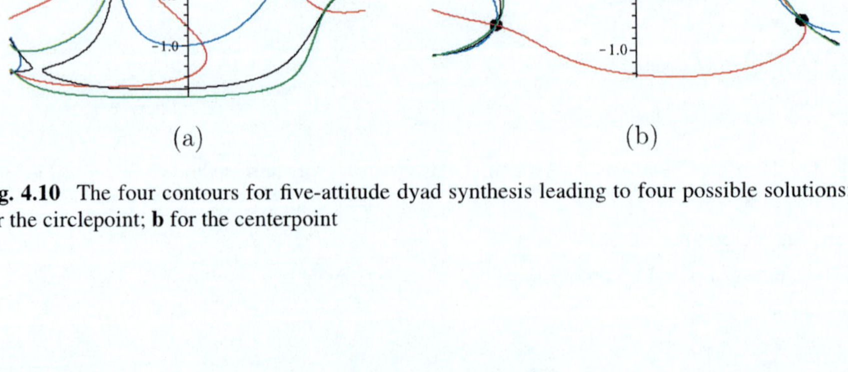

Fig. 4.10 The four contours for five-attitude dyad synthesis leading to four possible solutions: **a** for the circlepoint; **b** for the centerpoint

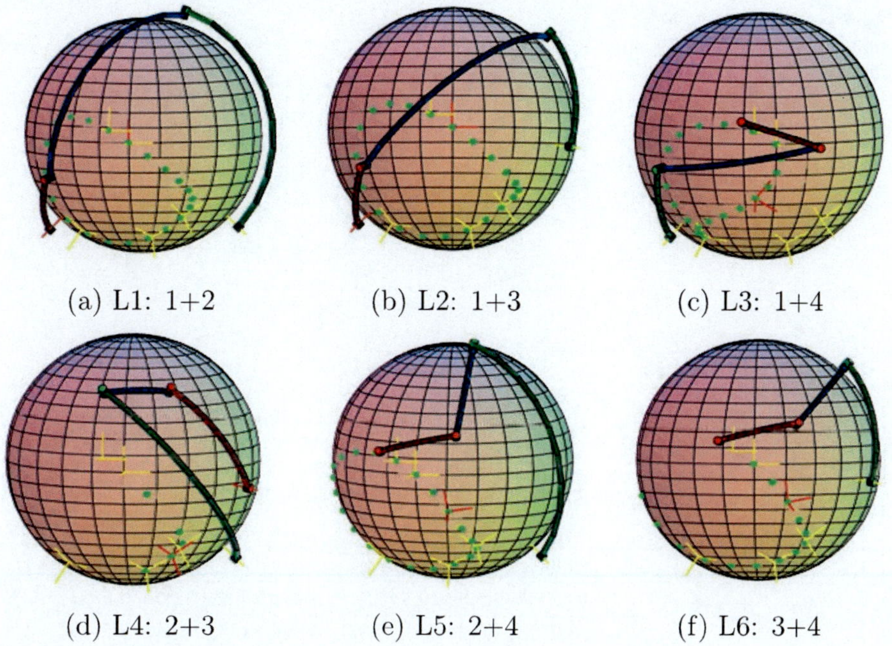

(a) L1: 1+2 (b) L2: 1+3 (c) L3: 1+4

(d) L4: 2+3 (e) L5: 2+4 (f) L6: 3+4

Fig. 4.11 Six synthesized mechanisms, shown together with all task orientations in yellow. Solid dots show traces of a point of the coupler link

Table 4.4 Five attitudes for the synthesis of a PR spherical dyad

ϕ_j [rad]	$\mathbf{e}^T$
0	[0,0,1]
0.2563	[−0.2280, −0.4553, −0.8606]
1.1307	[−0.0578, 0.2370, −0.9697]
1.1938	[0.5049, 0.8505, −0.1468]
1.3665	[0.7119, 0.6601, 0.2393]

found for $\mathbf{a}_0^*$ and $\mathbf{b}^*$, which restates that a P joint is a special case of the general spherical RR dyad. All results are listed in Table 4.5.

Altogether, we have three possible four-bar linkages containing one P joint, out of the given solutions, one being shown in Fig. 4.13. This is a branching-free linkage, as made apparent via animation. The spherical version of a P joint is implemented in mechanical equipment by means a circular guideway, as shown in Fig. 4.13.

As a matter of fact, circular guideways are common joints in machinery, available from manufacturers. Figure 4.14 shows one circular guideway from Germany-based

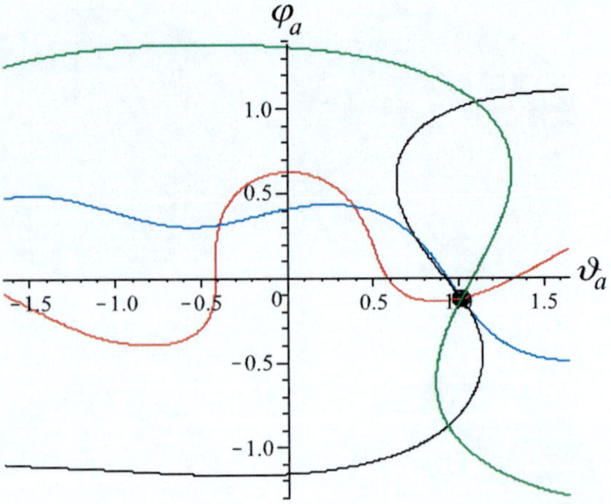

Fig. 4.12 Contour plot to find position vectors for the unit vector of the P joint

Table 4.5 Solutions of Example 3

	$\mathbf{a}_0^T$	$\mathbf{b}^T$
# 1	[0.2845, 0.3863, 0.8773]	[0.1219, −0.7089, −0.6946]
# 2	[0.7226, 0.5295, 0.4442]	[0.2309, 0.4566, 0.8591]
# 3	[0.9573, −0.2433, 0.1555]	[0.8134, 0.1643, 0.5579]
	$\mathbf{a}_0^{*T}$	$\mathbf{b}^{*T}$
# 4	[0.5221, 0.8442, −0.1208]	[0.0655, 0.1015, 0.9926]

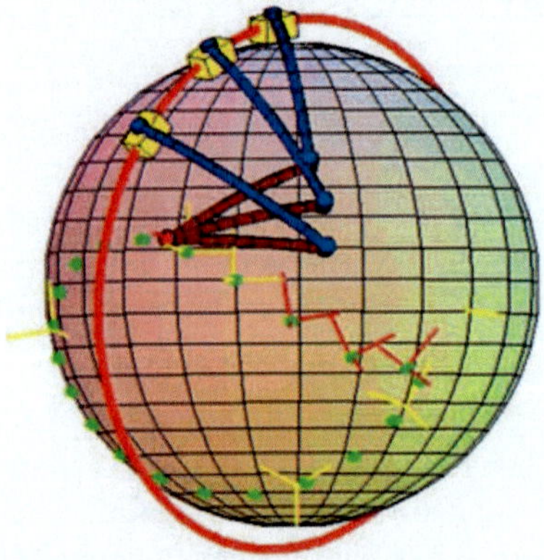

Fig. 4.13 Spherical four-bar linkage with a P joint

(a) (b)

Fig. 4.14 **a** An instance of an industrial circular guideway (*Source* Schaeffler), **b** a design with circular guideways

Schaeffler AG[21] along with a design of an integrated spherical motion generator (Li et al., 2018), in which circular guideways are used.

Approximate Synthesis of Spherical dyads

For approximate synthesis of spherical dyads, reader are recommended to read (Léger and Angeles, 2015) plus the pertinent references therein.

4.4 Spatial Four-Bar Linkages

In this section we study the Burmester problem in the realm of spatial linkages. We will focus on the problem of synthesis of two spatial four-bar linkages, namely, the CCCC and RCCC chains, for rigid-body guidance. The formulation of the synthesis equations and their robust solution is given due attention.[22]

A generic spatial four-bar linkage of the RCCC type is depicted in Fig. 4.15. As the coupler link AA^* moves, while visiting m given poses, the moving axes Z_3 and Z_4, represented by the dual vectors $\hat{\mathbf{a}}_0$ and $\hat{\mathbf{a}}_0^*$ at the reference pose, attain m *locations*. A line being defined by a point and a direction, the line is short of two dimensions to occupy a rigid-body pose. Henceforth, a *line location* will be understood as given by four independent parameters, grouped in a six-dimensional array of Plücker coordinates (Pottmann and Wallner, 2001). The location of a rigid body, known as the body *pose*, requires six independent parameters. A displacement undergone by a line is also known as an *incompletely specified rigid-body displacement* (Tsai and Roth,

[21] http://www.schaeffler.com/content.schaeffler.com/en/divisions/industrial/industrial.jsp.

[22] This section is based on earlier work reported by the authors (Bai and Angeles, 2012, 2015b).

1973). The m locations of the Z_3-axis are represented by the dual vectors $\hat{\mathbf{a}}_1, \ldots, \hat{\mathbf{a}}_m$, those of the Z_4-axis by $\hat{\mathbf{a}}_1^*, \ldots, \hat{\mathbf{a}}_m^*$.

Likewise, the dual vectors of the fixed axes Z_1 and Z_2 are represented by the dual vectors $\hat{\mathbf{b}}$ and $\hat{\mathbf{b}}^*$, respectively. With the foregoing model, the spatial Burmester problem is stated as:

> Find a spatial four-bar linkage that will conduct its coupler link through a set S of m poses, given by (i) the orthogonal matrices $\{\mathbf{Q}_j\}_1^m$ and (ii) the corresponding position vectors $\{\mathbf{r}_j\}_1^m$ of points $\{R_j\}_1^m$ of the coupler link, both vectors and matrices defined with respect to a reference pose, given by $\mathbf{r}_0 = \mathbf{0}$ and $\mathbf{Q}_0 = \mathbf{1}$.

4.4.1 Geometric Constraints of *CC* and *RC* dyads

With reference to the spatial linkage of Fig. 4.15, joint axes Z_3 and Z_4, which define uniquely the coupler link, undergo spatial motion while visiting the m given poses. The spatial displacement of the moving axis Z_3 in terms of dual vectors can be described by the relation

$$\hat{\mathbf{a}}_j = \hat{\mathbf{Q}}_j \hat{\mathbf{a}}_0, \quad j = 1, \ldots, m \tag{4.101}$$

where $\hat{\mathbf{a}}_0$ is the dual representation of line Z_3 at the reference pose of link 3, while $\hat{\mathbf{Q}}_j$ is the *dual orthogonal matrix* representing the spatial displacement of Z_3, namely,

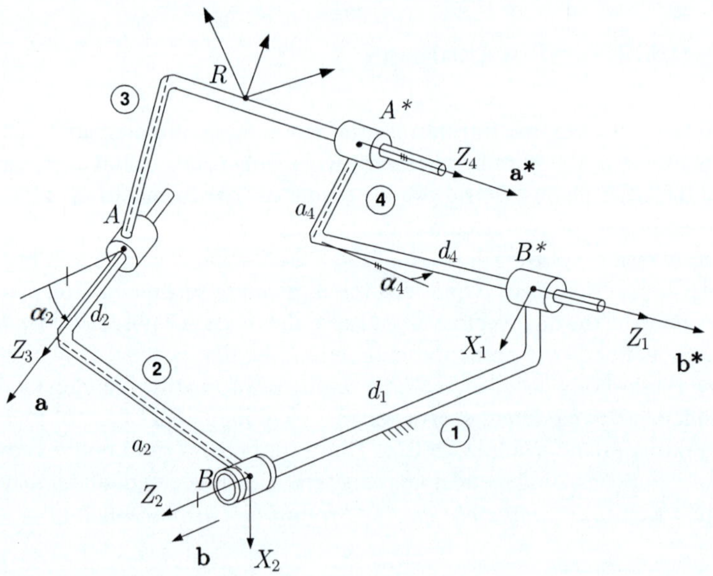

Fig. 4.15 The RCCC linkage

$$\hat{\mathbf{a}}_0 = \mathbf{a}_0 + \epsilon\bar{\mathbf{a}}_0, \quad \hat{\mathbf{Q}}_j = \mathbf{Q}_j + \epsilon\bar{\mathbf{Q}}_j \tag{4.102}$$

with $\mathbf{a}_0$ denoting the unit vector parallel to Z_3 at its reference pose and $\bar{\mathbf{a}}_0$ the *moment* of the same axis with respect to the origin O_1 of frame $\{X_1, Y_1, Z_1\}$. Moreover, $\mathbf{Q}_j$ is the proper orthogonal matrix that rotates the coupler link from its 0th to its jth attitude. In addition, $\bar{\mathbf{Q}}_j$ is the product of a *translation matrix* $\mathbf{R}_j$ by $\mathbf{Q}_j$, i.e.,

$$\bar{\mathbf{Q}}_j = \mathbf{R}_j\mathbf{Q}_j, \quad j = 1, \ldots, m \tag{4.103a}$$

$$\mathbf{R}_j = \text{CPM}(\mathbf{r}_j) = \partial(\mathbf{r}_j \times \mathbf{v})/\partial(\mathbf{v}) \tag{4.103b}$$

with $\mathbf{R}_j$ defined as the *cross-product matrix* of vector $\mathbf{r}_j$.

Given that the location of Z_3 with respect to Z_2 remains constant throughout the motion of the coupler link, the dual angle between the two axes remains constant as well, i.e.,

$$\hat{\mathbf{a}}_j^T\hat{\mathbf{b}} = \cos(\alpha_2 + \epsilon a_2) \equiv \cos\alpha_2 - \epsilon a_2 \sin\alpha_2, \quad j = 0, \ldots, m \tag{4.104}$$

where

$$\hat{\mathbf{b}} = \mathbf{b} + \epsilon\bar{\mathbf{b}} \tag{4.105}$$

with a_2, α_2 and $\mathbf{b}$ shown in Fig. 4.15, while vector $\bar{\mathbf{b}}$ denotes the moment of Z_2 with respect to O_1. Moreover, a_2 is measured along the common perpendicular to X_2 and Z_3, indicated as a dashed line in Fig. 4.15. Subtracting the 0th equation from the m remaining equations (4.104) leads to

$$(\hat{\mathbf{a}}_j - \hat{\mathbf{a}}_0)^T\hat{\mathbf{b}} = 0, \quad j = 1, \ldots, m \tag{4.106}$$

which, upon expansion into its primal and dual parts, for $j = 1, \ldots, m$, yield

$$\mathbf{a}_0^T(\mathbf{Q}_j^T - \mathbf{1})\mathbf{b} = 0, \quad j = 1, \ldots, m \tag{4.107a}$$

$$\mathbf{a}_0^T\bar{\mathbf{Q}}_j^T\mathbf{b} + \mathbf{a}_0^T(\mathbf{Q}_j^T - \mathbf{1})\bar{\mathbf{b}} + \bar{\mathbf{a}}_0^T(\mathbf{Q}_j^T - \mathbf{1})\mathbf{b} = 0, \quad j = 1, \ldots, m \tag{4.107b}$$

its unknowns, $\mathbf{a}_0$, $\bar{\mathbf{a}}_0$, $\mathbf{b}$ and $\bar{\mathbf{b}}$, being subject to the constraints

$$\mathbf{a}_0^T\bar{\mathbf{a}}_0 = 0; \quad \mathbf{b}^T\bar{\mathbf{b}} = 0 \tag{4.108a}$$

$$\|\mathbf{a}_0\|^2 = 1; \quad \|\mathbf{b}\|^2 = 1 \tag{4.108b}$$

We thus have two real sets of equations, (4.107a) for the primal and (4.107b) for the dual part. Of these, Eq. (4.107a) involves only unit vectors of the joint axes, while Eq. (4.107b) involves both unit and moment vectors. Therefore, while Eq. (4.107a) is dimensionless, Eq. (4.107b) has units of length. Following McCarthy and Soh (2011), we call the first set the *direction*, the second the *moment equations*. Both sets stand for the basic geometric constraints on different dyads, including CC, RC and RR. Furthermore, Eqs. (4.107a) are nothing but the synthesis equations for the spherical four-bar linkage, illustrated in Fig. 4.7, under motion generation.[23]

Equation (4.107b) can be cast into a different, more useful form. It is convenient to introduce the position vectors of points on the axes of rotation as design variables, namely,

$$\bar{\mathbf{a}}_0 = \mathbf{r}_{A0} \times \mathbf{a}_0; \quad \bar{\mathbf{b}} = \mathbf{r}_B \times \mathbf{b} \tag{4.109}$$

where $\mathbf{r}_{A0}$ and $\mathbf{r}_B$ are the position vectors of points A_0 and B on the two axes, in the $\{X_1, Y_1, Z_1\}$ frame. In light of Eqs. (4.109), (4.107b), after some manipulations, can be rewritten as

$$(\mathbf{r}_{Aj} - \mathbf{r}_B)^T(\mathbf{a}_j \times \mathbf{b}) - (\mathbf{r}_{A0} - \mathbf{r}_B)^T(\mathbf{a}_0 \times \mathbf{b}) = 0, \quad j = 1, \ldots, m \tag{4.110}$$

where $\mathbf{a}_j = \mathbf{Q}_j\mathbf{a}_0; \quad \mathbf{r}_{Aj} = \mathbf{r}_j + \mathbf{Q}_j\mathbf{r}_{A0}$. The above equation describes the *constancy* of the projection of the line segment defined by two points on the two axes onto their common normal.

4.4.2 *The Synthesis of the* CC *dyad*

The CC dyad under synthesis consists of a rigid link that is coupled to ground via one C joint and to the coupler link via a second C joint. The dyad can be geometrically regarded as a link composed of two skew lines, joined to each other by means of a third line, their common normal. A CC dyad is thus determined once the two skew lines are known. The problem of CC-dyad synthesis thus reduces to locating its two joint axes. The latter can be uniquely described by means of the dual vectors $\hat{\mathbf{a}}_0$ and $\hat{\mathbf{b}}$, which comprise a total of 12 scalar components. Notice that $\mathbf{a}_0$ and $\mathbf{b}$ are unit vectors, while a line can be regarded as a zero-pitch screw (McCarthy and Soh, 2011), as expressed by Eqs. (4.108a) and (4.108b). Those four scalar constraints thus reduce the number of independent variables to only eight.

Five-Pose Synthesis (m = 4)

Equations (4.107a) and (4.107b) lead to $2m$ constraint equations for m given poses. For the number of given poses to yield as many equations as unknowns, i.e., eight, one must have $2m = 8$. The system thus admits exact solutions for CC dyads in the case of $m = 4$, i.e., with five given poses.

---

[23] That is, if abstraction is made of the translation of the rigid body when formulating the problem stated in Sect. 5.3.1, then point R of Fig. 4.15 remains stationary, and the problem at hand becomes one of spherical synthesis.

Knowing that the direction equations are independent of the moment equations, we can find first the direction vectors through Eqs. (4.107a). These equations, identical to those of the Burmester problem for spherical linkages, admit at most six real solutions. With each of these solutions, the moment equations become linear in the moments, namely,

$$\mathbf{p}_j^T\bar{\mathbf{a}}_0 + \mathbf{q}_j^T\bar{\mathbf{b}} = w_j, \quad j = 1, \ldots, m \tag{4.111}$$

with

$$\mathbf{p}_j = (\mathbf{Q}_j^T - \mathbf{1})\mathbf{b}, \quad \mathbf{q}_j = (\mathbf{Q}_j - \mathbf{1})\mathbf{a}_0, \quad w_j = -\mathbf{a}_0^T\bar{\mathbf{Q}}_j^T\mathbf{b}, \quad j = 1, \ldots, m \tag{4.112}$$

In the case of $m = 4$, Eqs. (4.108a) and (4.111) amount to six linear equations for six unknowns, thus defining a unique pair of moments, $\bar{\mathbf{a}}_0$ and $\bar{\mathbf{b}}$, in the absence of singularities.

The six linear equations (4.108a) and (4.111) can be cast in the form

$$\mathbf{M}\mathbf{x} = \mathbf{w} \tag{4.113}$$

where

$$\mathbf{M} = \begin{bmatrix} \mathbf{p}_1^T & \mathbf{q}_1^T \\ \vdots & \vdots \\ \mathbf{p}_m^T & \mathbf{q}_m^T \\ \mathbf{a}_0^T & \mathbf{0}^T \\ \mathbf{0}^T & \mathbf{b}^T \end{bmatrix}, \quad \mathbf{x} = \begin{bmatrix} \bar{\mathbf{a}}_0 \\ \bar{\mathbf{b}} \end{bmatrix}, \quad \mathbf{w} = \begin{bmatrix} w_1 \\ \vdots \\ w_m \\ 0 \\ 0 \end{bmatrix} \tag{4.114}$$

and hence, $\mathbf{M}$ is a $(m + 2) \times 6$ matrix.

So far, we have reformulated the synthesis equations for CC dyads. Apparently, the spatial linkage synthesis with CC dyads is based essentially on the Burmester problem for spherical linkages. Notice that Eqs. (4.107a) also represent the synthesis equations for spherical four-bar linkages, which can be readily solved by semigraphical means (Angeles and Bai, 2010).

Four-Pose Synthesis (m = 3)

If only four poses are given, a total of six equations is obtained from Eqs. (4.107a) and (4.107b). In this case, the system of synthesis equations is underdetermined, and hence, infinitely many solutions are possible. These solutions can be regarded as sets of lines, called *congruences*, which define the moving and fixed axes. The generation of line congruences is described in Sect. 4.4.7.

4.4.3 The Synthesis of the RC dyad

A RC dyad is composed, of course, of one revolute and one cylindrical joint. The constraints for a CC spatial dyad also apply to a RC dyad. Moreover, compared with a CC dyad, a RC dyad is subject to one more constraint: the sliding s_j along the fixed axis, shown in Fig. 4.16, is zero. This constraint can be expressed in terms of the dual vector of the common normal.

According to Eq. (A.24), for the jth pose, the common normal to $\mathcal{L}_a$ and $\mathcal{L}_b$ is given by

$$\hat{\mathbf{n}}_j = \mathbf{n}_j + \epsilon\bar{\mathbf{n}}_j, \quad j = 0, \ldots, m \tag{4.115}$$

with

$$\mathbf{n}_j = \frac{1}{\sin\alpha_{2j}}(\mathbf{a}_j \times \mathbf{b}) \tag{4.116a}$$

$$\bar{\mathbf{n}}_j = \frac{1}{\sin\alpha_{2j}}(\mathbf{a}_j \times \bar{\mathbf{b}} + \bar{\mathbf{a}}_j \times \mathbf{b}) - \frac{a_{2j}\cos\alpha_{2j}}{\sin^2\alpha_{2j}}(\mathbf{a}_j \times \mathbf{b}) \tag{4.116b}$$

where a_{2j} and α_{2j} are the distance and the angle between $\mathcal{L}_a$ and $\mathcal{L}_b$, respectively, at the jth given pose. Now, by virtue of the link rigidity, $a_{2j} = a_{20} \equiv a_2$ and $\alpha_{2j} = \alpha_{20} \equiv \alpha_2$, the two above equations then becoming

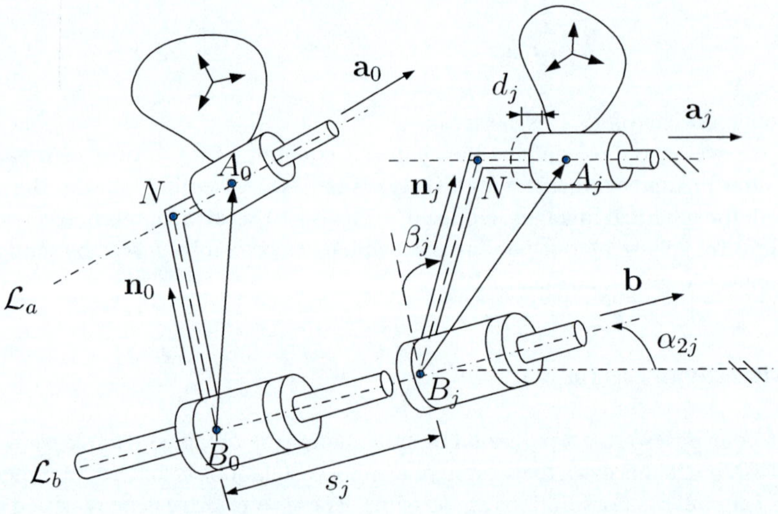

Fig. 4.16 A CC dyad, which becomes a RC dyad if the sliding s_j vanishes

$$\mathbf{n}_j = \frac{1}{\sin\alpha_2}(\mathbf{a} \times \mathbf{b}) \tag{4.117a}$$

$$\bar{\mathbf{n}}_j = \frac{1}{\sin\alpha_2}(\mathbf{a}_j \times \bar{\mathbf{b}} + \bar{\mathbf{a}}_j \times \mathbf{b}) - \frac{a_2\cos\alpha_2}{\sin^2\alpha_2}(\mathbf{a}_j \times \mathbf{b}) \tag{4.117b}$$

Furthermore, the dual angle between two poses of the common normals, $\hat{\mathbf{n}}_j$ and $\hat{\mathbf{n}}_0$, is given by

$$\cos\hat{\beta}_j = \hat{\mathbf{n}}_j^T\hat{\mathbf{n}}_0, \quad j = 1, \ldots, m \tag{4.118}$$

where $\hat{\beta}_j = \beta_j + \epsilon s_j$, with β_j and s_j denoting the jth angle of rotation and the jth sliding associated with the joint, respectively. Equating the dual parts of the expanded equation (4.118) leads to

$$-s_j\sin\beta_j = \frac{1}{\sin^3\alpha_2}(A_j a_2\cos\alpha_2 - B_j\sin\alpha_2) \tag{4.119}$$

with

$$A_j = 2\,(\mathbf{a}_0 \times \mathbf{b})^T(\mathbf{a}_j \times \mathbf{b}) \tag{4.120}$$

$$B_j = (\mathbf{a}_0 \times \mathbf{b})^T(\mathbf{a}_j \times \bar{\mathbf{b}} + \bar{\mathbf{a}}_j \times \mathbf{b}) + (\mathbf{a}_j \times \mathbf{b})^T(\mathbf{a}_0 \times \bar{\mathbf{b}} + \bar{\mathbf{a}}_0 \times \mathbf{b}) \tag{4.121}$$

For RC dyads, $s_j = 0$, while, for spatial dyads, α_2 does not vanish, Eq. (4.119) then leading to

$$A_j a_2\cos\alpha_2 - B_j\sin\alpha_2 = 0, \quad j = 1, \ldots, m \tag{4.122}$$

which is the set of constraint equations needed to guarantee the vanishing of the sliding on the fixed axis.

For a RC dyad, we have constraint equations (4.122), (4.107a) and (4.107b), which amount to $3m$ equations for m poses. The problem thus bears eight independent unknowns, which means that a RC-dyad synthesis problem admits exact solutions for up to $m = 2$, in which case the number of variables exceeds that of constraint equations by two, infinitely many solutions then being available.

A RR dyad, consisting of two revolute joints, requires one additional constraint, on top of those of the RC dyad. The additional equations pertain to the vanishing of the second sliding, namely, d_j of Fig. 4.16. Alternatively, the constraint equations can be derived from the geometric relationship between two axes, as included in Appendix B for the interested reader. The formulation involves $4m + 2$ equations for m poses. On the other hand, the formulation includes 10 independent variables, with four direction and six position variables. In summary, for $m = 2$, i.e., for the three prescribed poses, a RR spatial dyad admits exact solutions.

4.4.4 Synthesis of Four-Bar Linkages

A two-degree-of-freedom CCCC linkage can be built with two CC dyads. As shown in Sect. 4.4.2, each dyad admits exact solutions for $m = 4$, i.e., for five poses. A CCCC linkage can thus guide its coupler link through five poses, but such a linkage requires *two independent inputs*.

A RCCC linkage differs from a CCCC linkage, in that the former has one dof, the latter two. Moreover, a RCCC linkage has a *non-symmetrical topology*. As discussed in Sect. 4.4.1, a RC dyad admits exact solutions only for three given poses. Hence, no exact solutions for the RCCC linkage synthesis are possible when $m > 2$.

4.4.5 A Semigraphical Solution of the Direction Equations

The direction equations (4.107a) can be solved by semigraphical means as introduced in connection with the synthesis of spherical four-bar linkages (Angeles and Bai, 2010). In this regard, *spherical coordinates* on the unit sphere, namely, *longitude* and *latitude*, are used to describe the unit vectors of all four directions. Let, then, θ_a and ϕ_a denote the longitude and the latitude of a point on the unit sphere defining $\mathcal{L}_a$, θ_b and ϕ_b the counterpart spherical coordinates for the point defining $\mathcal{L}_b$. Hence,

$$\mathbf{a}_0 = \begin{bmatrix} \cos\phi_a \cos\theta_a \\ \cos\phi_a \sin\theta_a \\ \sin\phi_a \end{bmatrix}, \quad \mathbf{b} = \begin{bmatrix} \cos\phi_b \cos\theta_b \\ \cos\phi_b \sin\theta_b \\ \sin\phi_b \end{bmatrix} \tag{4.123}$$

Further, the direction equations are rewritten as

$$\mathbf{c}_j^T \mathbf{b} = 0, \quad j = 1, \ldots, 4 \tag{4.124}$$

or, upon assembling all four equations in array form,

$$\mathbf{C}\mathbf{b} = \mathbf{0} \in \mathbb{R}^4, \quad \mathbf{C} = \begin{bmatrix} \mathbf{c}_1^T \\ \mathbf{c}_2^T \\ \mathbf{c}_3^T \\ \mathbf{c}_4^T \end{bmatrix} \in \mathbb{R}^{3\times 4} \tag{4.125}$$

where

$$\mathbf{c}_j = (\mathbf{Q}_j - \mathbf{1})\mathbf{a}_0 \tag{4.126}$$

As vector $\mathbf{b}$ is of unit magnitude, it cannot vanish, and hence, $\mathbf{C}$ must be rank-deficient; therefore, its four 3×3 subdeterminants must all vanish:

$$\Delta_j(\mathbf{a}_0) \equiv \det(\mathbf{C}_j) = 0, \quad j = 1, \ldots, 4 \tag{4.127}$$

with $\mathbf{C}_j$ denoting the 3×3 matrix obtained upon deleting the jth row from $\mathbf{C}$. Now the four determinant equations (4.127) in $\mathbf{a}_0$ become equations in the trigonometric functions of ϕ_a and θ_a. They define four contours $\mathcal{C}_j$ in the ϕ_a-θ_a plane. If the four contours are plotted in the $-\pi/2 \leq \phi_a \leq \pi/2$, $-\pi/2 \leq \theta_a \leq \pi/2$ square, then the intersections of the four contours yield *all the real solutions* sought.[24] There can be six, four, two or no real solutions, in general (Chiang, 1988; McCarthy and Soh, 2011). Two special cases, zero and infinitely many solutions, are noteworthy. In the case of zero solutions, no common intersection of the four contours appears in the superimposed plots. At the other end of the spectrum, infinitely many solutions are possible when either ϕ_a or ϕ_b is identical to $\pm\pi/2$, which thus yields points at a pole of the unit sphere; poles are known to admit any possible longitude—at the poles of the Earth, any time is good, as all time zones converge there!

The intersections are estimated by inspection on the four contours. These estimates can then be refined if used as initial guesses of an iterative procedure to find the roots of *any* pair of the four foregoing equations. As there are as many as six such possible pairs, the choice of the two equations depends on the arbitrariness of the designer. If the choice turns out to be lucky, i.e., if the two contours intersect at "sharp enough" angles, then the solution will be reliable enough; else, the solution may be either unreliable—contours intersect at almost-coincident contour segments—or impossible to find—contours are tangent to each other at the intersection point. To help the designer, a least-square approximation of the overdetermined system of four nonlinear equations in only two unknowns is recommended here: in fact, the four contours will not intersect at one common point because of roundoff error; however, nonlinear least squares will return the most likely, and hence, the most reliable estimate of the solution sought. This is the same philosophy behind surveying, which is based on *triangulation*.

Once $\mathbf{a}_0$ is known, $\mathbf{c}_j$, for $j = 1, \ldots, 4$, can be computed from Eqs. (4.126), and hence, the 4×3 matrix $\mathbf{C}$ is available. Vector $\mathbf{b}$ can then be found as the unit vector spanning the nullspace of $\mathbf{C}$, which can be determined using, e.g., the QR-decomposition of $\mathbf{C}$ (Nash and Sofer, 1996). While this approach is sound, it entails one drawback: any roundoff error incurred in computing $\mathbf{C}$ is propagated into the computation of $\mathbf{b}$. A robust solution consists in computing $\mathbf{b}$ independent from the computation of $\mathbf{a}_0$. In this vein, the foregoing procedure is paraphrased, as applicable to the other end of the dyad, which is done by rewriting Eq. (4.107a) as

$$\mathbf{D}\mathbf{a}_0 = \mathbf{0}, \quad \mathbf{D} = \begin{bmatrix} \mathbf{d}_1^T \\ \mathbf{d}_2^T \\ \mathbf{d}_3^T \\ \mathbf{d}_4^T \end{bmatrix} \in \mathbb{R}^{3\times 4} \tag{4.128}$$

the determinant equations to find θ_b and ϕ_b then being derived as

$$\Delta_j(\mathbf{b}) \equiv \det(\mathbf{D}_j) = 0, \quad j = 1, \ldots, 4 \tag{4.129}$$

[24] Here, of course, the *antipodal* solutions of those lying in the square of side-length π are excluded.

with $\mathbf{D}_j$ denoting the 3×3 matrix obtained upon deleting the jth row from $\mathbf{D}$. Notice that $\mathbf{d}_j = (\mathbf{Q}_j^T - \mathbf{1})\mathbf{b}$.

4.4.6 Solving the Linear Equations of the Moment Variables

Once the unit vectors $\mathbf{a}_0$ and $\mathbf{b}$ are known, a system of six linear equations in six unknowns can be obtained from Eq. (4.113) for the moments $\bar{\mathbf{a}}_0$ and $\bar{\mathbf{b}}$. Now, as Eq. (4.113) is a system of $m + 2$ linear equations for six moment variables, a unique set of solutions for $\bar{\mathbf{a}}_0$ and $\bar{\mathbf{b}}$ can be found for $m = 4$.

4.4.7 Congruences of the Fixed and the Moving Axes

In the synthesis of *planar four-bar linkages* with four given poses, circlepoint- and centerpoint curves can be generated to select pivoting points for moving and fixed-joint centers, respectively. In analogy with the planar case, sets of lines for the moving and fixed axes can be determined in the spatial case. These sets, called *congruences*, define the moving and fixed axes of the CC dyads that guide a rigid body through four given poses.

We include below a method for defining and generating congruences of lines, to be used as fixed and moving axes. The method takes advantage of the three constraint equations derived below.

Referring to Eq. (4.125) with $m = 3$, $\mathbf{C}$ becomes a 3×3 matrix, which must be singular. This means that, under the four given poses,

$$F(\mathbf{a}_0) \equiv \det(\mathbf{C}) \equiv (\mathbf{c}_1 \times \mathbf{c}_2)^T \mathbf{c}_3 = 0 \tag{4.130}$$

Equation (4.130) defines a conic surface in Cartesian space. Its intersection with the unit sphere yields a spherical curve, called the *circlepoint curve* (McCarthy and Soh, 2011). Each point on the curve defines one unit vector giving the direction of the moving axis.

Likewise, the equation for the unit vector of the fixed axis bears the form

$$G(\mathbf{b}) \equiv \det(\mathbf{D}) = (\mathbf{d}_1 \times \mathbf{d}_2)^T \mathbf{d}_3 = 0, \quad \mathbf{d}_i \in \mathbb{R}^3, \quad i = 1, 2, 3 \tag{4.131}$$

The intersection of the cubic surface described by the above equation with the unit sphere yields the *spherical centerpoint curve*. Each point on the curve corresponds to a candidate unit vector for the fixed axis.

Notice that the centerpoint and the circlepoint curves are linked through the direction constraint equation (4.107a). This means that each point on one of the two curves has its unique matching point on the other curve.

The dual-vector parts of the two axes, i.e., the moments $\bar{\mathbf{a}}_0$ and $\bar{\mathbf{b}}$, can be found from Eq. (4.113). However, for $m = 3$, $\mathbf{M}$ of Eq. (4.113) is of 5×6, and the system is *underdetermined*. In order to face this indeterminacy, a sixth row $\mathbf{m}_6^T$ is added to matrix $\mathbf{M}$, this row being defined as

$$\mathbf{m}_6^T = [1, 0, 0, 0,\ 0,\ 0] \tag{4.132}$$

and, correspondingly, the right-hand side of Eq. (4.113) becomes

$$\mathbf{w} = [w_1, \ldots, w_m,\ 0,\ 0,\ \lambda]^T \tag{4.133}$$

which means that the x-component $(\bar{\mathbf{a}}_0)_x$ of $\bar{\mathbf{a}}_0$ is assigned a variable value λ. If the choice of $\mathbf{m}_6$ happens to render $\mathbf{M}$ either singular or ill-conditioned, then another component of $\bar{\mathbf{a}}_0$ should be chosen instead.

Because of the algebraic coupling between $\bar{\mathbf{a}}_0$ and $\bar{\mathbf{b}}$, moreover, the latter also turns out to be a function of parameter λ. Equation (4.113) thus yields, in this case, the moments sought as linear functions of parameter λ. We thus obtain solution axes in the form of Plücker coordinates $[\mathbf{a}_0^T\ \bar{\mathbf{a}}_0^T(\lambda)]^T$ and $[\mathbf{b}^T\ \bar{\mathbf{b}}^T(\lambda)]^T$, each standing for a set of parallel lines lying in the same plane.

Once vectors $\mathbf{a}_0$, $\bar{\mathbf{a}}_0$, $\mathbf{b}$ and $\bar{\mathbf{b}}$ are available, the position vectors $\mathbf{r}_{A0}$ and $\mathbf{r}_B$, as introduced in Eqs. (4.109), can be computed from the feet of the common perpendicular of the pairs of skew lines (Z_2 and Z_3) and (Z_3 and Z_4), respectively.

4.4.8 Examples

We include here examples—solved with computer algebra—of spatial-linkage synthesis to demonstrate the foregoing procedure. The data are taken from the literature (Bai and Angeles, 2012). The given poses are displayed in Table4.6, where the orientation is described with *natural invariants* (Angeles, 2014), i.e., the unit vector $\mathbf{e}_j$ of the axis of rotation and the angle of rotation φ_j, at the jth pose. The rotation matrix at this pose then takes the form Angeles (2014)

$$\mathbf{Q}_j = \mathbf{1} + \mathrm{s}\varphi_j\mathbf{E}_j + (1 - \mathrm{c}\varphi_j)\mathbf{E}_j^2 \tag{4.134}$$

where $\mathbf{E}_j$ denotes the cross-product matrix of $\mathbf{e}_j$, while $\mathrm{c}\varphi_j$ and $\mathrm{s}\varphi_j$ stand for $\cos\varphi_j$ and $\sin\varphi_j$, respectively.

Five-Pose Case

For starters, the direction unit vectors of the axes are found. The four trigonometric functions, produced from Eq. (4.129) for vector $\mathbf{b}$ of the fixed axis, are

Table 4.6 Five given poses

#	φ_j[rad]	$\mathbf{e}_j^T$	$\mathbf{r}_j^T$
1	0	[0,0,1]	[0,0,0]
2	0.2275	[−0.5516, −0.4597, 0.6959]	[8.5499, 4.2276, −11.3658]
3	0.3361	[−0.7508, −0.5362, 0.3855]	[28.3322, 12.7149, −13.0406]
4	0.3288	[−0.7629, −0.6357, −0.1169]	[41.4694, 23.2247, −1.1371]
5	0.2519	[−0.4171, −0.4950, −0.7621]	[34.2001, 29.1295, 9.0739]

$$\begin{aligned}\Delta_1(\mathbf{b}) = & \, 0.7954\, \mathrm{c}\theta_b \mathrm{c}^2\phi_b \mathrm{s}\theta_b \mathrm{s}\phi_b + 0.2216\, \mathrm{s}^3\phi_b - 0.5388\, \mathrm{c}^3\theta_b \mathrm{c}^3\phi_b \\ & + 0.8001\, \mathrm{s}^3\theta_b \mathrm{c}^3\phi_b - 1.1618\, \mathrm{c}^2\theta_b \mathrm{c}^2\phi_b \mathrm{s}\phi_b + 0.1403\, \mathrm{c}\theta_b \mathrm{c}^3\phi_b \mathrm{s}^2\theta_b \\ & + 0.0827\, \mathrm{c}\theta_b \mathrm{c}\phi_b \mathrm{s}^2\phi_b + 0.1425\, \mathrm{s}^2\theta_b \mathrm{c}^2\phi_b \mathrm{s}\phi_b - 0.6723\, \mathrm{s}\theta_b \mathrm{c}\phi_b \mathrm{s}^2\phi_b \\ & + 0.0601\, \mathrm{c}^2\theta_b \mathrm{c}^3\phi_b \mathrm{s}\theta_b \end{aligned} \tag{4.135a}$$

$$\begin{aligned}\Delta_2(\mathbf{b}) = & \, 0.3726\, \mathrm{c}\theta_b \mathrm{c}^2\phi_b \mathrm{s}\theta_b \mathrm{s}\phi_b + 0.2385\, \mathrm{s}^3\phi_b - 0.6712\, \mathrm{c}^3\theta_b \mathrm{c}^3\phi_b \\ & + 0.6292\, \mathrm{s}^3\theta_b \mathrm{c}^3\phi_b - 0.5949\, \mathrm{c}^2\theta_b \mathrm{c}^2\phi_b \mathrm{s}\phi_b + 0.6866\, \mathrm{c}\theta_b \mathrm{c}^3\phi_b \mathrm{s}^2\theta_b \\ & + 0.8044\, \mathrm{c}\theta_b \mathrm{c}\phi_b \mathrm{s}^2\phi_b - 0.2671\, \mathrm{s}^2\theta_b \mathrm{c}^2\phi_b \mathrm{s}\phi_b - 1.1069\, \mathrm{s}\theta_b \mathrm{c}\phi_b \mathrm{s}^2\phi_b \\ & - 0.1149\, \mathrm{c}^2\theta_b \mathrm{c}^3\phi_b \mathrm{s}\theta_b \end{aligned} \tag{4.135b}$$

$$\begin{aligned}\Delta_3(\mathbf{b}) = & - 0.1721\, \mathrm{c}\theta_b \mathrm{c}^2\phi_b \mathrm{s}\theta_b \mathrm{s}\phi_b + 0.2086\, \mathrm{s}^3\phi_b - 0.3794\, \mathrm{c}^3\theta_b \mathrm{c}^3\phi_b \\ & + 0.1339\, \mathrm{s}^3\theta_b \mathrm{c}^3\phi_b + 0.4073\, \mathrm{c}^2\theta_b \mathrm{c}^2\phi_b \mathrm{s}\phi_b + 0.8991\, \mathrm{c}\theta_b \mathrm{c}^3\phi_b \mathrm{s}^2\theta_b \\ & + 0.8365\, \mathrm{c}\theta_b \mathrm{c}\phi_b \mathrm{s}^2\phi_b - 0.6039\, \mathrm{s}^2\theta_b \mathrm{c}^2\phi_b \mathrm{s}\phi_b - 0.9426\, \mathrm{s}\theta_b \mathrm{c}\phi_b \mathrm{s}^2\phi_b \\ & - 0.2169\, \mathrm{c}^2\theta_b \mathrm{c}^3\phi_b \mathrm{s}\theta_b \end{aligned} \tag{4.135c}$$

$$\begin{aligned}\Delta_4(\mathbf{b}) = & -0.4783\, \mathrm{c}\theta_b \mathrm{c}^2\phi_b \mathrm{s}\theta_b \mathrm{s}\phi_b + 0.1969\, \mathrm{s}^3\phi_b - 0.0680\, \mathrm{c}^3\theta_b \mathrm{c}^3\phi_b \\ & - 0.1912\, \mathrm{s}^3\theta_b \mathrm{c}^3\phi_b + 0.6195\, \mathrm{c}^2\theta_b \mathrm{c}^2\phi_b \mathrm{s}\phi_b + 0.5308\, \mathrm{c}\theta_b \mathrm{c}^3\phi_b \mathrm{s}^2\theta_b \\ & + 0.4241\, \mathrm{c}\theta_b \mathrm{c}\phi_b \mathrm{s}^2\phi_b - 0.5335\, \mathrm{s}^2\theta_b \mathrm{c}^2\phi_b \mathrm{s}\phi_b - 0.3676\, \mathrm{s}\theta_b \mathrm{c}\phi_b \mathrm{s}^2\phi_b \\ & - 0.2046\, \mathrm{c}^2\theta_b \mathrm{c}^3\phi_b \mathrm{s}\theta_b \end{aligned} \tag{4.135d}$$

The foregoing equations involve *sextic bivariate expressions* of the harmonic functions of θ_b and ϕ_b. In this light, if the tan-half identities (Wampler et al., 1990) are introduced, Eqs. (4.135a)–(4.135d) become bivariate polynomial equations of the 12th degree, or, dodecic. This means that the Bezout number of any pair of these is $12^2 = 144$, the resolvent polynomial of the two thus being univariate, but of the 144th degree! It is well known (Forsythe, 1970) that polynomial-root finding is inherently an ill-conditioned problem, i.e., extremely sensitive to round-off error in the coefficients, as the polynomial degree becomes moderately large, of 10 or higher.

The semigraphical method proposed here comes to the rescue. The four contours stemming from Eqs. (4.135a)–(4.135d) are shown in Fig. 4.17a. Four intersections can be identified, which suggests four real solutions, for which accurate values are found with a nonlinear least-square solver—as per the discussion in Sect. 4.4.5 in connection with the solution of Eq. (4.127). The contour plots for vector $\mathbf{a}_0$ are shown in Fig. 4.17b. Identifying the intersection around $\phi_a = \pm\pi/2$, where all four contours are superimposed along the line $\phi_a = \pm\pi/2$, requires special attention. While there are infinitely many intersections, all correspond to one unique solution, $\mathbf{a}_0 = [0, 0, -1]^T$, no matter what value θ_a takes. This is the special case discussed in Sect. 4.4.5, which shows that the *semigraphical method* is capable of coping with special cases.

Four real solutions are found for the unit vectors of the moving- and the fixed-joint axes. These solutions are matched with the aid of Eq. (4.107a) into four sets, each being a CC-dyad solution. From these solutions, the moment components of the two axes of the CC dyad are found from Eq. (4.113), with

$$\mathbf{M} = \begin{bmatrix} 0.0491 & -0.0883 & -0.0194 & -0.0009 & 0.0013 & 0.0001 \\ 0.0164 & -0.0258 & -0.0041 & 0.0715 & -0.0981 & 0.0027 \\ -0.0867 & 0.1041 & -0.0003 & 0.1555 & -0.1943 & 0.0416 \\ -0.1622 & 0.1666 & -0.0194 & 0.1545 & -0.1932 & 0.0409 \\ 0.5566 & 0.4613 & -0.6909 & 0 & 0 & 0 \\ 0 & 0 & 0 & 0.7752 & 0.5565 & -0.2987 \end{bmatrix} \tag{4.136a}$$

$$\mathbf{w} = [-1.1049, -2.4925, -4.7123, -2.8470, 0, 0]^T \tag{4.136b}$$

Equations (4.136a) and (4.136b) display matrix $\mathbf{M}$ and vector $\mathbf{w}$ produced from the direction solution # 1. The corresponding moments can be directly obtained from $\mathbf{x} = \mathbf{M}^{-1}\mathbf{w}$. All dual vectors of the moving and fixed axes are listed in Table 4.7. A total of six CCCC mechanisms can be synthesized from the four sets of solutions. A mechanism generated from solutions # 3 and # 4 is shown in Fig. 4.18. Notice that this example is based on the example reported by Jamalov et al. (1984). The mechanism shown in Fig. 4.18, identical to the linkage in the foregoing reference, validates the method developed in this work.

Four-Pose Case

The synthesis of CCCC linkages with four poses was conducted using as data the first four rows of Table 4.6. For the four given poses, the pertinent centerpoint- and circlepoint curves are displayed in Fig. 4.19, on the unit sphere. Of these, the circlepoint curve is selected to generate the congruences of lines. Figure 4.20 shows the congruences of the fixed and moving axes, where solid edges of each plane indicate the direction of the set of lines. Notice that 24 points are selected from the centerpoint curve to generate the congruences.

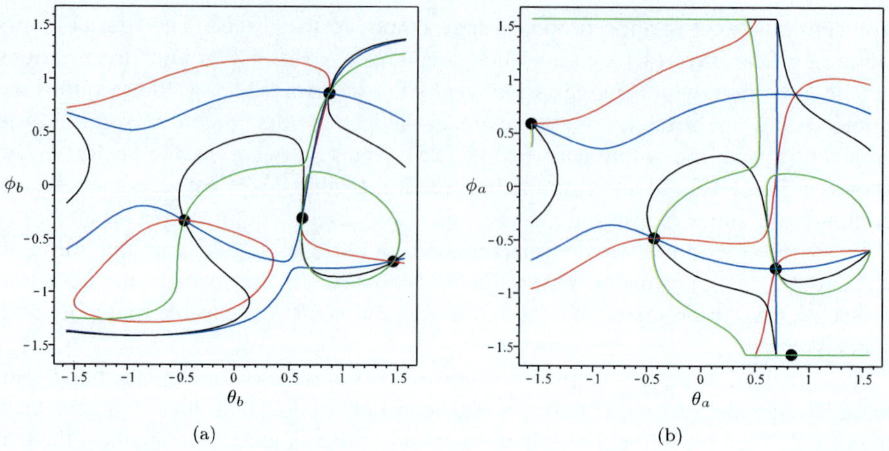

Fig. 4.17 Contour plots for: **a** angles θ_b and ϕ_b, and **b** angles θ_a and ϕ_a

Table 4.7 CCCC linkage solutions

	$\mathbf{a}_0$	$\mathbf{b}$
# 1	[0.557, 0.461, –0.691]	[0.775, 0.557, –0.299]
# 2	[0.805, –0.370, –0.463]	[0.843, –0.430, –0.324]
# 3	[0.000, 0.000, –1.000]	[0.417, 0.495, 0.762]
# 4	[0.000, –0.831, 0.557]	[0.083, 0.756, –0.649]
	$\bar{\mathbf{a}}_0$	$\bar{\mathbf{b}}$
# 1	[–16.724, 5.525, –9.784]	[–24.798, 1.756, –61.078]
# 2	[–27.809, 39.395, –79.918]	[–22.408, 35.386, –105.146]
# 3	[0.000, 0.000, 0.000]	[–24.109, –19.553, 25.894]
# 4	[–24.925, 29.496, 44.034]	[19.108, –22.451, –23.685]

4.4.9 Summary

Synthesis equations were derived in this chapter for three different types of spatial dyads, CC, CR and RR. A synthesis method of spatial linkages applicable to CCCC and RCCC linkages was proposed. The method consists of two steps: direction-equation solving and moment-equation solving, as demonstrated by means of an example. It is noted that the synthesis problem can involve none or infinitely many solutions.

Equation-solving was achieved with a semigraphical method, which provides a robust means of obtaining all possible solutions, including the case of CCCC linkages with four given poses. Other methods are equally applicable.

We thus conclude that: (i) the CC dyad can meet up to five given poses; (ii) the RC dyad can meet up to three given poses; and (iii) the spatial counterpart of

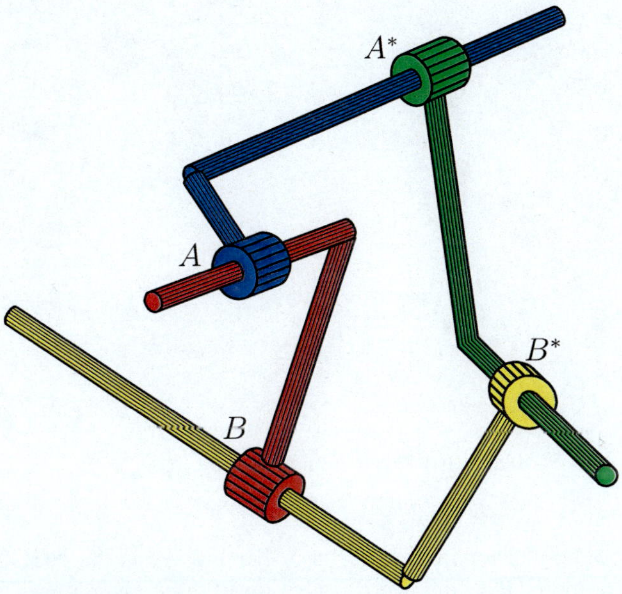

Fig. 4.18 A CCCC linkage for five given poses

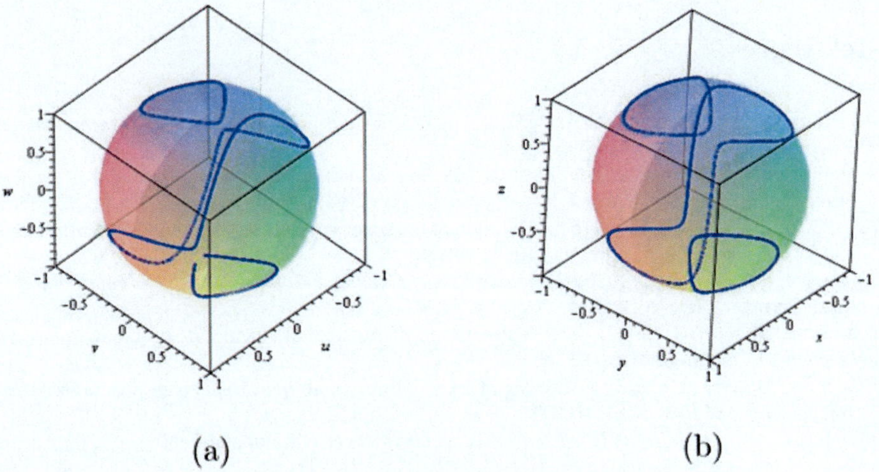

Fig. 4.19 **a** Centerpoint and **b** circlepoint spherical curves

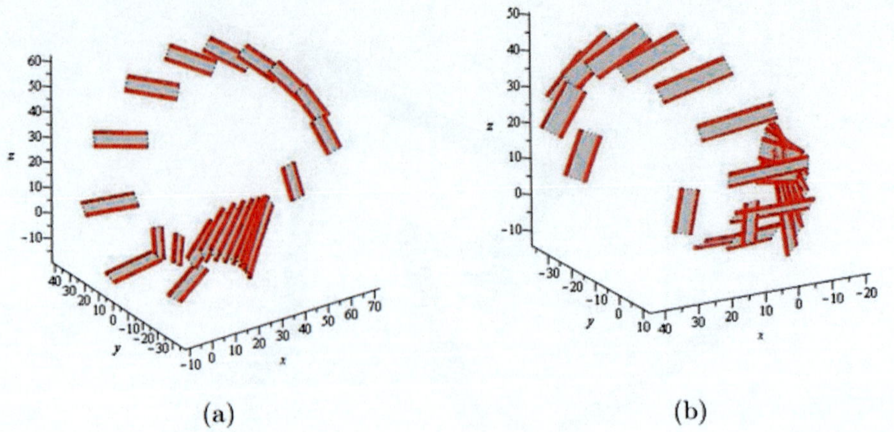

Fig. 4.20 Congruences of: **a** the fixed axis; and **b** the moving axis

the planar, or spherical for that matter, four-bar linkage is the RCCC linkage. In summary, then, the RCCC linkage can visit up to three given poses of its coupler link.

References

Angeles, J. (2006). Is there a characteristic length of a rigid-body displacement? *Mechanism and Machine Theory,41*(8), 884–896.

Angeles, J. (2014). *Fundamentals of robotic mechanical systems: Theory, methods, and algorithms* (4th ed.). Springer.

Angeles, J., & Bai, S. (2010). A robust solution of the spherical Burmester problem. In *Proceedings of ASME DETC2010*. Montreal, #MECH–28189.

Angeles, J. and Pasini, D. (2020). *Fundamentals of Geometry Construction: The Math Behind the CAD*. Springer, New York.

Bai, S., & Angeles, J. (2012). A robust solution of the spatial Burmester problem. *Journal of Mechanisms and Robotics,4*, 031003.

Bai, S., & Angeles, J. (2015). Synthesis of RCCC linkages to visit four given poses. *Journal of Mechanisms and Robotics,7*, 031004.

Bottema, O. and Roth, B. (1978). *Theoretical Kinematics*. North-Holland Publishers Co., Amsterdam.

Chang, C. Y., Angeles, J., and Gonzalez-Palacios, M. A. (1991). A semigraphical method for the solution of the Burmester problem. In *Proc. International Design Engineering Technical Conferences and Computers and Information in Engineering Conference*, Miami, Florida, USA, #DETC1991–0134, pp. 321–326

Chen, C., & Angeles, J. (2008). A novel family of linkages for advanced motion synthesis. *Mechanism and Machine Theory,43*(7), 882–890.

Chen, C., Bai, S., & Angeles, J. (2008). A comprehensive solution of the classic Burmester problem. *Transactions of the Canadian Society for Mechanical Engineering,32*(2), 137–154.

Chiang, C. H. (1988). *Kinematics of spherical mechanism*. Cambridge University Press.

Forsythe, G. E. (1970). Pitfalls in computation, or why a math book isn't enough. *American Mathematical Monthly,27*, 931–956.

Hartenberg, R., & Denavit, J. (1964). *Kinematic synthesis of linkages*. McGraw-Hill Book Company.

Hillier, S., & Liebermann, G. J. (1995). *Introduction to mathematical programming*. McGraw-Hill Inc.

Jamalov, R. I., Litvin, F. L., & Roth, B. (1984). Analysis and design of RCCC linkages. *Mechanism and Machine Theory,19*(4–5), 397–407.

Koetsier, T. (2010). Ludwig Burmester (1840-1927), Distinguished Figures in Mechanism and Machine Science, Their contributions and legacies, Part 2, M. Ceccarelli (editor), pp. 43–64, Springer, Cham.

Léger, J., & Angeles, J. (2014). A solution to the approximate spherical Burmester problem. In M. Ceccarelli & E. E. Hernández-Martínez (Eds.), *Proceedings of MUSME Conference on Multibody Mechatronic Systems* (pp. 521–529). Springer International Publishing.

Léger, J., & Angeles, J. (2015). A redundancy-resolution algorithm for five-degree-of-freedom tasks via sequential quadratic programming. In *TrC-IFToMM symposium on theory of machines and mechanisms*.

Li, X., Liu, J., Chen, W., & Bai, S. (2018). Integrated design, modeling and analysis of a novel spherical motion generator driven by electromagnetic principle. *Robotics and Autonomous Systems,106*, 69–81.

Liu, Z., & Angeles, J. (1992a). The constrained least-square optimization of spherical four-bar linkages for rigid-body guidance. *Transactions of the Canadian Society for Mechanical Engineering,16*(1), 47–60.

McCarthy, J. M., & Soh, G. S. (2011). *Geometric design of linkages*. Springer.

Nash, S. G., & Sofer, A. (1996). *Linear and nonlinear programming*. The McGraw-Hill Companies Inc.

Pottmann, H., & Wallner, J. (2001). *Computational line geometry*. Springer.

Roth, B. (1967). On the screw axes and other special lines associated with spatial displacements of a rigid body. *ASME Journal of Engineering for Industry,89*(1), 102–110.

Tsai, L. W., & Roth, B. (1973). Incompletely specified displacements: Geometry and spatial linkage synthesis. *ASME Journal of Engineering for Industry,95B*, 603–611.

Wampler, C. W., Morgan, A. P., & Sommese, A. J. (1990). Numerical continuation methods for solving polynomial systems arising in kinematics. *ASME Journal of Mechanical Design,112*(1), 59–68.

Chapter 5
Trajectory Generation

A recurrent problem in mechanical engineering design is the tracing of a continuous point trajectory by means of a mechanism. Examples abound in practice: cranes to upload and download containers from ships, a task that requires horizontal segments to ease the load on the motor; guiding laser beams to cut a profile from a metal plate; guiding the center of the wheel in the landing gear of small aircraft through a planar curve; and so on. The foregoing examples pertain to planar linkages. Other tasks call for *spherical curves*, i.e., curves lying on the surface of a sphere, which are most simply implemented by means of *spherical linkages*. An example of this task is the linkage needed to guide the focal axis of a solar collector in the form of a paraboloid of revolution. The axis must aim at the center of the Sun as this moves on the sky throughout the day, with daily adaptations of the linkage parameters, of course. One more class of tasks involves the tracing of the path generated by a line, namely a *ruled surface*, which is to be implemented with a spatial linkage. An application example here is the design of the landing gear (LG) of large aircraft, in which case the generating line is the axis of one of the wheels of the LG, and large aircraft carry several wheels on each of their two sides.

While the foregoing operations can be realized by means of robots, these become impractical when the operation involves endless repetitions through the same path. A single-dof linkage is the solution here, not only because of its low cost in terms of production, maintenance, and servicing, but also because of *reliability* and *repeatability*. *A robot cannot compete with a linkage in terms of the two foregoing demands*. Other applications include the synthesis of *dwell* in production lines. For example, the gluing of labels on or the filling of a bottle, presented to the pertinent mechanism of a packaging system, calls for contact of a mechanism link with the bottle during a finite amount of time. From the results of Sect. 3.3.1, it is apparent that a four-bar linkage cannot produce dwell, which requires that output velocity and acceleration *vanish simultaneously* during a finite time interval. This then calls for a multiloop linkage, e.g., a six-bar linkage with two kinematic loops. The synthesis of a dwell

J. Angeles and S. Bai, *Kinematics of Mechanical Systems*, Mathematical Engineering,
https://doi.org/10.1007/978-3-031-09544-3_5

mechanism thus requires the addition of an extra triad[1] to a four-bar linkage. The triad can be of two types, RRR or RPR, with the extreme R joints coupled, one, to the machine frame and, the other, to the coupler link of the four-bar linkage, at a designated point P. For triads of the first type, dwell is obtained by choosing P so that it traces, during a certain finite interval, a coupler curve that *locally* approximates a circle of radius r to a third order, meaning that the curvature of the curve is $1/r$ at the linkage posture at which the curvature becomes *stationary* with respect to the input angle of the linkage. Points of the coupler curve with the *stationarity property* are known (Hartenberg and Denavit, 1964b) to lie on a *cubic curve* fixed to the coupler link, this curve being quite appropriately known as the *cubic of stationary curvature* (CSC). For triads of the second type, P is chosen so that its coupler curve *locally* approximates a line segment to a second order, meaning that the curvature of the coupler curve traced by P vanishes at a given posture of the linkage. It is known (Hartenberg and Denavit, 1964b) that the locus of points of the coupler link of a planar four-bar linkage with a vanishing curvature is a circle, which is rightfully known as the *inflection circle*. Both loci are unique at a given linkage posture, meaning that these loci, fixed on the coupler link, change as the linkage moves from posture to posture.

The balance of the chapter includes the methodology behind the synthesis of planar, spherical, and spatial four-bar linkages with the property that one point of their coupler link will visit a prescribed discrete set of points.

5.1 Planar Linkages

The problem of interest here is formulated as follows:

Problem 5.1.1 Synthesize a planar four-bar linkage, as shown in Fig. 5.1, whose coupler point P will attain a set of positions $\{P_j\}_0^m$ defined in the fixed coordinate frame $\mathcal{F}$, as the linkage is driven by its input link.

Two types of path-synthesis problems are described in this section, one for *path generation with prescribed timing*, the subject of Sect. 5.1.1, and the other for *plain* coupler-curve synthesis, as studied in Sect. 5.1.2.

5.1.1 Planar Path Generation with Prescribed Timing

One class of path-synthesis problems consists in the *synchronization* of the points $\{P_j\}_0^m$ with the value of the input angle, namely that made by the crank BA with the x axis, the angle denoted $\{\psi_j\}_0^m$ at the prescribed poses. This is known as a problem

[1] Similar to a dyad, a triad is a two-link chain, with two LKPs at its free ends and one third pair coupling the two given links.

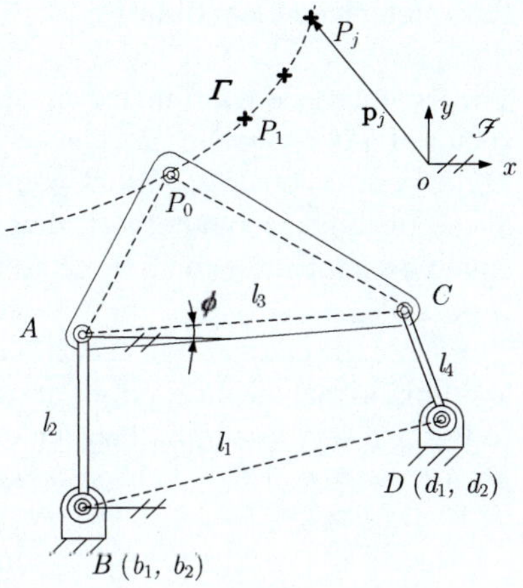

Fig. 5.1 A generic four-bar linkage

of *path generation with prescribed timing*. The problem occurs when one may need, for example, to have points $\{P_j\}_0^m$ laid down on a line with equal spacing between consecutive points, for equal increments of the input angle. In this case, the line segment is to be traced with a uniform speed as the driving joint turns at a uniform angular speed. The data are thus given as $\{\psi_j, \mathbf{p}_j\}_0^m$, with $\mathbf{p}_j$ denoting the position vector of P_j.

It is known from Chap. 4 that motion synthesis admits exact solutions for $m = 4$, i.e., with five poses. For path synthesis with prescribed timing, the problem remains nearly the same, except that the crank angle is prescribed, instead of the coupler orientation angle. The problem at hand thus admits exact solutions for $m = 4$ as well.

Synthesis Equations

We recall below the equations of motion synthesis:

$$\|(\mathbf{p}_j - \mathbf{b}) + \mathbf{Q}\mathbf{a}_0\|^2 = l_2^2; \quad \|(\mathbf{p}_j - \mathbf{d}) + \mathbf{Q}\mathbf{c}_0\|^2 = l_4^2 \tag{5.1}$$

Notice that this formulation relies on the assumption that the orientation angle ϕ is prescribed, along with $\mathbf{p}_j$. For the problem at hand, it is assumed that the crank rotation ψ is prescribed as a function of time. If the above formulation is used, we have to determine ϕ as a function of ψ, which will end up with a cumbersome formulation.

Instead of using ϕ in the formulation, we adopt alternative coordinates, namely parametrized coordinates, for point C, which streamline the formulation.

Parameterized Coordinates

The parametrized coordinates are applicable to ternary links, for which the position vector of a point on the link can be obtained as a linear combination of the position vectors of two other *landmark* points of the same link. This formulation was introduced for linkage synthesis (Bai et al., 2016), whereby a unified form of synthesis equations applicable to all three Stephenson linkages, planar, spherical, and spatial, is possible.

We use the coupler link APC to describe the parametrized coordinates, under the assumption that the three points at stake, A, P, and C, are *not aligned*. With reference to Fig. 5.2, and assuming that the coordinates of points A_j and P_j are known, the position vector of C_j can be expressed as

$$\mathbf{c}_j = \mathbf{p}_j + \mu\mathbf{u}_j + \nu\mathbf{v}_j, \quad j = 1, \ldots, m \tag{5.2}$$

where $\mathbf{u}_j$ and $\mathbf{v}_j$ denote orthogonal vectors attached to the coupler link, which define an orthogonal basis for $\mathbb{R}^2$. These vectors are not necessarily of unit magnitude, while μ and ν are real dimensionless parameters. Thus, let

$$\mathbf{u}_j = \mathbf{a}_j - \mathbf{p}_j; \quad \mathbf{v}_j = \mathbf{E}(\mathbf{a}_j - \mathbf{p}_j) \tag{5.3}$$

$\mathbf{p}_j$ denoting the position vector of P_j, while $\mathbf{E}$, as usual, is the 2×2 rotation matrix through 90° CCW, as defined in previous chapters. Equation (5.2) thus becomes

$$\mathbf{c}_j = \mathbf{p}_j + \mu(\mathbf{a}_j - \mathbf{p}_j) + \nu\mathbf{E}(\mathbf{a}_j - \mathbf{p}_j) \tag{5.4}$$

That is,

$$\mathbf{c}_j = (\mu\mathbf{1} + \nu\mathbf{E})\mathbf{a}_j + [(1 - \mu)\mathbf{1} - \nu\mathbf{E}]\,\mathbf{p}_j \tag{5.5}$$

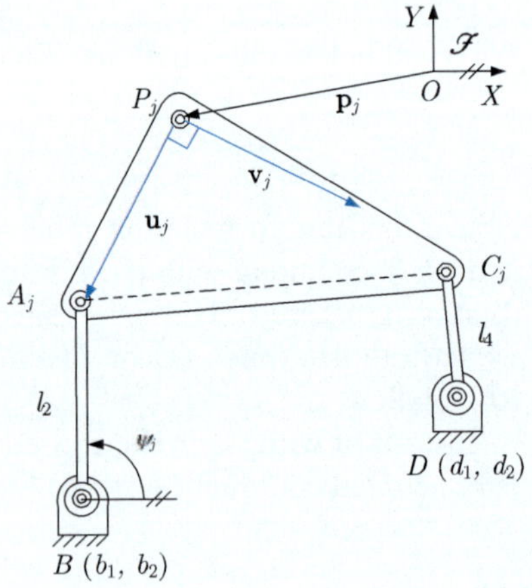

Fig. 5.2 Illustration of parametrized coordinates, whereby the position vector of point C_j is to be expressed as a linear combination of those of A_j and P_j

which is expressed in compact form as

$$\mathbf{c}_j = \mathbf{M}\mathbf{a}_j + \mathbf{N}\mathbf{p}_j \tag{5.6}$$

with

$$\mathbf{M} = \mu\mathbf{1} + \nu\mathbf{E}; \quad \mathbf{N} = (1-\mu)\mathbf{1} - \nu\mathbf{E} \tag{5.7}$$

M and **N** thus obeying the relation

$$\mathbf{N} = \mathbf{1} - \mathbf{M} \tag{5.8}$$

We have thus expressed the position vector of C as a linear combination of the position vectors of two other points on the same link. As shown in the formulation below, the parametrized position vectors simplify significantly the constraint equations.

In the formulation by means of loop-closure equations, the coordinates are expressed as functions of the angle of rotation of the driving link. As this angle is configuration-dependent, such an approach leads to as many intermediate variables as the number of prescribed configurations. In our example, the synthesis equation, based on loop-closure equations, requires using five angles of rotation of the coupler link as intermediate variables. Given that the newly introduced formulation of parametrized coordinates needs only two parameters, μ, ν, the new formulation is concise.

Moreover, **M** and **N** obey the identities

$$\mathbf{M}^T\mathbf{M} = (\mu^2 + \nu^2)\mathbf{1}, \quad \mathbf{N}^T\mathbf{N} = [(1-\mu)^2 + \nu^2]\mathbf{1} \tag{5.9}$$

from which it is apparent that **M** and **N** are *isotropic*, i.e., the two foregoing products are proportional to the 2×2 identity matrix. Moreover,

$$\mathbf{M}^T\mathbf{N} = (\mu - \mu^2 - \nu^2)\mathbf{1} - \nu\mathbf{E}, \quad \mathbf{N}^T\mathbf{M} = (\mu - \mu^2 - \nu^2)\mathbf{1} + \nu\mathbf{E} \tag{5.10}$$

which will lead to *numerical robustness*.

Alternative Constraint Equations

With the above-parametrized coordinates of point C, a new set of constraints can be formulated, as shown below:

Once ψ is prescribed, namely as ψ_j, the position vector of point A_j is expressed as a function of ψ_j:

$$\mathbf{a}_j = \mathbf{b} + \mathbf{R}_j\mathbf{s}_0 \tag{5.11}$$

where $\mathbf{b}$ is the position vector of B in the frame given in Fig. 5.2, $\mathbf{R}_j$ the rotation matrix through an angle $\beta_j = \psi_j - \psi_0$, and $\mathbf{s}_0 = \mathbf{a}_0 - \mathbf{b}$.

Further, two constraint equations are formulated. In doing so, and for simplicity, we assume point P_0 coincides with the origin, and hence, $\mathbf{p}_0 = \mathbf{0}$, thereby obtaining

$$\|\mathbf{p}_j - \mathbf{a}_j\|^2 = \mathbf{a}_0^T \mathbf{a}_0, \quad j = 1, \ldots, m \tag{5.12a}$$

$$\|\mathbf{c}_j - \mathbf{d}\|^2 = (\mathbf{c}_0 - \mathbf{d})^T (\mathbf{c}_0 - \mathbf{d}) \tag{5.12b}$$

With $\mathbf{a}_j$ defined as in Eq. (5.11), Eq. (5.12a) is rewritten as

$$\|\mathbf{p}_j - \mathbf{R}_j(\mathbf{a}_0 - \mathbf{b}) - \mathbf{b}\|^2 = \mathbf{a}_0^T \mathbf{a}_0$$

Upon expansion and simplification, the above relation leads to

$$\mathbf{p}_j^T \mathbf{p}_j - 2\mathbf{p}_j^T [\mathbf{R}_j \mathbf{a}_0 - (\mathbf{R}_j - \mathbf{1})\mathbf{b}] - 2\mathbf{a}_0^T (\mathbf{1} - \mathbf{R}_j^T)\mathbf{b} + 2\mathbf{b}^T (\mathbf{1} - \mathbf{R})\mathbf{b} = 0 \tag{5.13}$$

Apparently, this equation is *quadratic* in $\mathbf{b}$ and *linear* in $\mathbf{a}_0$.

We now look at Eq. (5.12b), if rewritten as

$$\|\mathbf{M}\mathbf{a}_j + \mathbf{N}\mathbf{p}_j - \mathbf{d}\|^2 = \|\mathbf{M}\mathbf{a}_0 - \mathbf{d}\|^2, \quad j = 1, \ldots, m$$

Expanding and simplifying the above equation lead to

$$\begin{aligned} &\mathbf{a}_j^T \mathbf{M}^T \mathbf{M} \mathbf{a}_j + \mathbf{p}_j^T \mathbf{N}^T \mathbf{N} \mathbf{p}_j - \mathbf{a}_0^T \mathbf{M}^T \mathbf{M} \mathbf{a}_0 \\ &+2\,(\mathbf{a}_j^T \mathbf{M}^T \mathbf{N} \mathbf{p}_j - \mathbf{d}^T \mathbf{N} \mathbf{p}_j - \mathbf{d}^T \mathbf{M} \mathbf{a}_j + \mathbf{d}^T \mathbf{M} \mathbf{a}_0) = 0, \quad j = 1, \ldots, m \end{aligned} \tag{5.14}$$

or

$$\begin{aligned} &(\mu^2 + \nu^2)\mathbf{a}_j^T \mathbf{a}_j + [(1-\mu)^2 + \nu^2]\mathbf{p}_j^T \mathbf{p}_j - (\mu^2 + \nu^2)\mathbf{a}_0^T \mathbf{a}_0 \\ &+ 2\,[(\mu - \mu^2 - \nu^2)\mathbf{a}_j^T \mathbf{p}_j - \nu \mathbf{a}_j^T \mathbf{E} \mathbf{a}_j - \mathbf{d}^T \mathbf{N} \mathbf{p}_j - \mathbf{d}^T \mathbf{M} \mathbf{a}_j + \mathbf{d}^T \mathbf{M} \mathbf{a}_0] \\ &= 0, \quad j = 1, \ldots, m \end{aligned} \tag{5.15}$$

which is linear in $\mathbf{d}$ and quadratic in (μ, ν).

Solutions for Path Synthesis

We can now readily find solutions for path synthesis. To this end, Eq. (5.13) is rewritten in a compact form as

$$\mathbf{G}\mathbf{m} = \mathbf{0} \tag{5.16}$$

with

$$\mathbf{G} = \begin{bmatrix} \mathbf{g}_1 & e_1 \\ \mathbf{g}_2 & e_2 \\ \mathbf{g}_3 & e_3 \\ \mathbf{g}_4 & e_4 \end{bmatrix} \in \mathbb{R}^{4\times 3}, \quad \mathbf{m} = \begin{bmatrix} \mathbf{a}_0 \\ 1 \end{bmatrix} \in \mathbb{R}^3 \tag{5.17}$$

and

$$\mathbf{g}_j = -2\mathbf{p}_j^T \mathbf{R}_j - 2\mathbf{b}^T(\mathbf{1} - \mathbf{R}_j) \tag{5.18}$$

$$e_j = \mathbf{p}_j^T \mathbf{p}_j + 2\mathbf{p}_j^T(\mathbf{R}_j - \mathbf{1})\mathbf{b} - 2\mathbf{b}^T \mathbf{R} \mathbf{b} + 2\mathbf{b}^T \mathbf{b}, \quad j = 1, \ldots, 4 \tag{5.19}$$

For the 4×3 matrix $\mathbf{G}$ to have a non-trivial null space, which is needed in light of the above definition of $\mathbf{m}$, $\mathbf{G}$ must be rank-deficient. This means that every 3×3 submatrix of $\mathbf{G}$ is bound to be singular. We can thus derive four bivariate polynomial equations in the Cartesian coordinates of point B, namely

$$\det(\mathbf{G}_j) = 0, \quad \text{for } j = 1, \ldots, 4 \tag{5.20}$$

where the 3×3 submatrix $\mathbf{G}_j$ is obtained by deleting the jth row from $\mathbf{G}$. Moreover, each of the four Eqs. (5.20) can be solved graphically for $\mathbf{b}$ from the intersections of the contours of the four equations.

For each real solution of $\mathbf{b}$ thus obtained, Eq. (5.16) is rewritten as

$$\underline{\mathbf{G}}\mathbf{a}_0 = \mathbf{e}, \quad \underline{\mathbf{G}} \in \mathbb{R}^{4\times 2} \tag{5.21}$$

where $\underline{\mathbf{G}} = [\mathbf{g}_1^T, \mathbf{g}_2^T, \mathbf{g}_3^T, \mathbf{g}_4^T]^T$, $\mathbf{e} = [e_1, e_2, e_3, e_4]^T$. Equation (5.21) is thus an overdetermined system of four equations in the two components of $\mathbf{a}_0$. In this light, we look for the *least-square approximation* $\mathbf{a}_0$ to the above system. In fact, since the system of equations (5.16) is consistent, its least-square approximation is its *solution*. While, in principle, *any* pair of the four Eqs. (5.20) is a solution, roundoff error, especially in the case of an ill-conditioned 3×3 submatrix of $\mathbf{G}$, is bound to yield distinct numerical values thereof. As a matter of fact, if only one triad of subequation (5.16) is ill-conditioned, an *unlucky* selection will lead to *catastrophic* results. Our approach here, as well as in the balance of the book, is to aim at a *robust* solution, and hence, we regard this system as "overdetermined," hence aiming at its "least-square approximation".

Once solutions for $\mathbf{a}_0$ and $\mathbf{b}$ are obtained, the solutions for $\mathbf{d}$ and the two alternative coordinates μ and ν can be found likewise. First, Eq. (5.15) is rewritten in the form

$$\mathbf{H}\mathbf{n} = \mathbf{0} \tag{5.22}$$

with

$$\mathbf{H} = \begin{bmatrix} \mathbf{h}_1 & f_1 \\ \mathbf{h}_2 & f_2 \\ \mathbf{h}_3 & f_3 \\ \mathbf{h}_4 & f_4 \end{bmatrix} \in \mathbb{R}^{4\times 3}; \quad \mathbf{n} = \begin{bmatrix} \mathbf{d} \\ 1 \end{bmatrix} \in \mathbb{R}^3 \tag{5.23}$$

where

$$\mathbf{h}_j = -2(\mathbf{M}\mathbf{a}_j)^T - 2(\mathbf{N}\mathbf{p}_j)^T + 2(\mathbf{M}\mathbf{a}_0)^T \tag{5.24}$$

$$f_j = (\mu^2 + \nu^2)\mathbf{a}_j^T\mathbf{a}_j + [(1-\mu)^2 + \nu^2]\mathbf{p}_j^T\mathbf{p}_j + 2[(\mu - \mu^2 - \nu^2)\mathbf{a}_j^T\mathbf{p}_j$$
$$- \nu\mathbf{a}_j^T\mathbf{E}\mathbf{a}_j - \mathbf{d}^T\mathbf{M}\mathbf{a}_0] - (\mu^2 + \nu^2)\mathbf{a}_0^T\mathbf{a}_0, \quad j = 1, \ldots, 4 \tag{5.25}$$

Once again, we obtain a 4×3 matrix, $\mathbf{H}$. Hence, the determinants of all its 3×3 submatrices $\mathbf{H}_j$ must vanish for Eq. (5.22) to admit a non-trivial solution, as required by $\mathbf{n}$ in Eq. (5.23), i.e.,

$$\det(\mathbf{H}_j) = 0, \quad \text{for } j = 1, \ldots, 4 \tag{5.26}$$

submatrix $\mathbf{H}_j$ being obtained upon deleting the jth row of $\mathbf{H}$. The system of equations (5.26) is solved graphically, which yields numerical approximations of μ and ν. Further, we rewrite Eq. (5.22) in the form

$$\underline{\mathbf{H}}\mathbf{d} = \mathbf{f} \tag{5.27}$$

where $\underline{\mathbf{H}} = [\mathbf{h}_1^T, \mathbf{h}_2^T, \mathbf{h}_3^T, \mathbf{h}_4^T]^T \in \mathbb{R}^{4\times3}$ and $\mathbf{f} = [f_1, f_2, f_3, f_4]^T$. Furthermore, Eq. (5.21) is an overdetermined system of four equations in the three components of $\mathbf{d}$. The system is thus solved, again, by means of least squares, thereby guaranteeing, again, a *robust solution* via a "least-square approximation," the quotation marks stressing that overdeterminacy is only *formal*.

Example

We demonstrate below the foregoing procedure. The prescribed position and input angles are given in Table 5.1. Following the procedure, we obtain the *four* contour plots from Eq. (5.20), as displayed in Fig. 5.3. Two intersections are obtained, as per the plots in Fig. 5.3a. The two intersections correspond to real solutions, i.e., the coordinates of point B. Accurate solutions can be found by resorting to any nonlinear-equation solver with an initial guess around each of these two points. Afterwards, the unique solution, corresponding to $\mathbf{a}_0$, is found from Eq. (5.21).

Table 5.1 Prescribed positions and input angles

j	$[x_i, y_j]$	β_j
0	[0.0, 0.0]	0
1	[−0.1865, 0.01828]	$\pi/4$
2	[−0.3360, −0.07635]	$\pi/2$
3	[−0.4153, −0.2260]	$3\pi/4$
4	[−0.4105, −0.3767]	π

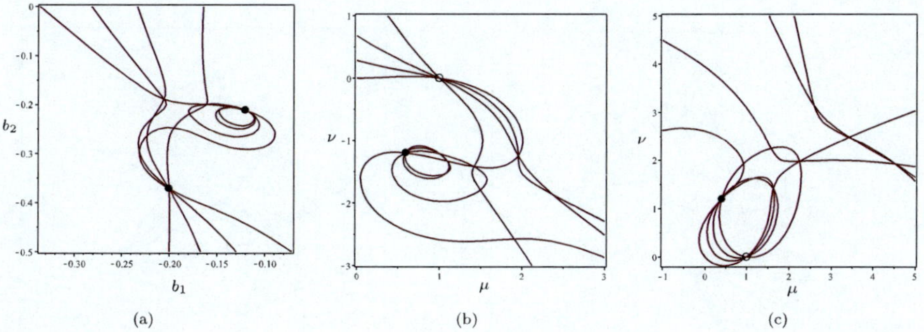

Fig. 5.3 Contour plots of **a** solutions for coordinates of point **b** and **b**, **c** corresponding solutions for (μ, ν)

Table 5.2 Linkage synthesis results (unit: m)

No.	$\mathbf{b}^T = [b_1, b_2]$	$\mathbf{a}_0^T = [a_1, a_2]$	$\mathbf{d}^T = [d_1, d_2]$	$[\mu, \nu]$	$\mathbf{c}_0^T = [c_1, c_2]$
1	[−0.2003, −0.3701]	[−0.1239, −0.1852]	[0.0385, −0.3709]	[0.4028, 1.200]	[0.1723, −0.2234]
2	[−0.1201, −0.2110]	[0.0540, −0.1389]	[0.0398, −0.3710]	[0.60, −1.2]	[−0.1342, −0.1482]

Once values of $\mathbf{a}_0$ and $\mathbf{b}$ are available, we can further find solutions for $\mathbf{d}$ and (μ, ν). For each set of solutions $\mathbf{a}_0$ and $\mathbf{b}$, contour plots are obtained from Eq. (5.26). Figure 5.3b, c shows plots associated with corresponding sets of solutions $\mathbf{a}_0$ and $\mathbf{b}$. In each of these figures, two contour intersections are identified. Of these intersections, the small circle stands for a *false* solution, as the point leads to coordinates of a point C coincident with point A. There is thus only one feasible solution in each of the two foregoing figures, indicated by a bullet. Accurate values at each point are found as in the case of $\mathbf{a}_0$, yielding solutions for (μ, ν). Finally, $\mathbf{d}$ is found through Eq. (5.27). The two linkages capable of visiting five points with prescribed timing, for this example, are listed in Table 5.2, their corresponding images displayed in Fig. 5.4.

5.1.2 Coupler-Curve Synthesis of Planar Four-Bar Linkages

Coupler-curve synthesis is a special path-synthesis problem, which consists in finding all the linkage parameters capable of tracing a given coupler curve. In this section, we describe how to find solutions to the *exact coupler-curve synthesis* problem.

The coupler curve is traced by one of the points on the coupler link of a four-bar linkage. The algebraic coupler curve of such a linkage is known to be a *sextic* bivariate polynomial (Cayley, 1875), namely

$$f(x, y) = \sum_{i,j=0}^{6} P_{ij} x^i y^j = 0, \quad i + j \leq 6 \tag{5.28}$$

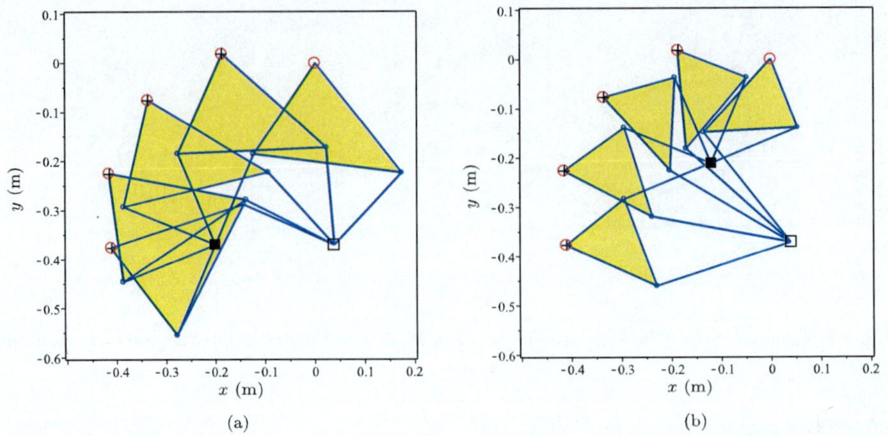

Fig. 5.4 The two linkages synthesized for **a** solution #1 and **b** solution #2. In each case, point P_0 is indicated by a circle, and other points P_j by encircled crosses. Locations of pivoting points B and D are indicated by small squares, black for B and white for D

where P_{ij} is a coefficient independent of x and y. It is also known that the four-bar linkage coupler curve is a *special form* of the sextic bivariate polynomial, with only 15 coefficients, as opposed to the 28 of its generic counterpart (Blechschmidt and Uicker, 1986).

The coupler-curve synthesis problem, up front, looks like an overdetermined problem, as its bivariate polynomial, Eq. (5.28), carries fifteen known coefficients, each establishing one equation, while the problem has nine unknown linkage parameters. As a matter of fact, the problem has been considered for a long time as *overdetermined*, for which only approximate solutions have been found, by means of either graphical or optimization methods (Cabrera et al., 2002; Bulatović and Djordjevic, 2004; Sancibrian et al., 2004).

Research progress in the twenty-first century has brought about a complete solution to the exact coupler-curve synthesis problem. In this solution, a *determined system*, i.e., a system with as many equations as unknowns, is first established (Bai, 2015a; Bai and Angeles, 2015a). Then, a univariate polynomial equation is derived, which yields all exact solutions (Bai, 2021b). We describe below the solution procedure in detail.

A generic four-bar linkage is illustrated in Fig. 5.5, whose coupler curve is Γ, as traced by a landmark point P of the coupler link. The linkage is generally defined via nine independent parameters: $\mathbf{x} = [m, h, b_1, b_2, d_1, d_2, l_2, l_3, l_4]^T$, including the link lengths $l_2, l_3, l_4 > 0$, of course. Length l_1 of the base link is calculated from the coordinates of points B and D, (b_1, b_2) and (d_1, d_2), respectively.

It is noted that parameters m and h, defining the location of P in the coupler link, are dimensionless, normalized by the link length l_3. It will be shown in the sequel that this definition of parameters eases the equation simplification.

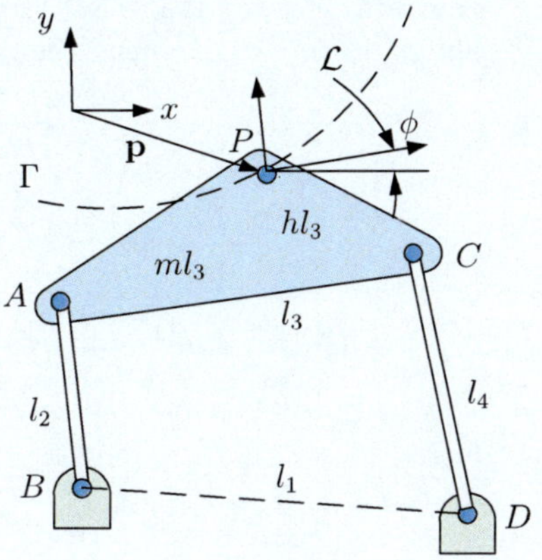

Fig. 5.5 A generic four-bar linkage with *nine* design parameters, including link lengths, l_2–l_4, the position of point P in the coupler link located by (ml_3, hl_3), and the coordinates of the two joint centers $B(b_1, b_2)$ and $D(d_1, d_2)$

With reference to Appendix B, the coupler curve of a four-bar linkage is expressed as

$$\begin{aligned} f(x, y) =&(x^2 + y^2)^3 + (k_1x + k_2y)(x^2 + y^2)^2 + (k_3x^2 + k_4xy + k_5y^2)(x^2 + y^2) \\ &+ k_6x^3 + k_7x^2y + k_8xy^2 + k_9y^3 + k_{10}x^2 + k_{11}xy + k_{12}y^2 \\ &+ k_{13}x + k_{14}y + k_{15} = 0 \end{aligned} \tag{5.29}$$

which is a polynomial bivariate equation of degree six. In total, we have 15 coefficients $\{k_i\}_1^{15}$, all functions of *nine independent parameters*.

The polynomial coefficients allow us to build a system of equations with a reduced number of unknowns through algebraic manipulations. In this vein, a six-dimensional array of design variables is defined as

$$\mathbf{y} = [m, h, b_1, b_2, d_1, d_2]^T \tag{5.30}$$

It is noteworthy that the three coefficients k_1, k_2, and k_4 of Eq. (5.29) are independent of l_2, l_3, and l_4. Their corresponding expressions, displayed in Eq. (B.16), can be expressed in terms of variable $\mathbf{y}$ as

$$k_j(\mathbf{y}) - k_j = 0, \quad j = 1, 2, 4 \tag{5.31a}$$

Lengths l_2, l_3, and l_4 can be eliminated from other coefficient equations. This is done by subtracting one from the other in each pair of the coefficient equations of (k_3, k_5), (k_6, k_8), and (k_7, k_9), which leads to three new equations

$$k_{ij}(\mathbf{y}) - (k_i - k_j) = 0, \quad (i, j) \in \{(3, 5), (6, 8), (7, 9)\} \tag{5.31b}$$

where $k_{ij}(\mathbf{y}) \equiv k_i(\mathbf{x}) - k_j(\mathbf{x})$, all being independent of l_2, l_3, and l_4.

Equations (5.31a) and (5.31b), upon expansion, lead to a *determined* system of *six equations* in *six unknowns*, the components of $\mathbf{y}$ in Eq. (5.30):

$$-2\,hb_2 + 2\,hd_2 + 2\,mb_1 - 2\,md_1 - k_1 - 4\,b_1 - 2\,d_1 = 0 \tag{5.32a}$$

$$2\,hb_1 - 2\,hd_1 + 2\,mb_2 - 2\,md_2 - k_2 - 4\,b_2 - 2\,d_2 = 0 \tag{5.32b}$$

$$\begin{aligned}-4\,hb_1^2 + 4\,hb_2^2 + 4\,hd_1^2 - 4\,hd_2^2 - 8\,mb_1b_2 + 8\,d_1md_2 + 8\,b_1b_2\\ +8\,b_1d_2 + 8\,b_2d_1 - k_4 = 0\end{aligned} \tag{5.32c}$$

$$\begin{aligned}-8\,hb_1b_2 + 8\,hd_1d_2 + 4\,mb_1^2 - 4\,mb_2^2 - 4\,md_1^2 + 4\,md_2^2 - 4\,b_1^2\\ -8\,b_1d_1 + 4\,b_2^2 + 8\,b_2d_2 + k_3 - k_5 = 0\end{aligned} \tag{5.32d}$$

$$\begin{aligned}8\,hb_1^2d_2 + 16\,hb_1b_2d_1 - 16\,hb_1d_1d_2 - 8\,hb_2^2d_2 - 8\,hb_2d_1^2 + 8\,hb_2d_2^2\\ -8\,mb_1^2d_1 + 16\,mb_1b_2d_2 + 8\,mb_1d_1^2 - 8\,mb_1d_2^2 + 8\,mb_2^2d_1\\ -16\,mb_2d_1d_2 + 8\,b_1^2d_1 - 16\,b_1b_2d_2 - 8\,b_2^2d_1 + k_6 - k_8 = 0\end{aligned} \tag{5.32e}$$

$$\begin{aligned}-8\,hb_1^2d_1 + 16\,hb_1b_2d_2 + 8\,hb_1d_1^2 - 8\,hb_1d_2^2 + 8\,hb_2^2d_1 - 16\,hb_2d_1d_2\\ -8\,mb_1^2d_2 - 16\,mb_1b_2d_1 + 16\,mb_1d_1d_2 + 8\,mb_2^2d_2 + 8\,mb_2d_1^2\\ -8\,mb_2d_2^2 + 8\,b_1^2d_2 + 16\,b_1b_2d_1 - 8\,b_2^2d_2 + k_7 - k_9 = 0\end{aligned} \tag{5.32f}$$

thereby obtaining a system of six *independent* equations for the six components of $\mathbf{y}$ of Eq. (5.30).

The remaining six coefficient equations for $k_{10}, \ldots, k_{15}$, contain only terms in l_2^4, $l_2^2l_4^2$, l_4^4, l_2^2, and l_4^2. These equations can be rewritten as

$$t_{i,1}l_2^4 + t_{i,2}l_2^2l_4^2 + t_{i,3}l_4^4 + t_{i,4}l_2^2 + t_{i,5}l_4^2 + t_{i,6} = 0; \quad i = 1, \ldots, 6 \tag{5.33}$$

which can be cast in linear homogenous form, namely

$$\mathbf{Tq} = \mathbf{0}, \quad \mathbf{q} = [l_2^4,\ l_2^2l_4^2,\ l_4^4,\ l_2^2,\ l_4^2,\ 1]^T \tag{5.34}$$

with $\mathbf{T}$ denoting a 6×6 matrix whose entries consist of the t_{ij} terms of Eq. (5.33). Moreover, Eq. (5.34) implies the singularity of $\mathbf{T}$, i.e.,

$$\det(\mathbf{T}) \equiv \Delta(\mathbf{y}, l_3) = 0 \tag{5.35}$$

for $\mathbf{y}$ defined in Eq. (5.30).

Thus far, a determined system of seven equations, Eqs. (5.32a)–(5.32f) and (5.35), with seven unknowns, has been established, which admits, in principle, exact solutions. The foregoing derivation proves a result on *coupler-curve synthesis* (Bai, 2021b):

If the coupler-curve equation of a four-bar linkage is known, a determined system of coefficient equations can be established, whose solutions yield the complete set of the parameters defining the linkage.

A Univariate Polynomial for One Pivoting Point

We show first how the coefficient equations can be reduced to a univariate polynomial. We start by eliminating parameters m and h from Eqs. (5.32a)–(5.32f), which we do upon solving for h and m from these equations. Indeed, Eqs. (5.32a) and (5.32b) lead to

$$h = -\frac{1}{2L_1}(k_1 b_2 - k_1 d_2 - k_2 b_1 + k_2 d_1 - 6 b_1 d_2 + 6 b_2 d_1) \tag{5.36a}$$

$$\begin{aligned} m =& \frac{1}{2L_1}(k_1 b_1 - k_1 d_1 + k_2 b_2 - k_2 d_2 \\ &+ 4(b_1^2 + b_2^2) - 2(b_1 d_1 + b_2 d_2 + d_1^2 + d_2^2)) \end{aligned} \tag{5.36b}$$

where

$$L_1 \equiv l_1^2 = b_1^2 - 2 b_1 d_1 + b_2^2 - 2 b_2 d_2 + d_1^2 + d_2^2 \tag{5.36c}$$

Substituting h and m into the remaining equations, after simplification, yields

$$\begin{aligned} k_1 b_2 + k_1 d_2 + k_2 b_1 + k_2 d_1 + 4 b_1 b_2 \\ +2 b_1 d_2 + 2 b_2 d_1 + 4 d_1 d_2 + 0.5 k_4 = 0 \end{aligned} \tag{5.37a}$$

$$\begin{aligned} k_1 b_1 + k_1 d_1 - k_2 b_2 - k_2 d_2 + 2 b_1^2 + 2 b_1 d_1 - 2 b_2^2 \\ -2 b_2 d_2 + 2 d_1^2 - 2 d_2^2 - 0.5 k_3 + 0.5 k_5 = 0 \end{aligned} \tag{5.37b}$$

$$\begin{aligned} -2 k_1 b_1 d_1 + 2 k_1 b_2 d_2 + 2 k_2 b_1 d_2 + 2 k_2 b_2 d_1 - 4 b_1^2 d_1 + 8 b_1 b_2 d_2 \\ -4 b_1 d_1^2 + 4 b_1 {d_2}^2 + 4 b_2^2 d_1 + 8 b_2 d_1 d_2 + 0.5 k_6 - 0.5 k_8 = 0 \end{aligned} \tag{5.37c}$$

$$\begin{aligned} -2 k_1 b_1 d_2 - 2 k_1 b_2 d_1 - 2 k_2 b_1 d_1 + 2 k_2 b_2 d_2 - 4 b_1^2 d_2 - 8 b_1 b_2 d_1 \\ -8 b_1 d_1 d_2 + 4 b_2^2 d_2 - 4 b_2 d_1^2 + 4 b_2 d_2^2 + 0.5 k_7 - 0.5 k_9 = 0 \end{aligned} \tag{5.37d}$$

The above system of four equations includes four unknowns, b_1, b_2, d_1, and d_2. A careful examination of these equations shows that products from coordinates of points B and D, such as $b_1 d_1$, $b_2 d_2$, $b_1 d_2$, and $b_2 d_1$, can be eliminated. In this light, the equations are further simplified as follows:

Step 1: (5.37a)$\cdot 2b_2$ + (5.37b)$\cdot 2b_1$ + (5.37c); (5.37a)$\cdot(-2b_1)$ + (5.37b)$\cdot 2b_2$ + (5.37d):

$$2k_1 b_1^2 - 2k_1 b_2^2 - 4k_2 b_1 b_2 + 4b_1^3 - 12 b_1 b_2^2 + k_3 b_1 \\ - k_4 b_2 - k_5 b_1 + 0.5(k_6 - k_8) = 0 \tag{5.38a}$$

$$4k_1 b_1 b_2 + 2k_2 b_1^2 - 2k_2 b_2^2 + 12 b_1^2 b_2 - 4b_2^3 + k_3 b_2 \\ + k_4 b_1 - k_5 b_2 + 0.5(k_7 - k_9) = 0 \tag{5.38b}$$

Step 2: (5.37a)$\cdot 2d_2$ + (5.37b)$\cdot 2d_1$ + (5.37c); (5.37a)$\cdot(-2d_1)$ + (5.37b)$\cdot 2d_2$ + (5.37d):

$$2k_1 d_1^2 - 2k_1 d_2^2 - 4k_2 d_1 d_2 + 4d_1^3 - 12 d_1 d_2^2 + k_3 d_1 \\ - k_4 d_2 - k_5 d_1 + 0.5(k_6 - k_8) = 0 \tag{5.38c}$$

$$4k_1 d_1 d_2 + 2k_2 d_1^2 - 2k_2 d_2^2 + 12 d_1^2 d_2 - 4d_2^3 + k_3 d_2 \\ + k_4 d_1 - k_5 d_2 + 0.5(k_7 - k_9) = 0 \tag{5.38d}$$

The above equations represent two subsystems, each carrying only the two coordinates of one pivoting point, either B or D. A close examination of the two foregoing subsystems shows that they are identical. This can be readily explained, as the pivoting points B and D can be swapped in designing a linkage. In this light, we need to only solve one set of equations, for example, Eqs. (5.38c) and (5.38d).

Next, we eliminate one variable from the selected subsystem. We choose, for example, d_1 to be eliminated. To this end, we first reduce the order of d_1 in Eq. (5.38c). This is achieved from the *linear combination* (5.38c)$\cdot(6d_2 + k_2)$ − (5.38d)$\cdot 2d_1$, thereby obtaining

$$4k_1 d_2 d_1^2 + 2k_1 k_2 d_1^2 - 2k_4 d_1^2 - 32 k_2 d_2^2 d_1 - 64 d_2^3 d_1 + k_2 k_3 d_1 \\ -4k_5 d_1 d_2 - k_2 k_5 d_1 - k_7 d_1 + k_9 d_1 + 4k_3 d_1 d_2 - 4k_2^2 d_1 d_2 - k_2 k_4 d_2 \\ +3k_6 d_2 - 12 k_1 d_2^3 - 6k_4 d_2^2 - 2k_1 k_2 d_2^2 + 0.5 k_2 k_6 - 3k_8 d_2 - 0.5 k_2 k_8 = 0 \tag{5.39}$$

which is a quadratic equation in d_1. Next, combining the above equation with Eq. (5.38d), we obtain a system of two quadratic equations in d_1, which is rewritten as

$$E_1 d_1^2 + F_1 d_1 + G_1 = 0 \tag{5.40}$$

$$E_2 d_1^2 + F_2 d_1 + G_2 = 0 \tag{5.41}$$

The two equations above must have common roots, in spite of their different coefficients. The conditions under which the two foregoing equations admit common roots can be found from root analysis. Here, we take another approach by virtue of the *Sylvester matrix*, introduced by Laidacker (1969). In this light, a new equation can be established:

$$\mathbf{M}\mathbf{d}_1 = \mathbf{0} \tag{5.42}$$

with $\mathbf{d}_1 = [d_1^3, d_1^2, d_1, 1]^T$ and

$$\mathbf{M} = \begin{bmatrix} E_1 & F_1 & G_1 & 0 \\ E_2 & F_2 & G_2 & 0 \\ 0 & E_1 & F_1 & G_1 \\ 0 & E_2 & F_2 & G_2 \end{bmatrix} \tag{5.43}$$

where

$$E_1 = 2\,k_2 + 12\,d_2 \tag{5.44a}$$
$$F_1 = 4\,k_1\,d_2 + k_4 \tag{5.44b}$$
$$G_1 = -\,2\,k_2\,d_2^2 - 4\,d_2^3 + k_3\,d_2 - k_5\,d_2 + 0.5(k_7 - k_9) \tag{5.44c}$$
$$E_2 = 2\,k_1\,k_2 + 4\,k_1\,d_2 - 2\,k_4 \tag{5.44d}$$
$$F_2 = -\,4\,k_2^2 d_2 - 32\,k_2\,d_2^2 - 64\,d_2^3 + k_2\,k_3 - k_2\,k_5 + 4\,k_3\,d_2 - 4\,k_5\,d_2 - k_7 + k_9 \tag{5.44e}$$
$$G_2 = -\,2\,k_1\,k_2\,d_2^2 - 12\,k_1\,d_2^3 - k_2\,k_4\,d_2 - 6\,k_4\,d_2^2 + 0.5k_2\,k_6 - 0.5k_2\,k_8 + 3\,k_6\,d_2 - 3\,k_8\,d_2 \tag{5.44f}$$

Moreover, Eq. (5.42) implies

$$\det(\mathbf{M}) = 0 \tag{5.45}$$

which is a 10th-degree polynomial equation in d_2. On the other hand, the system of Eqs. (5.38c) and (5.38d), according to Bezout's theorem, admits a maximum of nine solutions. This means that Eq. (5.45) admits one extra solution. Notice that, in the elimination procedure, we multiply Eq. (5.38c) by $(6d_2 + k_2)$, which brings about one extra root of the system. In this light, we divide both sides of Eq. (5.45) by $(6d_2 + k_2)$, to obtain the final, single equation

$$g = \sum_k p_k\,d_2^k = 0, \quad k = 0, \ldots, 9 \tag{5.46}$$

with all coefficients p_k, except for p_9, which is a known constant, being functions of k_i, $i = 1, \cdots, 9$. Moreover, all are available, as they can be calculated directly. The top four coefficients are displayed below, and the others, included in the Appendix, being skipped here for brevity:

$$\begin{aligned} p_6 &= 4096\,k_1^2 k_2 + 8192\,k_2^3 + 1024\,k_1\,k_4 - 13312\,k_2\,k_3 \\ &\quad + 13312\,k_2\,k_5 - 3072\,k_7 + 3072\,k_9 \\ p_7 &= 4096\,k_1^2 + 28672\,k_2^2 - 12288\,k_3 + 12288\,k_5 \\ p_8 &= 49152\,k_2 \\ p_9 &= 32768 \end{aligned} \tag{5.47}$$

The above formulation leads to one more result (Bai, 2021b):

If the coupler-curve equation of a four-bar linkage is known, a univariate nonic polynomial equation in one coordinate of its pivoting point can be uniquely established from the nine coefficients of the known coupler curve.

Exact Coupler-Curve Synthesis

Upon the derivation of the univariate polynomial of Eq. (5.46), the exact coupler-curve synthesis problem has been established, its solution described below.

The value of d_2 is first found from Eq. (5.46). According to the *Roberts–Chebyshev Theorem* (Roberts, 1869), this equation admits three real solutions for d_2. If the solutions are distinct, we can utilize a linear equation in d_1, obtained from Eqs. (5.38c) and (5.38d), to calculate one value of d_1 for each value of d_2. Otherwise, either Eq. (5.38c) or (5.38d) is used to find d_1, depending on whether d_2 admits a double or, even, a triple solution.

The solutions for the coordinates of point D provide those of point B as well. Once the coordinates of points B and D are available, values of m and h are calculated using Eqs. (5.36b). Furthermore, l_2, l_3, and l_4 can be uniquely determined. From Properties 2 and 3, Eqs. (B.16), for $i = 3, 5, 6, 7, 8, 9$, are all linear in l_2^2, l_3^2, and l_4^2. We can thus take one equation out of each pair $(3, 5)$, $(6, 8)$, $(7, 9)$, $i = 5, 6, 7$, for example, thereby formulating a system of three linear equations in three unknowns, from which all three unknowns are available.

The block diagram of Fig. 5.6 summarizes the steps of the foregoing synthesis procedure. In general, we need only nine coefficients, $k_1, \ldots, k_9$, to determine all linkage parameters. However, for some special cases, as mentioned above, some coefficients disappear; we then need to select alternative coefficients. A special case is included in the examples below.

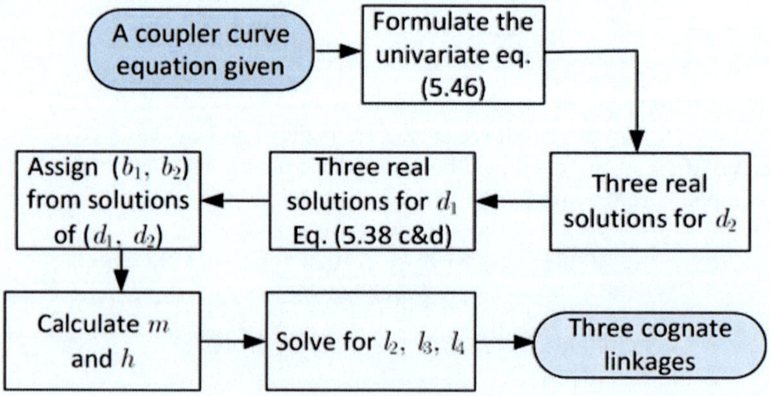

Fig. 5.6 Flowchart for exact coupler-curve synthesis

In the univariate polynomial equation (5.46), all coefficients p_k are functions of the given coefficients of an algebraic curve. From this equation, it can be determined whether real solutions exist for a given coupler curve. Several methods are available to this end, e.g., Descartes' rules of signs (Curtiss, 1918) or Sturm's Theorem (Basu et al., 2006).

Examples

We include below two examples to demonstrate the coupler-curve synthesis procedure. The examples are implemented in computer-algebra code, available in a `GoogleDrv` shared file.[2]

Example I

In the first example, the coupler curve is given in the form of a sextic polynomial equation, namely

$$\begin{aligned} f =& x^6 + 3x^4y^2 + 3x^2y^4 + y^6 + 0.1875x^5 + 0.5375x^4y + 0.375x^3y^2 \\ &+ 1.075x^2y^3 + 0.1875xy^4 + 0.5375y^5 - 0.125859x^4 + 0.16625x^3y \\ &- 0.017968x^2y^2 + 0.16625xy^3 + 0.10789y^4 - 0.010304x^3 - 0.007859x^2y \\ &+ 0.006695xy^2 - 0.035859y^3 + 0.0112623x^2 + 0.006071xy - 0.007966y^2 \\ &+ 0.00255779x + 0.00141746y + 0.00017165 = 0 \end{aligned} \tag{5.48}$$

With the foregoing procedure, a univariate polynomial equation in d_2 is obtained from Eq. (5.46):

$$\begin{aligned} g =& 32768d_2^9 + 1228.8d_2^8 + 479.36d_2^7 - 93.8234d_2^6 - 16.81892d_2^5 \\ &+ 5.91408d_2^4 - 0.75498d_2^3 + 0.05946d_2^2 - 0.010376d_2 + 0.0004337 = 0 \end{aligned} \tag{5.49}$$

Three real solutions for d_2 are readily found, as displayed in Fig. 5.7. Using the foregoing procedure, three solutions for d_1 are also calculated, thereby obtaining three sets of coordinates for the pivoting point D, as listed in Table 5.3. As discussed above, these results stand for the coordinates of B as well. In this vein, we simply assign a different set of coordinate values to B, thereby ending up with a total of six solutions. From the pivoting-point coordinates, other unknowns are found accordingly. The entire linkage parameters obtained for this example are listed in Table 5.3. The six solutions lead to three linkages, noted M1 to M3, which are *cognate* of the given coupler curve. The two solutions for each linkage represent two modes of assembly. Figure 5.7 displays the three linkages, along with the given coupler curve.

The current formulation allows one to also obtain and display the solutions graphically in the $d_1 - d_2$ plane. Figure 5.8 shows a density plot by treating the left-hand sides of Eqs. (5.38c) and (5.38d) as functions of d_1 and d_2, similar to the domain

[2] https://drive.google.com/file/d/1MOL3qUCqkSfF2DNaa7aU_jLKSKFFXbVK/view?usp=sharing.

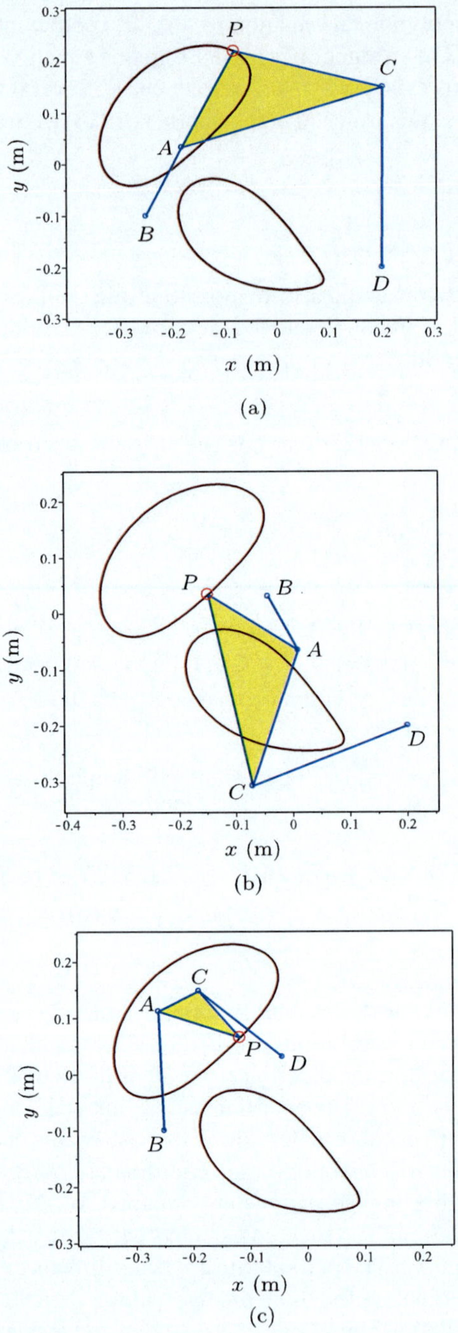

Fig. 5.7 Three cognate linkages shown together with their coupler curves: **a** Linkage M1; **b** Linkage M2; **c** Linkage M3

Table 5.3 Linkage synthesis results for Example I

No.	$[d_1, d_2](m)$	$[b_1, b_2](m)$	m	h	$l_2(m)$	$l_3(m)$	$l_4(m)$	Note
1	[0.2, –0.2]	[–0.25, –0.1]	0.375	0.375	0.15	0.4	0.35	M1
2	[0.2, –0.2]	[–0.0437, 0.0312]	–0.1764	–0.7059	0.1093	0.2551	0.2915	M2
3	[–0.25, –0.1]	[–0.0437, 0.0312]	–0.3333	1.3333	0.1856	0.0795	0.2121	M3
4	[–0.25, –0.1]	[0.2, –0.2]	0.625	–0.375	0.35	0.4	0.15	M1
5	[–0.0437, 0.0312]	[0.2, –0.2]	1.1764	0.7058	0.2915	0.2551	0.1093	M2
6	[–0.0437, 0.0312]	[–0.25, –0.1]	1.3333	–1.3333	0.2121	0.0795	0.1856	M3

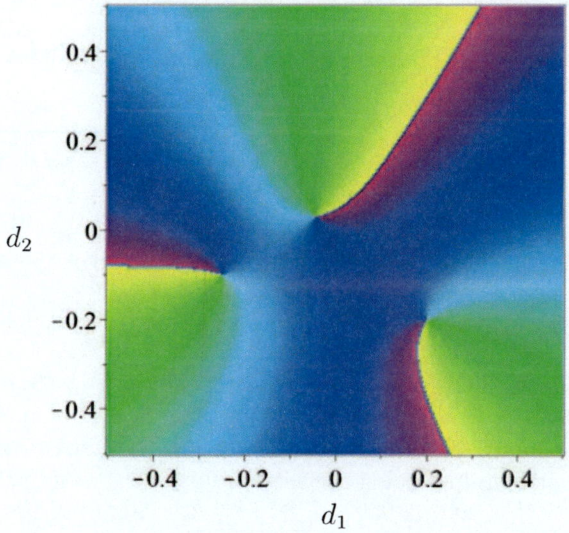

Fig. 5.8 Density plot to locate the pivoting point, the three darkest points indicating the location of point D

coloring plot. Three darkest points are distinguished, which are the plausible locations of the pivoting point D. For point B, the density plot will be the same, as this is obtained from identical functions, as per Eqs. (5.38a) and (5.38b).

The univariate polynomial (5.49) can be used to detect the existence of real solutions prior to equation-solving. To this end, let ν be the number of sign changes of the coefficients, n_p and n_n, denoting the numbers of positive and negative solutions, respectively. For the problem at hand, we assume a positive d_2; then we can show that (1) $\nu = 6$ for $g(d_2)$ and (2) $\nu = 3$ for $g(-d_2)$. According to Descartes' rules of signs (Curtiss, 1918), case (1) means that $n_p \leq 6$ and $(6 - n_p)$ is an even integer. In other words, we have 6, 4, 2, or 0 positive solutions. Furthermore, case (2) means that $n_n \leq 3$ and $3 - n_n$ is an even number, which implies that we have either 3 or 1 negative solutions. Combining both cases allows us to conclude that real solutions exist.

Table 5.4 Linkage synthesis results for Example II

No.	$[d_1, d_2](m)$	$[b_1, b_2](m)$	$r(m)$	m	h	$l_2(m)$	$l_3(m)$	$l_4(m)$	Note
1	[0.5, 0.0]	[0.25, 0.0]	0.075	−1.0	0.0	0.15	0.15	0.11	L1
2	[0.5, 0.0]	[0.0, 0.0]	0.1	0.5	0	0.3	0.2	0.3	L2
3	[0.25, 0.0]	[0.0, 0.0]	0.075	2.0	0.0	0.1	0.15	0.15	L3
4	[0.25,0.0]	[0.5, 0.0]	0.075	2.0	0	0.1	0.15	0.15	L1
5	[0.0, 0.0]	[0.25, 0.0]	0.075	0.25	0	0.15	0.15	0.1	L3
6	[0.0, 0.0]	[0.5, 0.0]	0.1	0.5	0	0.3	0.2	0.3	L2

Example II

In a second example, the coupler curve is given as

$$\begin{aligned} f =& x^6 + 3x^4y^2 + 3x^2y^4 + y^6 - 1.5x^5 - 3x^3y^2 - 1.5xy^4 + 0.6525x^4 + 0.805x^2y^2 \\ &+ 0.1525y^4 - 0.0275x^3 - 0.0275xy^2 - 0.027975x^2 - 0.010475y^2 + 0.0018x \\ &+ 0.0004 = 0 \end{aligned} \tag{5.50}$$

This curve, upon plotting, turns out to be what is known as a *Watt curve*, which is a *lemniscate*, i.e., an ∞-shaped curve[3]

This curve stems from a special case of Eq. (B.10), whereby all six coefficients, $k_2, k_4, k_7, k_9, k_{11}$, and k_{13} vanish. A univariate polynomial is obtained as

$$g = (32768\,d_2^6 + 3072\,d_2^4 + 72\,d_2^2 + 0.5)d_2^3 = 0 \tag{5.51}$$

which admits zero as a triple root. Using Eq. (5.38c), three distinct solutions, $d_1 = 0,\ 0.25,\ 0.5$, are found. Other parameters are obtained accordingly, all being listed in Table 5.4. It is noted that, in finding l_2, l_3, l_4, both k_7 and k_9 vanish, additional equations thus being needed. For this case, one more equation is introduced:

$$k_{10}(\mathbf{x}) - k_{12}(\mathbf{x}) - (k_{10} - k_{12}) = 0 \tag{5.52}$$

in addition to the coefficient equations for k_5 and k_6. The above equation is also linear in $\{l_2^2,\ l_3^2,\ l_4^2\}$. In total, the six sets of solutions thus obtained lead to three linkages, as displayed in Fig. 5.9.

[3] The curve, best known by its 90°-turned version, was so named by J. Bernoulli, from Latin *lemniscus*, meaning "a pendant ribbon" (https://mathworld.wolfram.com/Lemniscate.html).

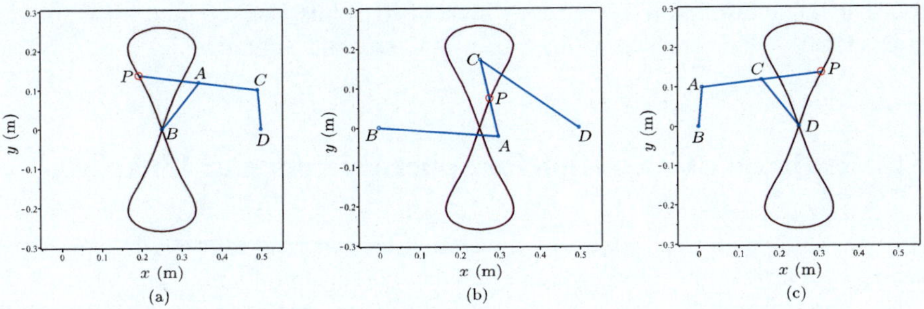

Fig. 5.9 Three cognate Watt linkages shown together with their coupler curve: **a** L1; **b** L2; **c** L3

Cognates and the Roberts–Chebyshev Theorem

In the two foregoing examples, we found three *cognate linkages* capable of visiting the given points on a coupler curve. This is exactly what the *Roberts–Chebyshev Theorem* states (Nolle, 1974):

For a given coupler curve, there exist three four-bar linkages, which will generate the same path.

If one revolute joint is replaced by a prismatic joint, e.g., the joint at point D, a crank-slider mechanism is obtained. In such a case, only two cognates exist that visit exactly a given couple curve. An example is available in the literature (Bai, 2015b).

Summary

In this section, we introduced a *complete and exact* solution to the classic coupler-curve synthesis problem of planar linkages. Upon revealing the algebraic properties of all coefficient polynomials and the hidden relations among all coefficient equations, we converted successfully the system of eighth-order polynomials of unknown coefficients into one univariate polynomial equation of known coefficients, which yields all real solutions to the problem.

Exact coupler-curve synthesis is a *fundamental problem in linkage synthesis*. The solution described in this chapter is applicable to other types of synthesis problems, the discrete path-synthesis problem, among others. Moreover, the univariate polynomial equation reveals the conditions for the existence of real solutions leading to four-bar linkages for a given algebraic curve, based on the analysis of the roots of univariate polynomials.

One more typical path-synthesis problem, namely the synthesis for a discrete set of points, was not covered in this section. As shown in Eq. (B.10), the underlying sextic is a function of nine linkage parameters. The path synthesis with discrete points thus admits exact solutions for nine prescribed trajectory points. The formulation is trivial for this problem, but the main challenge lies in finding its solutions. A complete four-bar linkage solution to visit nine prescribed points was reported by Wampler et al. (1990), which ends up with a total of 4326 distinct solutions!

5.2 Trajectory Generation for Spherical Four-Bar Linkages

Now we study path-, i.e., trajectory generation for spherical four-bar linkages consisting of four revolute joints, which is another classical problem.

In spherical four-bar linkages, the motion of every link is a *pure rotation* about one given point, labeled O. Trajectory generation, in the case at hand, consists in determining the linkage parameters that allow a given point of the mechanism coupler link to follow a prescribed path on the surface of the unit sphere, with the center at O.

Furthermore, a method of path synthesis for spherical four-bar linkages is developed. Synthesis examples, including both *discrete-point synthesis* and *great-arc synthesis*, are provided.

A spherical four-bar linkage for trajectory synthesis is depicted in Fig. 5.10, its four linkage dimensions being $\{\alpha_j\}_1^4$. The two fixed revolute joints are labeled B and D, the points at which their axes intersect the unit sphere. The two moving revolute joints, likewise, are labeled A and C. The axes of rotation of the fixed dyads are thus segments $\overline{OA}$ and $\overline{OB}$; the position vectors of A and B are **a** and **b**, respectively, both of unit magnitude, i.e.,

$$\|\mathbf{a}\| = 1, \quad \|\mathbf{b}\| = 1 \tag{5.53}$$

Likewise, the position vectors of points C and D are denoted by the unit vectors **c** and **d**. Moreover, a coupler point is defined on the coupler link, namely point P of

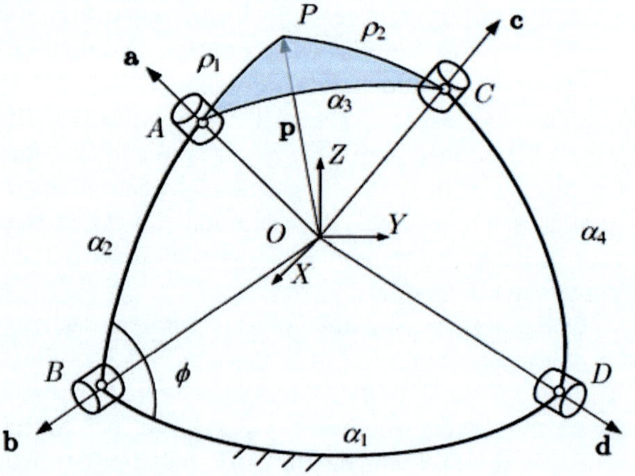

Fig. 5.10 Illustration of a spherical four-bar linkage for trajectory synthesis

Fig. 5.10, of position vector **p**, likewise of unit magnitude. The coupler link is defined by arcs ρ_1, ρ_2, and α_3.

The derivation of the coupler-curve equation for the linkage under study is provided in Appendix B. Referring to the spherical four-bar linkage shown in Fig. 5.10, the position vector $\mathbf{p} = [x, y, z]^T$ of the coupler point P is directed along line OP. The coupler curve is described by two equations, namely the equation of the unit sphere

$$g(x, y, z) = x^2 + y^2 + z^2 - 1 = 0 \tag{5.54}$$

and that of the spherical curve:

$$\begin{aligned} f(x, y, z) =& A_1^2C_2^2 - 2\,A_1C_2A_2C_1 + A_2^2C_1^2 + B_1C_2A_1B_2 - B_1^2C_2A_2 \\ & - C_1B_2^2A_1 + C_1B_2A_2B_1 - A_1^2B_2^2 + 2\,A_1B_2A_2B_1 - A_2^2B_1^2 = 0 \end{aligned} \tag{5.55}$$

where

$$A_1 = \mathbf{p}^T(\mathbf{1} - \mathbf{b}\mathbf{b}^T)\mathbf{a}_0 \tag{5.56a}$$
$$B_1 = \mathbf{p}^T\mathbf{B}\mathbf{a}_0 \tag{5.56b}$$
$$C_1 = \mathbf{p}^T\mathbf{b}\mathbf{b}^T\mathbf{a}_0 - \cos\rho_1 \tag{5.56c}$$
$$A_2 = \mathbf{d}^T(\mu\mathbf{1} + \nu\mathbf{P})(\mathbf{1} - \mathbf{b}\mathbf{b}^T)\mathbf{a}_0 \tag{5.56d}$$
$$B_2 = \mathbf{d}^T(\mu\mathbf{1} + \nu\mathbf{P})\mathbf{B}\mathbf{a}_0 \tag{5.56e}$$
$$C_2 = \mathbf{d}^T(\mu\mathbf{1} + \nu\mathbf{P})\mathbf{b}\mathbf{b}^T\mathbf{a}_0 + \lambda\mathbf{d}^T\mathbf{p} - \cos\alpha_4 \tag{5.56f}$$

with **B** and **P** denoting the CPMs of vectors **b** and **p**, respectively. The above equation is a *trivariate quartic polynomial*, of the form

$$f(x, y, z) = \sum_{i,j,k=0}^{4} K_{ijk}x^iy^jz^k = 0, \quad i + j + k \le 4 \tag{5.57}$$

whose coefficients K_{ijk} are functions of the linkage dimensions. The foregoing equation defines a surface in three-dimensional space passing through the origin, its *intersection* with the unit sphere, Eq. (5.54), generating the spherical coupler curve of interest.

5.2.1 *Path Synthesis with Discrete Positions*

The coupler-curve equation (5.57) enables path synthesis with a prescribed set of discrete positions on the surface of the unit sphere. The unknowns are the unit vectors **b** and **d**, along with the geometric parameters $\alpha_1, \alpha_2, \alpha_3, \rho_1, \rho_2$, which amount to a total of *nine* unknowns.[4] The path-synthesis problem for spherical four-bar linkages thus allows for a maximum of *nine prescribed points on the unit sphere for exact synthesis*, as in the planar case.

5.2.2 Path Synthesis with Prescribed Timing

For the problem at hand, the jth position of point P, noted P_j, to be visited by the coupler link of the mechanism is given by vector $\mathbf{p}_j$, corresponding to the angle of rotation ϕ_j, for $j = 0, \ldots, m$, of link AB. Constraint equations (B.28) and (B.33), upon identifying the pose number with j, then become

$$\mathbf{p}_j^T \underbrace{[\mathbf{b}\mathbf{b}^T + s_j\mathbf{B} + c_j(\mathbf{1} - \mathbf{b}\mathbf{b}^T]}_{\mathbf{R}_j}\mathbf{a}_0 = \cos\rho_1 = \mathbf{p}_0^T\mathbf{a}_0 \tag{5.58a}$$

$$\mathbf{d}^T\mathbf{M}_j\mathbf{c}' = \cos\alpha_4 = \mathbf{d}^T\mathbf{c}_0 \tag{5.58b}$$

where $s_j = \sin\beta_j$ and $c_j = \cos\beta_j$, with $\beta_j \equiv \phi_j - \phi_0$. Moreover,

$$\mathbf{M}_j = [\mathbf{p}_j,\ \mathbf{R}_j\mathbf{a}_0,\ \mathbf{B}_j^T\mathbf{R}_j\mathbf{a}_0] \in \mathbb{R}^{3\times 3}, \quad j = 0, ..., m \tag{5.59}$$

Furthermore, $\mathbf{c}'$ is a vector containing parametrized coordinates of point C as introduced in Eq. (B.29) of Appendix B, which is found as

$$\mathbf{c}' = \mathbf{M}_0^{-1}\mathbf{c}_0 \tag{5.60}$$

Synthesis equations (5.58a) and (5.58b) are displayed below in homogenous form:

$$\mathbf{p}_j^T[\mathbf{b}\mathbf{b}^T + s_j\mathbf{B} + c_j(\mathbf{1} - \mathbf{b}\mathbf{b}^T)]\mathbf{a}_0 - \mathbf{p}_0^T\mathbf{a}_0 = 0 \tag{5.61a}$$

$$\mathbf{d}^T(\mathbf{M}_j\mathbf{M}_0^{-1} - \mathbf{1})\mathbf{c}_0 = 0, \ j = 0, \ldots, m \tag{5.61b}$$

whose unknowns are vectors $\mathbf{a}_0$, $\mathbf{b}$, $\mathbf{c}_0$, and $\mathbf{d}$. As these are three-dimensional vectors subject to unit-length conditions, the total number of unknowns at hand is only eight. For $m = 4$, the system of equations is thus determined and hence admits to *exact solutions*. Equations (5.61a) and (5.61b) are apparently quadratic in $\mathbf{b}$ and linear in $\mathbf{a}_0$. A solution to this type of system has been reported (Angeles and Bai, 2010). Once solutions for $\mathbf{a}_0$ and $\mathbf{b}$ are available, the solutions for $\mathbf{d}$ and $\mathbf{c}_0$ can be found likewise.

---

[4] $\mathbf{b}$ and $\mathbf{d}$, being unit vectors, each entails two independent components.

5.2.3 *Examples*

Two examples are included to demonstrate the foregoing procedure. The first example illustrates the spherical-linkage synthesis for a discrete set of prescribed points, the second example addressing the same problem, but with prescribed timing.

Path Synthesis for a Set of Discrete Points

Equation (5.57) allows for a maximum of nine prescribed points for exact synthesis. Considering the numerical challenges with nine-point path synthesis, we include an example for the synthesis with only five points, similar to the case of planar four-bar linkages (Huang et al., 2008).

In the case at hand, two pivoting points on the grounded link are specified as $\mathbf{b} = [1, 0, 0]^T$ and $\mathbf{d} = [0.25, 0.9682, 0]^T$, the five points on the sphere being listed in Table 5.5. Upon substituting all numerical values into Eq. (5.55), a system of five synthesis equations in five unknowns is obtained. This system, solved with computer algebra, led to seven real solutions. It is noteworthy that no constraints were introduced upon equation-solving, other than specifying all dimensions to lie in the $(0, 2\pi)$ range. These solutions were further examined graphically. Two of them, listed in Table 5.6, are feasible and free of branching- and order-defects. The synthesis errors, defined as the rms values of the residuals of Eq. (5.55), are calculated for all solutions. The residuals in the third column, labeled "error," of Table 5.6 indicate that a fairly accurate synthesis solution was achieved (Fig. 5.11).

Table 5.5 Data for five-point synthesis

i	$\mathbf{p}_i^T$
1	[0.94724, −0.16168, 0.27674]
2	[0.90678, −0.31278, 0.28269]
3	[0.60792, −0.33632, 0.71924]
4	[0.53010, −0.25980, 0.80715]
5	[0.63011, −0.07979, 0.77238]

Table 5.6 Defect-free solutions for five-point path synthesis

Linkage	$[\alpha_2, \alpha_3, \alpha_4, \rho_1, \rho_2]$ (rad)	Error
L1	[0.349063, 1.04731, 1.27573, 0.673500, 0.570403]	3.01982×10^{-10}
L2	[0.394362, 1.91134, 1.00221, 0.660250, 3.79991]	2.72647×10^{-23}

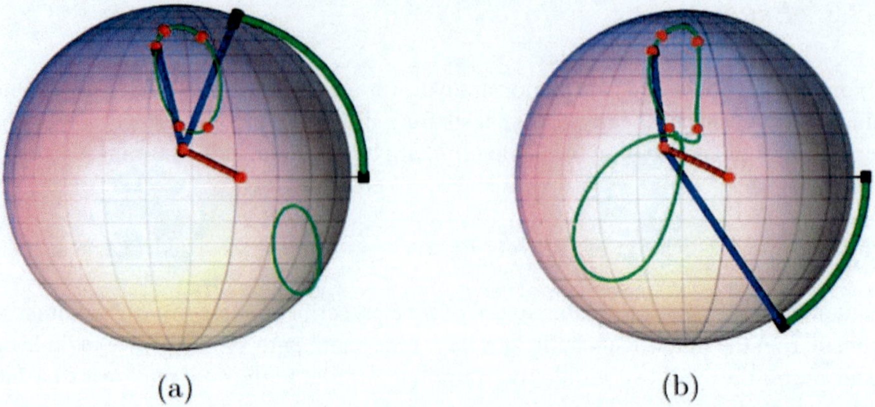

Fig. 5.11 Two linkages synthesized to visit five prescribed point positions: **a** linkage L1 and **b** linkage L2, all five locations being displayed as red dots

Path Synthesis with Prescribed Timing

Now a spherical four-bar linkage is to be synthesized to visit five points located on the unit sphere with *prescribed timing*. The latter is specified via the angles of rotation of the input link, as given in Table 5.7.

Using Eqs. (5.61a) and (5.61b), four equations are obtained for $\mathbf{a}_0$ and $\mathbf{b}$. As the number of independent variables is equal to four, we are able to solve the system for $\mathbf{a}_0$ and $\mathbf{b}$ *exactly* in the absence of singularities. These solutions are then substituted into Eq. (5.61b), thereby obtaining another system of four equations for $\mathbf{c}_0$ and $\mathbf{d}$. Solutions to the latter are found likewise. Table 5.8 lists two solutions, obtained with computer algebra. The spherical four-bar linkages corresponding to these solutions are displayed in Figs. 5.12 and 5.13, where green dots indicate the five given points. Notice that the problem may admit a large number of real solutions. All should be further explored to filter out *unfeasible solutions*.

Table 5.7 Five prescribed positions with timing

i	ψ_i (rad)	$\mathbf{p}_i^T$
0	0	[1,0,0]
1	$3\pi/10$	[0.91465, −0.40138, −0.04787]
2	$3\pi/5$	[0.74750, −0.57010, −0.34089]
3	$9\pi/10$	[0.66712, −0.35876, −0.65285]
4	$6\pi/5$	[0.62057, 0.24057, −0.74633]

Table 5.8 Results of path synthesis with timing

	Solution 1	Solution 2
$\mathbf{a}_0$	$[0.9866, -0.14685, 0.070568]^T$	$[0.452785, 0.661978, -0.597303]^T$
$\mathbf{b}$	$[0.88459, -0.20248, -0.42010]^T$	$[0.670596, 0.739318, -0.060944]^T$
$\mathbf{c}_0$	$[0.86986, 0.01238, -0.49313]^T$	$[0.671804, -0.732535, 0.10990]^T$
$\mathbf{d}$	$[0.43747, -0.13095, -0.88964]^T$	$[0.40946, 0.30681, 0.85918]^T$
Error	1.42855×10^{-11}	4.60493×10^{-11}

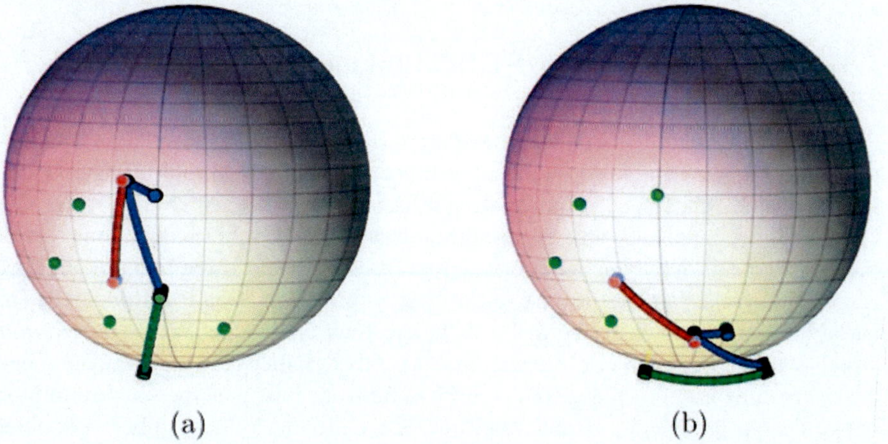

Fig. 5.12 Solution 1 of the linkages synthesized to visit the prescribed positions: **a** the first posture; **b** the fifth posture

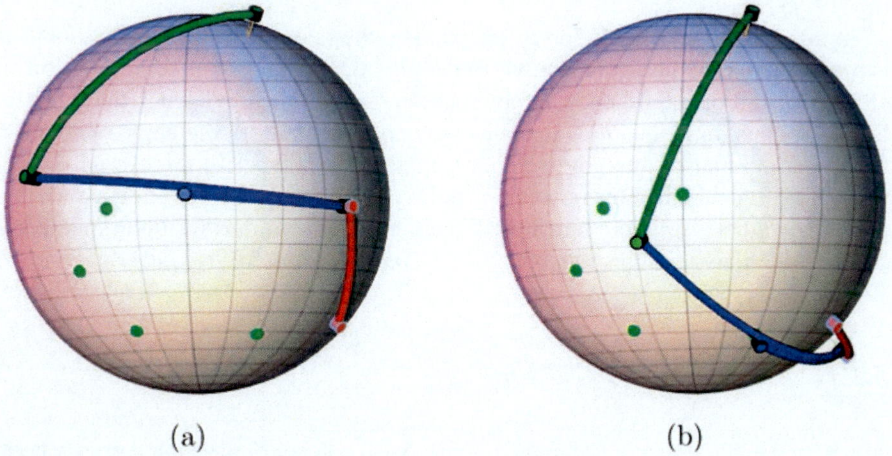

Fig. 5.13 Solution 2 of the synthesized linkages to visit the prescribed positions: **a** the first posture; **b** the fifth posture

5.2.4 *Summary*

A procedure of path synthesis for spherical four-bar linkages was introduced. The procedure is based on the coupler curve of spherical four-bar linkages, which is an implicit, trivariate quartic function, together with the implicit function of the unit sphere. Examples of synthesis with prescribed discrete positions, including five-point synthesis, approximate great-circle path generation, and path synthesis with prescribed timing, were included to demonstrate the procedure of synthesis.

5.3 Path Generation for RCCC Linkages

Path synthesis for RCCC linkages, compared with their planar and spherical counterparts, has been barely considered in the literature, except for approximate synthesis (Chu and Sun, 2009), exact path synthesis remaining nearly untouched. The challenges are manifold. One is the formulation of the synthesis equations. Unlike motion synthesis, in which both position and orientation of the coupler link are specified, path synthesis pertains only to position, while orientation is unspecified. That is, additional unknowns have to be included, then *eliminated in the formulation*, which complicates the model substantially. Another challenge is the large number of design parameters. For planar and spherical four-bar linkages, the maximum number of design parameters is nine, and for RCCC linkages, the number is doubled, as shown in the sequel. The synthesis equations thus lead to a *high-order multivariate polynomial system*, which poses significant challenges, even to state-of-the-art polynomial-equation solvers (Baskar and Bandyopadhyay, 2019; Plecnik and Fearing, 2017).

In this section, a formulation of the path-synthesis problem for RCCC linkages is proposed, based on recent results (Bai et al., 2021). The formulation relies on an alternative coordinate frame, whereby rotation matrices are obviated in the formulation. The constraint equations thus involve only the input driving angle, which can be further eliminated, thereby yielding synthesis equations that involve only functions of design parameters and coupler-point positions. The formulation shows that the path-synthesis problem under study admits exact solutions for up to nine-point positions, *exactly the same number as in the planar and spherical cases*.

5.3.1 *A Generic RCCC Linkage*

A generic RCCC linkage is depicted in Fig. 5.14, whose coupler link carries a point P to attain a set of m positions $\{P_j\}_1^m$. Each link is defined, moreover, by two parameters, namely the twist angle α_i and the distance a_i, for $i = 1, \ldots 4$. Moreover, the displacement of each joint is described by rotation θ_i and translation d_i, with θ_2

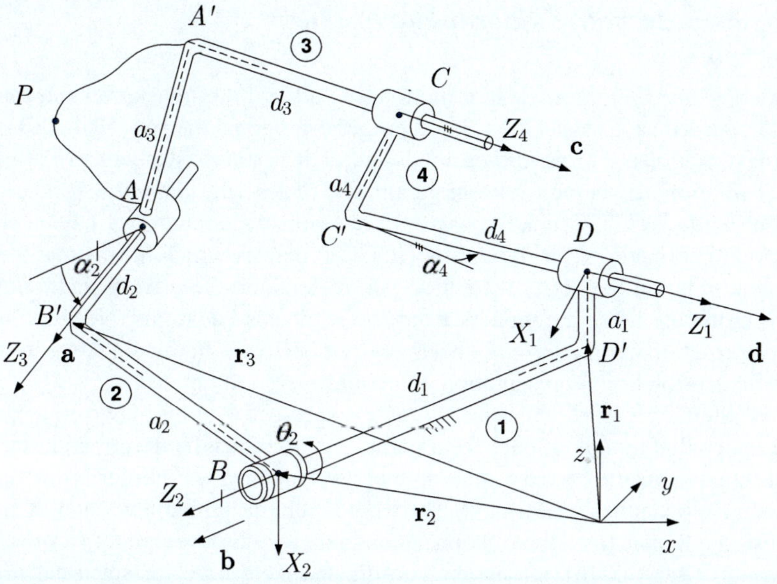

Fig. 5.14 The generic RCCC linkage with all local frames and linkage parameters

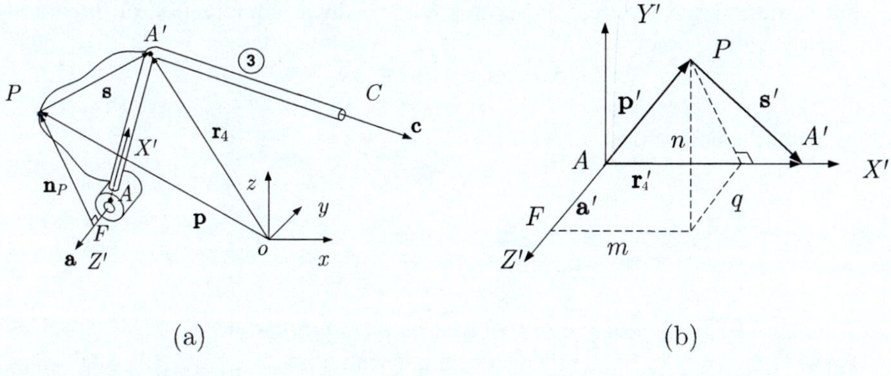

Fig. 5.15 The coupler link, showing the coupler point in **a** the global coordinate system and **b** the local coordinate system

denoting the designated driving input. In addition to the global coordinate system $\{O, x, y, z\}$, four local coordinate systems $\{O_i, X_i, Y_i, Z_i\}_{i=1}^4$ are established for each body, as per the D-H notation (Denavit and Hartenberg, 1955).

The position of the coupler point P in the global coordinate system is denoted $\mathbf{p} = [x, y, z]^T$ in the local system, as shown in Fig. 5.15a, $\mathbf{p}' = [m, n, q]^T$. As the coupler link AC moves, point P traces a path, namely the *coupler curve* of interest.

5.3.2 An Alternative Coordinate Frame

The coupler link undergoes spatial motion, including both translation and rotation, the latter described, as usual, by a 3×3 proper orthogonal matrix. While this matrix describes rotations in three-dimensional space, it is not convenient in the case at hand. A rotation matrix requires four parameters to describe *unambiguously* a spatial rotation, while a RCCC linkage has only one dof. Apparently, we have a surplus of rotation parameters over the dof of RCCC linkages, which means that we have to find the relation between all rotation parameters and the driving input angle to, finally, eliminate these parameters from the synthesis equations. Needless to say, using the rotation matrix *verbatim* becomes cumbersome. In this light, we introduce an alternative rotation representation, to avoid the rotation matrix.

In the alternative representation, a vector is expressed in terms of two other vectors. This implies that the coordinates of a point on a body can be described in terms of two designated vectors, fixed to the same body. This idea was earlier introduced for planar-linkage synthesis (Bai et al., 2016), whereby a unified approach of linkage synthesis applicable to all three Stephenson linkages was developed. In a subsequent publication (Bai, 2021a), alternative coordinates were used for spherical four-bar linkage analysis and synthesis. We extend now this formulation, as applicable to spatial linkages.

For starters, any vector can be expressed as a linear combination of three non-coplanar vectors, namely

$$\mathbf{x} = \lambda \mathbf{x}_1 + \mu \mathbf{x}_2 + \nu \mathbf{x}_3 \tag{5.62a}$$

or, in a more concise form,

$$\mathbf{x} = \mathbf{M}\mathbf{x}^{\star} \tag{5.62b}$$

with

$$\mathbf{M} = [\mathbf{x}_1,\ \mathbf{x}_2,\ \mathbf{x}_3]; \quad \mathbf{x}^{\star} = [\lambda,\ \mu,\ \nu]^T \tag{5.62c}$$

coefficients λ, μ, ν playing the role of *alternative coordinates*.

Fixed to the coupler link, a local coordinate frame $\{A, X', Y', Z'\}$ is established, as shown in Fig. 5.15. In the local frame, the position vector of point A', denoted $\mathbf{r}'_4$, is fully defined. Using the alternative representation, we can express vectors $\mathbf{s}$ and $\mathbf{c}$, the position vector of C, with reference to Fig. 5.15, in the local frame, as

$$\mathbf{s}' = \mathbf{N}\mathbf{s}^{\star}, \quad \mathbf{c}' = \mathbf{N}\mathbf{c}^{\star} \tag{5.63a}$$

where

$$\mathbf{s}' = \mathbf{r}'_4 - \mathbf{p}' \tag{5.63b}$$

$$\mathbf{N} = [\mathbf{a}', \mathbf{n}'_p, \mathbf{a}' \times \mathbf{n}'_p] \tag{5.63c}$$

The three vectors in matrix $\mathbf{N}$ being non-coplanar, $\mathbf{N}$ is invertible. The vectors $\mathbf{s}^\star$ and $\mathbf{c}^\star$ of alternative coordinates are then defined as

$$\mathbf{s}^\star = [\lambda_s, \mu_s, \nu_s]^T = \mathbf{N}^{-1}\mathbf{s}' \tag{5.63d}$$
$$\mathbf{c}^\star = [\lambda_c, \mu_c, \nu_c]^T = \mathbf{N}^{-1}\mathbf{c}' \tag{5.63e}$$

Equations (5.63a) are also valid in the global frame, i.e.,

$$\mathbf{s} = \mathbf{M}\mathbf{s}^\star, \quad \mathbf{c} = \mathbf{M}\mathbf{c}^\star \tag{5.64a}$$

where

$$\mathbf{M} = [\mathbf{a}\ \mathbf{n}_p\ \mathbf{a} \times \mathbf{n}_p] \tag{5.64b}$$

This can be simply proven upon multiplying both sides of each of Eqs. (5.63a) by a rotation matrix $\mathbf{Q}$. In the above formulation, $\mathbf{n}_p$ and $\mathbf{n}'_p$ denote the normal vector from point P to line Z_3, expressed in the global and in the local frames, respectively.

The global coordinates of point A', referring to Fig. 5.15a, are then found from the relation

$$\mathbf{r}_4 = \mathbf{p} + \mathbf{s} = \mathbf{p} + \mathbf{M}\mathbf{s}^\star \tag{5.65}$$

5.3.3 Constraint Equations of the CC and RC dyads

With the above preliminaries established, we can now formulate the constraint equations for the RCCC linkage.

RC dyad BAP

With reference to the RCCC linkage of Fig. 5.14, the displacement of the moving axis Z_3 in dual form is represented as

$$\hat{\mathbf{a}} = \mathbf{a} + \epsilon\bar{\mathbf{a}} \tag{5.66}$$

with $\mathbf{a}$ defined as the unit vector parallel to Z_3 and $\bar{\mathbf{a}}$ the *moment* of the same axis with respect to the origin O_1 of frame $\{X_1, Y_1, Z_1\}$, which can be expressed as

$$\bar{\mathbf{a}} = \mathbf{r}_3 \times \mathbf{a} \tag{5.67}$$

$\mathbf{r}_3$ denoting the position vector of point B', the orthogonal projection of point B on axis Z_3. Further, let $\mathbf{r}_1$ and $\mathbf{r}_2$ be the position vectors of points D' and B, respectively. As axis Z_3 rotates about Z_2, we have

$$\mathbf{a} = \mathbf{R}\mathbf{a}_0; \quad \mathbf{r}_3 = \mathbf{r}_2 + \mathbf{R}\mathbf{s}_{20}; \quad \mathbf{r}_2 = \mathbf{r}_1 + d_1\mathbf{b} \tag{5.68}$$

with $\mathbf{R}$ denoting the matrix rotating link 2 about Z_2 through an angle θ_2, while $\mathbf{a}_0$ is the direction vector parallel to line Z_3, $\mathbf{s}_{20}$ being the vector of the common perpendicular of Z_2 and Z_3 from the former to the latter, both expressed in frame $\{X_2, Y_2, Z_2\}$, namely

$$\mathbf{a}_0 = \mathbf{R}_x(\alpha_2); \quad \mathbf{s}_{20} = [a_2,\ 0,\ 0]^T \tag{5.69}$$

and $\mathbf{R}_x(\alpha_2)$ is the matrix of rotation about the X-axis through an angle α_2. Once the driving input is specified, line Z_3 is uniquely determined. Based on Eq. (A.41), a geometric constraint can thus be established:

$$(\mathbf{p} \times \mathbf{a} - \bar{\mathbf{a}})^T(\mathbf{p} \times \mathbf{a} - \bar{\mathbf{a}}) = \mathbf{n}_p^T\mathbf{n}_p \equiv l_p^2 \tag{5.70}$$

with $\mathbf{n}_P$ shown in Fig. 5.15a, and the distance from the coupler point P to axis Z_3, denoted l_p, remaining constant. This distance can be found from the local frame of the coupler link, as shown in Fig. 5.15. In the local frame, apparently, $\mathbf{p}' = [m, n, q]^T$, $\mathbf{a}' = [0, 0, 1]^T$, and $\bar{\mathbf{a}}' = [0, 0, 0]^T$. We thus have

$$l_p^2 = ||\mathbf{p}' \times \mathbf{a}'||^2 = m^2 + n^2 \tag{5.71}$$

CC dyad DCP

Given that the relative position of Z_4 with respect to Z_1 remains constant for all motions of the linkage, the dual angle between the two axes is constant as well, and hence,

$$\hat{\mathbf{c}}^T\hat{\mathbf{d}} = \cos(\alpha_4 + \epsilon a_4) \equiv \cos\alpha_4 - a_4\epsilon\sin\alpha_4 \tag{5.72}$$

where

$$\hat{\mathbf{d}} = \mathbf{d} + \epsilon\bar{\mathbf{d}}; \quad \hat{\mathbf{c}} = \mathbf{c} + \epsilon\bar{\mathbf{c}}; \tag{5.73}$$

with vectors $\mathbf{c}$ and $\mathbf{d}$, as well as a_4 and α_4, shown in Fig. 5.14.

The primal and the dual parts of Eq. (5.72) yield, correspondingly,

$$\mathbf{c}^T\mathbf{d} = \alpha_4 \tag{5.74a}$$

$$\mathbf{c}^T\bar{\mathbf{d}} + \bar{\mathbf{c}}^T\mathbf{d} = -a_4\sin\alpha_4 \tag{5.74b}$$

where $\mathbf{d}$ and $\bar{\mathbf{d}}$ are constant. By virtue of the alternative coordinates, $\mathbf{c}$ and $\bar{\mathbf{c}}$ can be readily expressed as functions of the coupler-point position, i.e.,

$$\mathbf{c} = \mathbf{M}\mathbf{c}^\star; \quad \bar{\mathbf{c}} = (\mathbf{p} + \mathbf{s}) \times \mathbf{c} \tag{5.75}$$

expressions for $\mathbf{s}$ and $\mathbf{M}$ being displayed in Eqs. (5.64a) and (5.64b), respectively.

5.3.4 N-point Path Synthesis

We now look at the problem of path synthesis for N prescribed positions. The jth location of the coupler link is indicated by the position vector $\mathbf{p}_j$ of point P_j, $j = 1, \ldots, N$. Moreover, let the moving axes Z_3 and Z_4 be represented by the dual vectors $\hat{\mathbf{a}}_j$ and $\hat{\mathbf{c}}_j$, corresponding to the j-th coupler-link pose. Likewise, the dual vectors of the fixed axes Z_1 and Z_2 are represented by $\hat{\mathbf{d}}$ and $\hat{\mathbf{b}}$, respectively. With reference to Fig. 5.14, the design parameters are included in the 22-dimensional vector array

$$\mathbf{x}_d = [\mathbf{b}^T, \bar{\mathbf{b}}^T, \mathbf{d}^T, \bar{\mathbf{d}}^T, a_2, \alpha_2, a_3, \alpha_3, a_4, \alpha_4, d_1, m, n, q]^T \tag{5.76}$$

whose four vectors contain only eight independent design parameters, which thus leads to a total of 18 unknowns.

With the constraint equations formulated, the synthesis equations for path generation are

$$(\mathbf{p}_j \times \mathbf{a}_j - \bar{\mathbf{a}}_j)^T(\mathbf{p}_j \times \mathbf{a}_j - \bar{\mathbf{a}}_j) = m^2 + n^2 \tag{5.77a}$$

$$\mathbf{c}_j^T\mathbf{d} = \cos\alpha_4 \tag{5.77b}$$

$$\mathbf{c}_j^T\bar{\mathbf{d}} + \bar{\mathbf{c}}_j^T\mathbf{d} = -a_4\sin\alpha_4 \tag{5.77c}$$

where, for $j = 1, \ldots, N$,

$$\mathbf{a}_j = \mathbf{R}_j\mathbf{a}_0 \tag{5.78a}$$

$$\bar{\mathbf{a}}_j = \mathbf{r}_3 \times \mathbf{a}_j \tag{5.78b}$$

$$\mathbf{c}_j = \mathbf{M}_j\mathbf{c}^\star \tag{5.78c}$$

$$\bar{\mathbf{c}}_j = (\mathbf{p}_j + \mathbf{M}_j\mathbf{s}^\star) \times \mathbf{c}_j \tag{5.78d}$$

with

$$\mathbf{M}_j = [\mathbf{a}_j,\ \mathbf{n}_{pj},\ \mathbf{a}_j \times \mathbf{n}_{pj}] \tag{5.78e}$$

and

$$\mathbf{n}_{pj} = \mathbf{r}_{pj} - (\mathbf{r}_{pj}^T\mathbf{a}_j)\mathbf{a}_j \quad \text{and} \quad \mathbf{r}_{pj} = (\mathbf{p}_j - \mathbf{r}_3) \tag{5.79}$$

Equations (5.77a)–(5.77c), after expansion, yield

$$A_{1j}c^2 + A_{2j}cs + A_{3j}c + A_{4j}s + A_{5j} = 0 \tag{5.80a}$$

$$B_{1j}c^2 + B_{2j}cs + B_{3j}c + B_{4j}s + B_{5j} = 0 \tag{5.80b}$$

$$C_{1j}c^3 + C_{2j}c^2 + C_{3j}cs + C_{4j}c + C_{5j}s + C_{6j} = 0 \tag{5.80c}$$

where $c \equiv \cos\theta_2$ and $s \equiv \sin\theta_2$. All coefficients in the above equations are functions of the design parameters. It is noteworthy that θ_2 varies for different coupler poses.

Using tan-half-angle substitutions for c and s, the foregoing equations become

$$D_{4j}u^4 + D_{3j}u^3 + D_{2j}u^2 + D_{1j}u + D_{0j} = 0 \tag{5.81a}$$

$$E_{4j}u^4 + E_{3j}u^3 + E_{2j}u^2 + E_{1j}u + E_{0j} = 0 \tag{5.81b}$$

$$F_{6j}u^6 + F_{5j}u^5 + F_{4j}u^4 + F_{3j}u^3 + F_{2j}u^2 + F_{1j}u + F_{0j} = 0 \tag{5.81c}$$

where $u = \tan(\theta_2/2)$. All coefficients in the above equations being extremely lengthy, they are not displayed here; they are available online[5] For this polynomial system, u can be eliminated using *resultants* (Gelfand et al., 2008), the synthesis equations thus becoming

$$f_j(\mathbf{x}_d, \mathbf{p}_j) = \det(\mathbf{M}_{1j}) = 0 \tag{5.82a}$$

$$g_j(\mathbf{x}_d, \mathbf{p}_j) = \det(\mathbf{M}_{2j}) = 0 \tag{5.82b}$$

$\mathbf{M}_{1j}$ and $\mathbf{M}_{2j}$ representing the *resultant matrices* (Gelfand et al., 2008) for Eqs. (5.81a) and (5.81b) and (5.81a) and (5.81c), respectively,

$$\mathbf{M}_{1j} = \begin{bmatrix} D_{4j} & D_{3j} & D_{2j} & D_{1j} & D_{0j} & 0 & 0 & 0 \\ E_{4j} & E_{3j} & E_{2j} & E_{1j} & E_{0j} & 0 & 0 & 0 \\ 0 & D_{4j} & D_{3j} & D_{2j} & D_{1j} & D_{0j} & 0 & 0 \\ 0 & E_{4j} & E_{3j} & E_{2j} & E_{1j} & E_{0j} & 0 & 0 \\ 0 & 0 & D_{4j} & D_{3j} & D_{2j} & D_{1j} & D_{0j} & 0 \\ 0 & 0 & E_{4j} & E_{3j} & E_{2j} & E_{1j} & E_{0j} & 0 \\ 0 & 0 & 0 & D_{4j} & D_{3j} & D_{2j} & D_{1j} & D_{0j} \\ 0 & 0 & 0 & E_{4j} & E_{3j} & E_{2j} & E_{1j} & E_{0j} \end{bmatrix} \tag{5.83}$$

and

$$\mathbf{M}_{2j} = \begin{bmatrix} D_{4j} & D_{3j} & D_{2j} & D_{1j} & D_{0j} & 0 & 0 & 0 & 0 & 0 \\ F_{6j} & F_{5j} & F_{4j} & F_{3j} & F_{2j} & F_{1j} & F_{0j} & 0 & 0 & 0 \\ 0 & D_{4j} & D_{3j} & D_{2j} & D_{1j} & D_{0j} & 0 & 0 & 0 & 0 \\ 0 & F_{6j} & F_{5j} & F_{4j} & F_{3j} & F_{2j} & F_{1j} & F_{0j} & 0 & 0 \\ 0 & 0 & D_{4j} & D_{3j} & D_{2j} & D_{1j} & D_{0j} & 0 & 0 & 0 \\ 0 & 0 & F_{6j} & F_{5j} & F_{4j} & F_{3j} & F_{2j} & F_{1j} & F_{0j} & 0 \\ 0 & 0 & 0 & D_{4j} & D_{3j} & D_{2j} & D_{1j} & D_{0j} & 0 & 0 \\ 0 & 0 & 0 & F_{6j} & F_{5j} & F_{4j} & F_{3j} & F_{2j} & F_{1j} & F_{0j} \\ 0 & 0 & 0 & 0 & D_{4j} & D_{3j} & D_{2j} & D_{1j} & D_{0j} & 0 \\ 0 & 0 & 0 & 0 & 0 & D_{4j} & D_{3j} & D_{2j} & D_{1j} & D_{0j} \end{bmatrix} \tag{5.84}$$

respectively.

---

[5] https://drive.google.com/file/d/1UzGjb1hRPZdIsWBa8AqOL1OLpPLpyzK4/view?usp=sharing.

Equations (5.82a) and (5.82b) allow for path synthesis with N prescribed discrete positions. For these positions, the equations lead to $2N$ constraint equations. Notice that a total of 18 unknowns are to be found. One must thus have $2N = 18$ equations for exact solutions. We thus conclude that *the path synthesis of RCCC linkages admits exact solutions for a maximum of nine prescribed positions, the same number in the case of planar and spherical linkages.*

On the path-synthesis problem, Burmester surmised that *the maximum number of points that can be prescribed for a four-bar linkage is nine* (Nolle, 1975). While this statement is true for planar and spherical linkages, to the authors' knowledge, it has not yet been reported for their spatial counterparts. We just showed that the statement holds for RCCC linkages as well, thereby closing an important chapter in the history of linkage synthesis.

We thus have an interesting result:

Theorem The synthesis of a four-bar linkage for path generation through n prescribed point positions admits exact solutions for $n = 9$ for all linkage types: *planar, spherical, and spatial.*

Equations (5.82a) and (5.82b) also stand for implicit functions of the coupler curve. This means that we can, in principle, generate a coupler curve for a RCCC linkage from two functions. However, it is noteworthy that spurious solutions are introduced upon variable elimination. As a matter of fact, the intersection of two function plots generates additional curves, other than the coupler curve. This is unavoidable, because of the algebraic operations involved, but must be taken into account.

5.3.5 Example

We include here an illustrative example to demonstrate the foregoing formulation and the corresponding path-synthesis procedure.

We assume that the ground link is specified, i.e., $\hat{\mathbf{b}}$ and $\hat{\mathbf{d}}$ are given, which means that axes Z_2 and Z_1, respectively, *are known*. We find below the linkage parameters of the three moving links, along with the locations of both the coupler point and the moving revolute joint. The number of unknowns amounts then to 10.

We resort to computer algebra for equation-solving. Moreover, two *alternative* systems of equations, one with Eqs. (5.82a) and (5.82b) and the other with Eqs. (5.77a)–(5.77c), are available, one of which is to be solved. Either alternative works, the latter leading to a faster solution.

Five positions of P_j are given in Table 5.9. With these data, one solution obtained is listed in Table 5.10, the corresponding linkage displayed in Fig. 5.16.

Table 5.9 Five prescribed positions

j	$\mathbf{p}_j^T$
1	[−2.76397, −0.39367, 4.16027]
2	[−3.37794, 1.07553, 4.34335]
3	[−4.24293, 1.87940, 4.64811]
4	[−5.28101, 2.15610, 4.59700]
5	[−6.34872, 2.10364, 3.80349]

Table 5.10 The RCCC linkage synthesized in the example

$\mathbf{b}^T$, $\bar{\mathbf{b}}^T$	$\mathbf{d}^T$, $\bar{\mathbf{d}}^T$	Specified
[0, 0, 1], [0, 0, 0]	[0, 0.866, 0.5], [−0.866, 2.50, −4.33]	
$[m, n, q]$, d_1	(a_2, a_3, a_4), $(\alpha_2, \alpha_3, \alpha_4)$	Synthesized
[4, 3, 5], 1.0	(2, 4, 3.5), (0.5235, 0.9599, 0.7856)	

Fig. 5.16 The synthesized RCCC linkage, with red dots indicating the given coupler-point positions to be visited by a point P of the coupler link

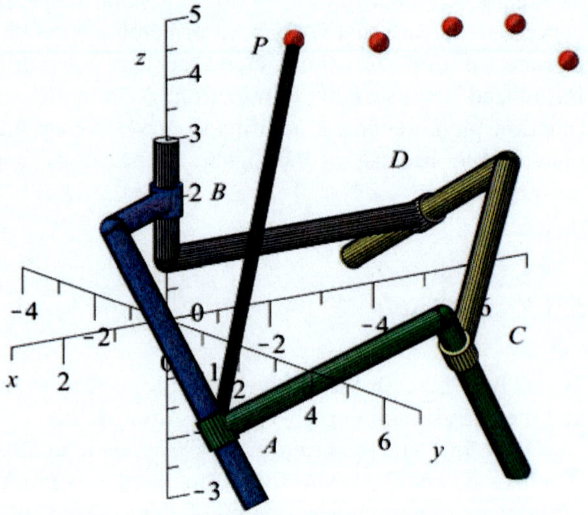

The three constraint equations (5.77a)–(5.77c) allow us to find the coupler curve using the angle of rotation at the input link as a parameter. For any given rotation of the input link, a coupler-curve point can be determined through three constraint equations. A coupler curve is thus readily obtained using this procedure. Figure 5.17 includes the coupler curve of the synthesized linkage, which shows that the curve has *two branches*. In the same figure, all prescribed points are also displayed. It is seen that the five prescribed points lie all on one and the same branch, the linkage thus being free of branch defect.

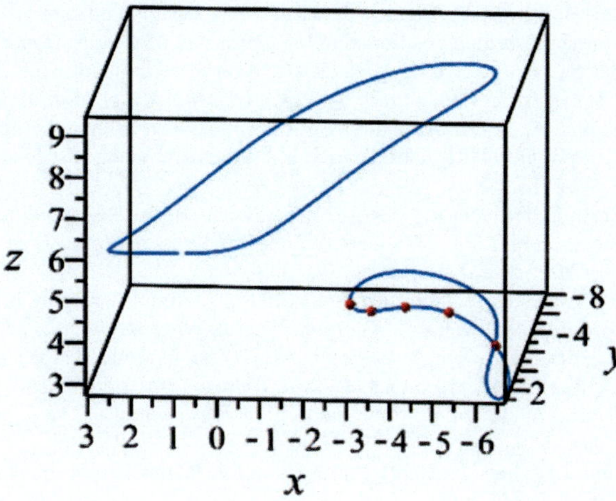

Fig. 5.17 Coupler curve of the synthesized RCCC linkage, shown together with the five prescribed points

5.3.6 Summary

The synthesis equations for path generation using a RCCC linkage were formulated, which reveal that the path synthesis of these linkages admits exact solutions for a maximum of nine prescribed path points. It is noteworthy that *this number is the same for planar and spherical four-bar linkages*. A numerical example was included to illustrate the use of the proposed synthesis procedure.

Moreover, with the foregoing formulation, it is possible to conduct exact path synthesis for RCCC linkages, but the numerical challenges still remain. In this chapter, we included only one example of a five-point synthesis. Examples of nine-point synthesis are available (Bai et al., 2021). Further development of efficient *nonlinear-equation solvers* should help us face the numerical challenges involved.

References

Angeles, J., & Bai, S. (2010). A robust solution of the spherical Burmester problem. In *Proceedings of ASME DETC2010*, Montreal, #MECH–28189.

Bai, S. (2015a). Determination of linkage parameters from coupler curve equations. In *Mechanisms, Transmissions and Applications: Proceedings of the Third MeTrApp Conference 2015* (pp. 49–57). Springer International Publishing.

Bai, S. (2015b). Exact synthesis of cognate linkages with algebraic coupler curves. In *Proceedings of 14th IFToMM World Congress* (pp. 210–216).

Bai, S. (2021). Algebraic coupler curve of spherical four-bar linkages and its applications. *Mechanism and Machine Theory, 158*, 104218.

Bai, S. (2021b). A note on the univariate nonic derived from the coupler curve of four-bar linkages. *Mechanism and Machine Theory, 162*.

Bai, S., & Angeles, J. (2015). Coupler-curve synthesis of four-bar linkages via a novel formulation. *Mechanism and Machine Theory, 94*, 177–187.

Bai, S., Li, Z., & Angeles, J. (2021). Exact path synthesis of RCCC linkages for a maximum of nine prescribed positions. *ASME Journal of Mechanisms and Robotics*.

Bai, S., Wang, D., & Dong, H. (2016). A unified formulation for dimensional synthesis of Stephenson linkages. *ASME Journal of Mechanisms and Robotics, 8*(4), 041009.

Baskar, A., & Bandyopadhyay, S. (2019). An algorithm to compute the finite roots of large systems of polynomial equations arising in kinematic synthesis. *Mechanism and Machine Theory, 133*, 493–513.

Basu, S., Pollack, R., & Roy, M.-F. (2006). *Algorithms in real algebraic geometry*. Springer.

Blechschmidt, J. L., & Uicker, J. J. (1986). Linkage synthesis using algebraic curves. *ASME Journal of Mechanical Design, 108*(4), 543–548.

Bulatović, R., & Djordjevic, S. (2004). Optimal synthesis of a four-bar linkage by method of controlled deviation. *Theoretical and Applied Mechanics, 31*, 265–280.

Cabrera, J., Simon, A., & Prado, M. (2002). Optimal synthesis of mechanisms with genetic algorithms. *Mechanism and Machine Theory, 37*(10), 1165–1177.

Cayley, A. (1875). On three-bar motion. *Proceedings of The London Mathematical Society, s1-7*(1), 136–166.

Chu, J., & Sun, J. (2009). Research on RCCC mechanism path generation by using numerical atlas method. *Zhongguo Jixie Gongcheng/China Mechanical Engineering, 20*, 1024–1028.

Curtiss, D. (1918). Recent extentions of Descartes' rule of signs. *Annals of Mathematics*, 251–278.

Denavit, J., & Hartenberg, R. (1955). A kinematic notation for lower pair mechanisms based on matrices. *ASME Journal of Applied Mechanics, 77*, 215–221.

Gelfand, I. M., Kapranov, M., & Zelevinsky, A. (2008). *Discriminants, resultants, and multidimensional determinants*. Springer Science & Business Media.

Hartenberg, R., & Denavit, J. (1964). *Kinematic synthesis of linkages*. McGraw-Hill.

Huang, X., Liao, Q., Wei, S., & Xu, Q. (2008). Five precision point-path synthesis of planar four-bar linkage using algebraic method. *Frontiers of Electrical and Electronic Engineering in China, 3*, 470–474.

Laidacker, M. A. (1969). Another theorem relating Sylvester's matrix and the greatest common divisor. *Mathematics Magazine, 42*(3), 126–128.

Nolle, H. (1974). Linkage coupler curve synthesis: A historical review—II. Developments after 1875. *Mechanism and Machine Theory, 9*(3-4), 325–348.

Nolle, H. (1975). Linkage coupler curve synthesis: A historical review—III. Spatial synthesis and optimization. *Mechanism and Machine Theory, 10*(1), 41–55.

Plecnik, M., & Fearing, R. (2017). Finding only finite roots to large kinematic synthesis systems. *Journal of Mechanisms and Robotics, 9*(2), 021005.

Roberts, S. (1869). On the motion of a plane under certain conditions. *Proceedings of the London Mathematical Society, s1-3*(1), 286–319.

Sancibrian, R., Viadero, F., Garciá, P., & Fernández, A. (2004). Gradient-based optimization of path synthesis problems in planar mechanisms. *Mechanism and Machine Theory, 39*(8), 839–856.

Wampler, C. W., Morgan, A. P., & Sommese, A. J. (1990). Numerical continuation methods for solving polynomial systems arising in kinematics. *ASME Journal of Mechanical Design, 112*(1), 59–68.

Chapter 6
Synthesis of Planar Complex Linkages

In previous chapters, we studied kinematic-synthesis methods applied mainly to four-bar linkages. In this chapter we extend the study to *complex linkages*, i.e., those with more than four links; in particular, we extend the procedure to six- and 10-bar linkages, still with one degree of freedom. However, we limit the chapter to planar linkages, which have been extensively studied in the archival literature.

As a matter of fact, the literature on multiloop spatial kinematic chains has been expanded, mainly in the realm of parallel robotics. Examples of these mechanisms have been included in Figs. 2.6, 2.17, 2.18, 2.19, 2.20, 2.21, 2.22 and 2.23.

6.1 Six-Bar Linkages

A six-bar linkage, as a one degree-of-freedom mechanism, can be used for motion generation, similar to its four-bar counterpart. Compared with the latter, six-bar linkages, in general, offer a *high synthesis flexibility*. This means that a six-bar linkage can be designed with constraints on both link dimensions and joint locations (Shen et al., 2008), or on velocity and accelerations as the case of dwell linkages (Dong and Wang, 2007; Cavic et al., 2019).

Some methods of synthesis of six-bar linkages are available in the literature. An early work on the subject was proposed by McLarnan (1963), who considered three mechanisms, one Watt linkage and two inversions of the Stephenson linkage. Design equations for Watt-I and Stephenson-I, II, and III path generators were reported by Kim et al. (1972), who proposed an extension of the synthesis of planar four-bar mechanisms. A method of six-bar linkage synthesis was developed by Soh and McCarthy (2008), who considered the linkages as constrained 3R chains.

In this chapter, we look at the motion-generation synthesis of the Stephenson linkage. Such a linkage comprises two ternary links that are connected by binary links. Depending on the grounded link, either one of the two ternary links or a binary

J. Angeles and S. Bai, *Kinematics of Mechanical Systems*, Mathematical Engineering,
https://doi.org/10.1007/978-3-031-09544-3_6

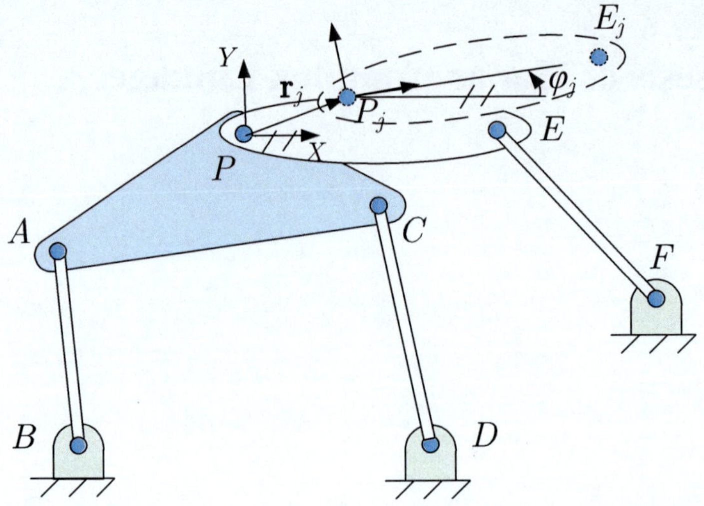

Fig. 6.1 A Stephenson-III linkage, with *grounded dyads* AB, CD, and EF, to be synthesized

link, Stephenson linkages are classified into three types. These linkages bear two kinematic loops, the number of links in each loop varying with the linkage type. In most works in the literature, the synthesis equations are formulated as loop-closure equations (Plecnik et al., 2014). In this book, a *unified* motion-synthesis procedure is proposed for the three types of Stephenson linkages. The method is based on the formulation introduced by Bai et al. (2016).

A link, shown shaded in Fig. 6.1, carries joints A, C and P; the link is to be guided through a discrete set of m poses, given by $\{\mathbf{r}_j, \varphi_j\}_0^m$, where $\mathbf{r}_j$ is the position vector of a landmark point P of the link at the jth pose, φ_j the angle that line P_jE_j of the link makes with line BD of the frame at this pose. The problem consists in finding the joint centers A_0 and B, a.k.a. the *circlepoint* and the *centerpoint*, respectively, that define the BA *dyad*, the same applying to dyads DC and FE.

The problem at hand thus consists in finding the three binary—revolute-revolute (RR)—dyads AB, CD, and EF of Fig. 6.1 that carry one joint fixed to the frame, to construct a six-bar linkage capable of visiting a set of prescribed poses of the *driven link* PE. The linkage in Fig. 6.1 is of the *Stephenson-III* type,[1] which is studied first, those of types I and II following suit.

In the linkage-synthesis problem defined above, the origin of the moving coordinate system is located at point P, as appearing in Fig. 6.1. The unknowns are the coordinates of points $A, B, \ldots, F$, via their position vectors $\mathbf{a}, \mathbf{b}, \ldots, \mathbf{f}$ in the X-Y frame. For a moving point, a subscript is attached to indicate the pose number. Moreover, we let $\mathbf{Q}_j$ and $\mathbf{R}_j$, $j = 1, \ldots, m$, denote the rotation matrices of links AP and PE at the jth linkage posture, respectively, with $\mathbf{Q}_0 = \mathbf{R}_0 = \mathbf{1}$.

[1] We follow the well-established nomenclature in the art (Norton, 2004).

6.2 Synthesis of the Stephenson-III Linkage

We start the procedure with a single RR dyad, namely, the BA dyad, as shown in Fig. 6.2.

Upon recalling the pertinent synthesis equations of four-bar linkages, we have

$$\mathbf{b}^T(\mathbf{1}-\mathbf{Q}_j)\mathbf{a}_0 + \mathbf{r}_j^T\mathbf{Q}_j\mathbf{a}_0 - \mathbf{r}_j^T\mathbf{b} + \frac{1}{2}\mathbf{r}_j^T\mathbf{r}_j = 0\,, \quad j = 1, \ldots, m \tag{6.1}$$

which are, now, the *synthesis equations* to be used to compute the design parameters of the linkage at hand, namely, vectors $\mathbf{a}_0$ and $\mathbf{b}$. It is noteworthy that, in linkage synthesis, the unknowns are, first and foremost, linkage parameters, linkage variables being motion dependent.

Likewise, the synthesis equations for the CD dyad in Fig. 6.1 are

$$\mathbf{d}^T(\mathbf{1}-\mathbf{Q}_j)\mathbf{c}_0 + \mathbf{r}_j^T\mathbf{Q}_j\mathbf{c}_0 - \mathbf{r}_j^T\mathbf{d} + \frac{1}{2}\mathbf{r}_j^T\mathbf{r}_j = 0, \quad j = 1, \ldots, m \tag{6.2}$$

while those for the EF dyad in the same figure are

$$\mathbf{f}^T(\mathbf{1}-\mathbf{R}_j)\mathbf{e}_0 + \mathbf{r}_j^T\mathbf{R}_j\mathbf{e}_0 - \mathbf{r}_j^T\mathbf{f} + \frac{1}{2}\mathbf{r}_j^T\mathbf{r}_j = 0, \quad j = 1, \ldots, m \tag{6.3}$$

with $\mathbf{R}_j$ denoting the rotation matrix through an angle φ_j for link PE, as shown in Fig. 6.1.

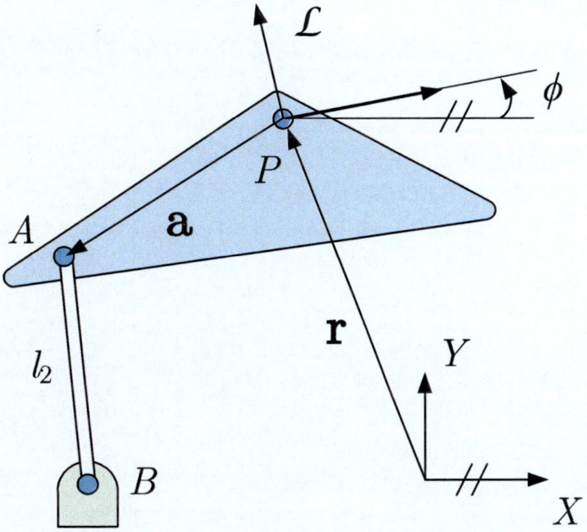

Fig. 6.2 Revolute-revolute (RR) dyad BA

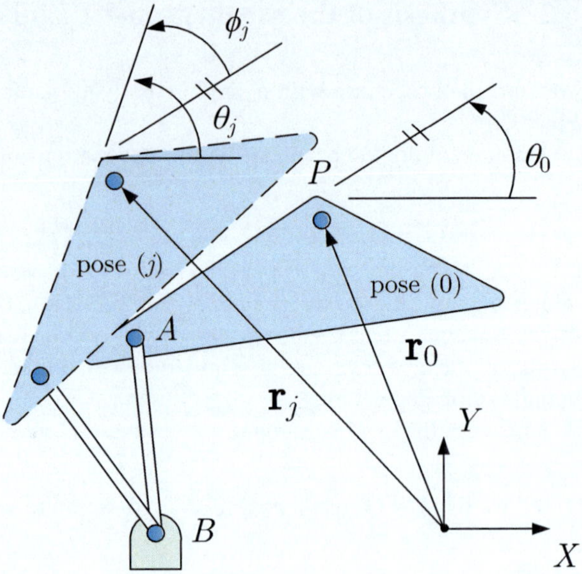

Fig. 6.3 Two finitely separated poses of the coupler link

Relations (6.1)–(6.3) constitute the system of synthesis equations for the Stephenson-III linkage, which include vectors denoting the positions of points A_0, B, C_0, D, E_0, and F, the orientation variable being ϕ_j, as shown in Fig. 6.3. The latter is a configuration-dependent variable, or *motion variable* for brevity, which is, hence, to be eliminated.

6.2.1 *Elimination of Motion Variables*

The orientation angle ϕ_j of link AP, illustrated in Fig. 6.3, is now eliminated, as shown below. This can be done upon recalling the synthesis equations of dyads AB and CD.

To begin with, all terms of Eq. (6.1) are expressed upon writing $\mathbf{Q}_j$ in the form $\mathbf{Q}_j = c_j\mathbf{1} + s_j\mathbf{E}$, in which $s_j \equiv \sin\phi_j$ and $c_j \equiv \cos\phi_j$, as usual. Moreover, $\mathbf{1}$ denotes the 2×2 identity matrix, while $\mathbf{E}$ is the 2×2 rotation matrix through an angle of $90°$. Hence, the first two terms of Eq. (6.1) are expanded as

$$\begin{aligned}\mathbf{b}^T(\mathbf{1} - \mathbf{Q}_j)\mathbf{a}_0 &= \mathbf{b}^T(\mathbf{1} - c_j\mathbf{1} - s_j\mathbf{E})\mathbf{a}_0 \\ &= \mathbf{b}^T\mathbf{a}_0 - c_j\mathbf{b}^T\mathbf{a}_0 - s_j\mathbf{b}^T\mathbf{E}\mathbf{a}_0, \quad j = 1, \ldots, m \qquad (6.4a)\end{aligned}$$

$$\begin{aligned}\mathbf{r}_j^T\mathbf{Q}_j\mathbf{a}_0 &= \mathbf{r}_j^T(c_j\mathbf{1} + s_j\mathbf{E})\mathbf{a}_0 \\ &= c_j\mathbf{r}_j^T\mathbf{a}_0 + s_j\mathbf{r}_j^T\mathbf{E}\mathbf{a}_0, \quad j = 1, \ldots, m \qquad (6.4b)\end{aligned}$$

Equation (6.1) can now be written, for dyad AB, in abbreviated form:

$$A_1 c_j + B_1 s_j + C_1 = 0 \tag{6.5a}$$

with coefficients defined as

$$A_1 = \mathbf{r}_j^T \mathbf{a}_0 - \mathbf{b}^T \mathbf{a}_0 \tag{6.5b}$$

$$B_1 = \mathbf{r}_j^T \mathbf{E}\mathbf{a}_0 - \mathbf{b}^T \mathbf{E}\mathbf{a}_0 \tag{6.5c}$$

$$C_1 = \mathbf{b}^T \mathbf{a}_0 - \mathbf{r}_j^T \mathbf{b} + \mathbf{r}_j^T \mathbf{r}_j / 2 \tag{6.5d}$$

Likewise, the synthesis equation for dyad CD leads to

$$A_2 c_j + B_2 s_j + C_2 = 0, \quad j = 1, \ldots, m \tag{6.6a}$$

with

$$A_2 = \mathbf{r}_j^T \mathbf{c}_0 - \mathbf{d}^T \mathbf{c}_0 \tag{6.6b}$$

$$B_2 = \mathbf{r}_j^T \mathbf{E}\mathbf{c}_0 - \mathbf{d}^T \mathbf{E}\mathbf{c}_0 \tag{6.6c}$$

$$C_2 = \mathbf{d}^T \mathbf{c}_0 - \mathbf{r}_j^T \mathbf{d} + \mathbf{r}_j^T \mathbf{r}_j / 2 \tag{6.6d}$$

Equations (6.5a) and (6.6a), with unknows c_j and s_j, then yield

$$c_j = \frac{B_1 C_2 - C_1 B_2}{A_1 B_2 - A_2 B_1}, \quad s_j = -\frac{A_1 C_2 - A_2 C_1}{A_1 B_2 - A_2 B_1}, \quad j = 1, \ldots, m \tag{6.7}$$

Finally, substitution of the above expressions for c_j and s_j into the identity $s_j^2 + c_j^2 = 1$ leads to

$$\begin{aligned} & A_1^2 C_2^2 - 2\, A_1 C_2 A_2 C_1 + A_2^2 C_1^2 + B_1 C_2 A_1 B_2 - B_1^2 C_2 A_2 \\ & - C_1 B_2^2 A_1 + C_1 B_2 A_2 B_1 - A_1^2 B_2^2 + 2\, A_1 B_2 A_2 B_1 - A_2^2 B_1^2 = 0 \end{aligned} \tag{6.8}$$

which is the equation for the four-bar linkage $ABCD$, *as applicable to pose* j. In this equation, the coefficients A_i, B_i, C_i, $i = 1, 2$, are configuration-dependent, but free of ϕ_j. Apparently, these equations are dependent only on the translations, but not on the rotations of link ACP of Fig. 6.1. In this light, the dyad-synthesis problem is converted into one of *path generation*, from one of motion generation, or rigid-body guidance. On the other hand, the synthesis of dyad EF is still a problem of *rigid-body guidance*.

6.3 Extension to Stephenson-I and II Linkages

The foregoing procedure, introduced for the synthesis of the Stephenson-III linkage, is now applied to the other two types of Stephenson linkages, as described presently. Given that the two linkages studied in this section bear topologies different from their

Stepenson-III counterpart, additional variables of linkage rotation are needed, if we follow exactly the same procedure. This will naturally increase the complexity of the problem. To avoid this issue, alternative coordinates are used in the formulation. Below we synthesize the Stephenson-II linkage to demonstrate the formulation.

6.3.1 Stephenson-II Linkage

The linkage of interest can be constructed based on two dyads, AB and EF, together with a third dyad, CD, that connects links AB and EP. For dyads AB and EF, the Burmester equation can be directly applied, which leads, correspondingly, to

$$AB \text{ dyad: } \mathbf{b}^T(\mathbf{1}-\mathbf{Q}_j)\mathbf{a}_0+\mathbf{r}_j^T\mathbf{Q}_j\mathbf{a}_0-\mathbf{r}_j^T\mathbf{b}+\frac{1}{2}\mathbf{r}_j^T\mathbf{r}_j=0\,, \quad j=1,\ldots,m \tag{6.9a}$$

$$EF \text{ dyad: } \mathbf{f}^T(\mathbf{1}-\mathbf{R}_j)\mathbf{e}_0+\mathbf{r}_j^T\mathbf{R}_j\mathbf{e}_0-\mathbf{r}_j^T\mathbf{f}+\frac{1}{2}\mathbf{r}_j^T\mathbf{r}_j=0, \quad j=1,\ldots,m \tag{6.9b}$$

where $\mathbf{Q}_j$ and $\mathbf{R}_j$ are the rotation matrices for links AP and EP, respectively.

Link CD provides additional kinematic constraints to the kinematic chain $BAPEF$, which lead to the desired *single-dof mechanism*. For link CD, the equation below applies

$$\|(\mathbf{c}_j-\mathbf{d}_j\|^2=\|\mathbf{c}_0-\mathbf{d}_0\|^2\,, \qquad \text{for } j=1,\ldots,m \tag{6.10}$$

where $\mathbf{d}_j=\mathbf{r}_j+\mathbf{R}_j\mathbf{d}_0$.

The jth position vector of point C, $\mathbf{c}_j$, at the jth pose, cannot be simply expressed in the same way as that of point D, as the orientation of link AB is unknown. However, since point C lies on the same rigid body AB, its coordinates can be parametrized in terms of those of A and B, as described presently.

6.3.2 Parameterized Coordinates of Point D

The parametrized coordinates are applicable to ternary links, for which the position vector of a point on the link can be obtained as a linear combination of those of any other two points of the same link. We take the ternary link ABC of Fig. 6.4 to describe the coordinate parametrization. Assuming that the coordinates of points A and B are known, as is the case when this is the input link, the position vector of point C_j can be expressed, at the jth linkage posture, as

$$\mathbf{c}_j=\mathbf{b}+\alpha\mathbf{u}_j+\beta\mathbf{v}_j \tag{6.11}$$

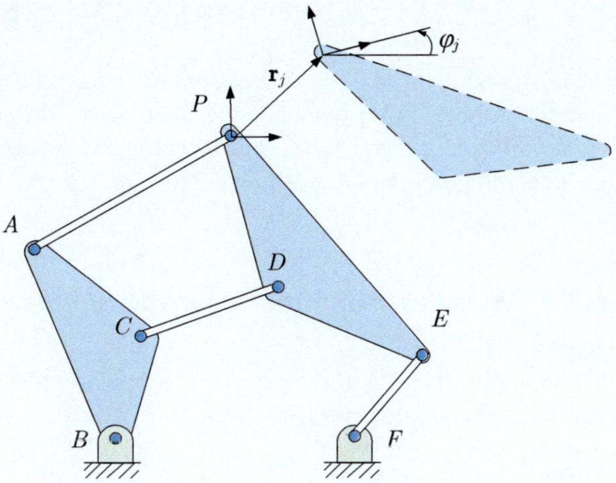

Fig. 6.4 The Stephenson-II linkage, where dyads AB, CD, and EF are to be synthesized

where $\mathbf{u}_j$ and $\mathbf{v}_j$ are orthogonal vectors attached to link ABC, thereby forming an orthogonal basis for the $\mathbb{R}^2$ space. However, these vectors *need not be of unit magnitude*. Moreover, α and β are dimensionless parameters. In this light, we define the two foregoing vectors as

$$\mathbf{u}_j = \mathbf{a}_j - \mathbf{b}; \quad \mathbf{v}_j = \mathbf{E}\mathbf{u}_j \tag{6.12}$$

Equation (6.11) thus becoming

$$\mathbf{c}_j = \mathbf{b} + \alpha(\mathbf{a}_j - \mathbf{b}) + \beta\mathbf{E}(\mathbf{a}_j - \mathbf{b}) \tag{6.13}$$

That is,

$$\mathbf{c}_j = (\alpha\mathbf{1} + \beta\mathbf{E})\mathbf{a}_j + [(1-\alpha)\mathbf{1} - \beta\mathbf{E}]\,\mathbf{b} \tag{6.14}$$

or, in compact form,

$$\mathbf{c}_j = \mathbf{M}\mathbf{a}_j + \mathbf{N}\mathbf{b} \tag{6.15}$$

with

$$\mathbf{M} = \alpha\mathbf{1} + \beta\mathbf{E}; \quad \mathbf{N} = (1-\alpha)\mathbf{1} - \beta\mathbf{E} \tag{6.16}$$

where, apparently,

$$\mathbf{M} + \mathbf{N} = \mathbf{1} \tag{6.17}$$

Moreover, notice that $\mathbf{a}_j = \mathbf{r}_j + \mathbf{Q}_j\mathbf{a}_0$, and hence,

$$\mathbf{c}_j = \mathbf{M}(\mathbf{r}_j + \mathbf{Q}_j\mathbf{a}_0) + \mathbf{N}\mathbf{b} \tag{6.18}$$

or

$$\mathbf{c}_j = \mathbf{M}\mathbf{Q}_j\mathbf{a}_0 + \mathbf{M}\mathbf{r}_j + \mathbf{N}\mathbf{b} \tag{6.19}$$

the position vector of C_j thus being expressed as a linear combination of the position vectors of three[2] other points on the same body. Moreover, of these points, A and C are *circular-*, B a *centerpoint*. The foregoing parametrization simplifies significantly the constraint equations, as shown below.

6.3.3 Synthesis Equations for the Stephenson-II Linkage

With the introduction of the parametrized position vector of point C_j, in Eqs. (6.19) and (6.10) is now rewritten as

$$\|\mathbf{M}\mathbf{Q}_j\mathbf{a}_0 + \mathbf{M}\mathbf{r}_j + \mathbf{N}\mathbf{b} - (\mathbf{r}_j + \mathbf{R}_j\mathbf{d}_0)\|^2 = \|\mathbf{M}\mathbf{a}_0 + \mathbf{N}\mathbf{b} - \mathbf{d}_0\|^2\,, \quad \text{for } j = 1, \ldots, m \tag{6.20}$$

Upon expansion and simplification, the above equation yields

$$\begin{aligned}(\mathbf{M}\mathbf{Q}_j\mathbf{a}_0)^T\left[\mathbf{N}(\mathbf{b}-\mathbf{r}_j) - \mathbf{R}_j\mathbf{d}_0\right] - \left[\mathbf{N}(\mathbf{b}-\mathbf{r}_j)\right]^T\mathbf{R}_j\mathbf{d}_0 - (\mathbf{N}\mathbf{b})^T\mathbf{N}\mathbf{r}_j \\ -(\mathbf{M}\mathbf{a}_0)^T\mathbf{N}\mathbf{b} + (\mathbf{M}\mathbf{a}_0)^T\mathbf{d}_0 + (\mathbf{N}\mathbf{b})^T\mathbf{d}_0 + (\mathbf{N}\mathbf{r}_j)^T(\mathbf{N}\mathbf{r}_j)/2 = 0\end{aligned} \tag{6.21}$$

Notice that Eq. (6.17) was used in the above derivation. Upon substitution of $\mathbf{Q}_j = c_j\mathbf{1} + s_j\mathbf{E}$ into Eq. (6.21), this equation leads to

$$A_2c_j + B_2s_j + C_2 = 0 \tag{6.22a}$$

with coefficients

$$A_2 = (\mathbf{M}\mathbf{a}_0)^T\left[\mathbf{N}(\mathbf{b}-\mathbf{r}_j) - \mathbf{R}_j\mathbf{d}_0\right] \tag{6.22b}$$

$$B_2 = (\mathbf{M}\mathbf{E}\mathbf{a}_0)^T\left[\mathbf{N}(\mathbf{b}-\mathbf{r}_j) - \mathbf{R}_j\mathbf{d}_0\right] \tag{6.22c}$$

$$\begin{aligned}C_2 = &-\left[\mathbf{N}(\mathbf{b}-\mathbf{r}_j)\right]^T\mathbf{R}_j\mathbf{d}_0 - (\mathbf{N}\mathbf{b})^T\mathbf{N}\mathbf{r}_j \\ &- (\mathbf{M}\mathbf{a}_0)^T\mathbf{N}\mathbf{b} + (\mathbf{M}\mathbf{a}_0)^T\mathbf{d}_0 + (\mathbf{N}\mathbf{b})^T\mathbf{d}_0 + (\mathbf{N}\mathbf{r}_j)^T(\mathbf{N}\mathbf{r}_j)/2\end{aligned} \tag{6.22d}$$

C_2 being further expanded as

$$\begin{aligned}C_2 = &-\left[\mathbf{N}(\mathbf{b}-\mathbf{r}_j)\right]^T\mathbf{R}_j\mathbf{d}_0 - (\mathbf{M}\mathbf{a}_0)^T\mathbf{N}\mathbf{b} + (\mathbf{M}\mathbf{a}_0)^T\mathbf{d}_0 + (\mathbf{N}\mathbf{b})^T\mathbf{d}_0 \\ &+ [(1-\alpha)^2 + \beta^2](-\mathbf{r}_j^T\mathbf{b} + \mathbf{r}_j^T\mathbf{r}_j/2)\end{aligned} \tag{6.23}$$

which has a form similar to that of the Stephenson-III linkage, Eq. (6.6a). Equations (6.3), (6.5a), and (6.22a) constitute the system of synthesis equations for the Stephenson-II linkage.

[2] Two points suffice in the plane, the use of three streamlines the formulation.

6.3.4 Stephenson-I Linkage

Compared with the Stephenson-II linkage, its Stephenson-I counterpart, shown in Fig. 6.5, differs only in that joint D is located on the turning link EF, rather than on the floating link EP. The position vector of point D needs to be parametrized as well. The corresponding analysis follows that of the Stephenson-II linkage. To this end, we express $\mathbf{d}_j$ as a *generalized*[3] linear combination of $\mathbf{e}_j$ and $\mathbf{f}$, namely,

$$\mathbf{d}_j = \mathbf{U}\mathbf{e}_j + \mathbf{V}\mathbf{f} \tag{6.24}$$

with the 2×2 matrices $\mathbf{U}$ and $\mathbf{V}$ defined as

$$\mathbf{U} \equiv \mu\mathbf{1} + \nu\mathbf{E}, \quad \mathbf{V} \equiv (1-\mu)\mathbf{1} - \nu\mathbf{E} \tag{6.25}$$

where μ and ν are dimensionless parameters. Notice that, $\mathbf{U} + \mathbf{V} = \mathbf{1}$, $\mathbf{d}_j$ of Eq. (6.24) then reducing to a simple *linear combination* of vectors $\mathbf{e}_j$ and $\mathbf{f}$.

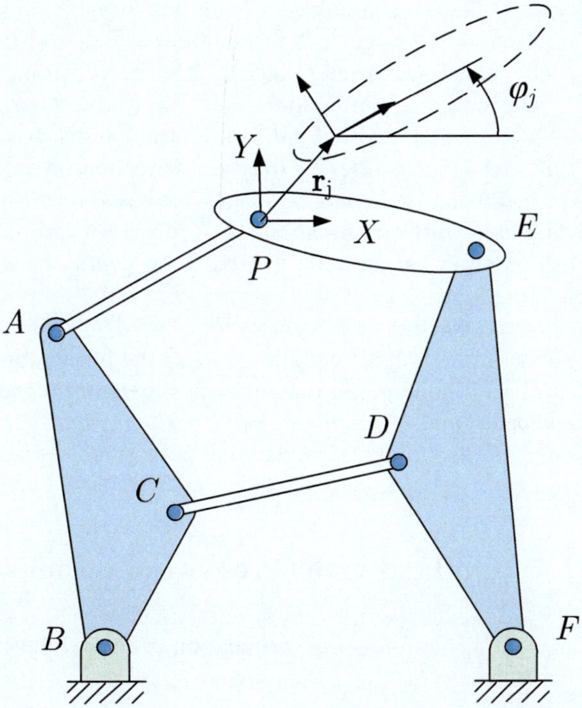

Fig. 6.5 The Stephenson-I linkage, where dyads AB, CD, EF, are to be synthesized

[3] In the usual case, $\mathbf{U}$ and $\mathbf{V}$ in Eq. (6.25) are replaced by different scalars, which can be interpreted as these two matrices are distinct multiples of the pertinent identity matrix.

Substituting Eqs. (6.15) and (6.24) into Eq. (6.10) leads to

$$\|\mathbf{MQ}_j\mathbf{a}_0 + \mathbf{Mr}_j + \mathbf{Nb} - (\mathbf{UR}_j\mathbf{e}_0 + \mathbf{Ur}_j + \mathbf{Vf})\|^2 = \|\mathbf{Ma}_0 + \mathbf{Nb} - \mathbf{Ue}_0 - \mathbf{Vf}\|^2 \tag{6.26}$$

After expanding the above equation, and simplifying the equation thus resulting, this equation attains the form of Eq. (6.22a), with the coefficients given below

$$A_2 = (\mathbf{Ma}_0)^T(\mathbf{Pr}_j + \mathbf{Nb} - \mathbf{UR}_j\mathbf{e}_0 - \mathbf{Vf}) \tag{6.27a}$$

$$B_2 = (\mathbf{MEa}_0)^T(\mathbf{Pr}_j + \mathbf{Nb} - \mathbf{UR}_j\mathbf{e}_0 - \mathbf{Vf}) \tag{6.27b}$$

$$\begin{aligned} C_2 = &(\mathbf{Nb} - \mathbf{UR}_j\mathbf{e}_0 - \mathbf{Vf})^T\mathbf{Pr}_j - (\mathbf{Ma}_0)^T(\mathbf{Nb} - \mathbf{Ue}_0 - \mathbf{Vf}) \\ &-(\mathbf{Vf})^T\mathbf{U}(\mathbf{1} - \mathbf{R}_j)\mathbf{e}_0 + (\mathbf{Nb})^T\mathbf{U}(\mathbf{1} - \mathbf{R}_j)\mathbf{e}_0 \\ &+[(\alpha - \mu)^2 + (\beta - \nu)]\mathbf{r}_j^T\mathbf{r}_j/2 \end{aligned} \tag{6.27c}$$

where $\mathbf{P} = \mathbf{M} - \mathbf{U}$.

Equations (6.3), (6.5a), and (6.22a), with the coefficients A_2, B_2, C_2 defined above, compose the system of synthesis equations for the Stephenson-I linkage.

So far, we have formulated a unified system of synthesis equations for the three types of Stephenson linkages. With this formulation, all the three types can be synthesized with the standard formulation of Eqs. (6.3), (6.5a) and (6.6a), where only coefficients A_2, B_2, and C_2 need to change accordingly.

The foregoing formulation offers some advantages over the formulation based on loop-closure equations. Firstly, our formulation is based always on three dyads, AB, CD, and EF, regardless of the type of Stephenson linkage at play. On the contrary, in the methods based on loop-closure one has to formulate the equations separately for each loop with varying numbers of links. For example, the Stephenson-III linkage has one four-bar and one five-bar loop, while the Stephenson-II linkage has two five-bar loops.

Secondly, the new formulation, based on parametrized coordinates, leads to polynomial synthesis equations, while the formulation with loop-closure equations involves trigonometric functions. *A system of polynomial equations usually offers computational advantages over its trigonometric counterpart.* For one thing, the number of solutions of a system of polynomial equations can be predicted!

6.4 Synthesis in the Presence of Incomplete Data Sets

In light of the foregoing formulation of the synthesis equations, we can draw two conclusions:

(a) The number of linkage parameters in three RR dyads, in terms of either relative or parametrized coordinates, amounts to 12.
(b) The total number of synthesis equations equals $2m$.

Apparently, the system admits exact solutions when $m = 6$, i.e., for seven prescribed poses. However, the synthesis equations for dyad EF involve only coordinates of points E_0 and F, which admit exact solutions with four equations, i.e., for at most $m = 4$. In this light, six-bar linkages can admit exact solutions for a maximum of $m = 4$, i.e., for five poses, the same number as in the case of four-bar linkages. On the other hand, for six-bar linkages we have 12 parameters. The six-bar linkage synthesis problem thus leads to an *underdetermined system* of equations. The problem of six-bar linkage synthesis thus admits infinitely many solutions for five given poses.

Underdeterminacy of a system of synthesis equations is always *welcome*, as it leads to *design flexibility*, which allows the designer to freely assign a few design parameters, while exact solutions are still possible. Different design parameters can be assigned. Some possibilities include:

1. Select one of the two dyads, either AB or CD, as given in the data. In this case, the problem requires only finding the other dyad, together with dyad EF. This is the case of synthesis with constrained 3R chains, as reported by Soh and McCarthy (2008).
2. Select and define two grounded points, i.e., centerpoints, B and F. The problem becomes one of finding two circlepoints, A_0 and C_0, together with the EF dyad.
3. Specify any four coordinates for the four points A_0, B, C_0, and D.

Other selection possibilities can be explored (Riva, 1981). Notice that an unlucky selection *might lead to no real solutions*. Thus, how to specify *robustly* additional conditions (constraints) is a challenging problem, that calls for further study.

Since the motion synthesis of a six-bar linkage does not lead to a determined system of synthesis equations, but to an underdetermined one, the exact synthesis of a six-bar linkage is essentially a problem with an *incomplete set of pose data*. As a matter of fact, this type of problem can also be found in other cases, where *topologically* non-symmetric[4] kinematic chains exist. For example, the synthesis of spatial RCCC linkages involves an incomplete data set too, as the RC (revolute-cylindrical) and CC (cylindrical-cylindrical) dyads require different numbers of prescribed poses for exact solutions. The synthesis with incomplete data stands as a special type of problem that calls for further research as well. Given the incompleteness of the data set, additional conditions should be included to obtain a solution, which implies flexibility in the design procedure, to freely choose some design parameters. The *robust selection* of parameters is a challenging problem, which is worth further study.

[4] The *symmetry* here refers *only* to the *type* of kinematic pairs, e.g., RCCC versus RCCR, independent of link dimensions.

Table 6.1 Five given poses

j	$\mathbf{r}_j$ (mm)	φ_j
0	$[0, 0]^T$	0
1	$[-92.0, -10.0]^T$	9.2°
2	$[-195.0, -35.0]^T$	18.8°
3	$[-336.0, -106.5]^T$	32.2°
4	$[-469.4, -344.8]^T$	58.9°

Table 6.2 Solution for the poses given in Table 6.1

Point	Coordinates (mm)	Point	Coordinates (mm)	Point	Coordinates (mm)
B	[−300, −500]*	A_0	[−450.0, −200.0]*	D	[50, −250.0]
F	[100.19, −500.08]	E_0	[500.08, −200.0]	$[\alpha, \beta](C)$	[0.467, −0.40]

* user-specified parameters

Table 6.3 Dimensions of the synthesized Stephenson-II linkage (lengths in mm)

Link	BA	BF	CD	BC	AC	EF	AP	EP	ED	PD
Length	335.4	400.0	304.1	206.3	223.5	500.0	492.4	538.5	452.7	254.9

6.4.1 A Design Example

We provide an example to illustrate the foregoing synthesis method.

Here we consider the synthesis of a Stephenson-II linkage. The set of prescribed poses is given in Table 6.1. As mentioned above, the design flexibility of a six-bar linkage allows us to specify pivoting points. In this example, two such points, B and D, of Fig. 6.5, are given. With Eqs. (6.3) and (6.8), the solution obtained is shown in Table 6.2, with a synthesis error of 4.9×10^{-6}. The corresponding link dimensions are shown in Table 6.3, which led to the linkage illustrated in Fig. 6.6. Notice that this solution was obtained with only points B and D given. Obviously, many feasible solutions are possible, as different pivoting points can be prescribed, which is what we mean by *design flexibility*.

6.5 A 10-Bar Linkage

A single-dof 10-bar linkage is shown in Fig. 6.7b, which derives from the 3-RRR planar linkage of Fig. 6.7a. The mechanism is mounted on a fixed base via revolute joints labeled B_i, for $i = 1, 2, 3$. The coupler link, playing the role of the mobile platform in a *parallel robot*, is a *ternary link*, $A_1A_2A_3$, coupled to the three limbs at corresponding revolute joints A_i, for $i = 1, 2, 3$, in the same figure. The middle joint of limb i is labeled C_i.

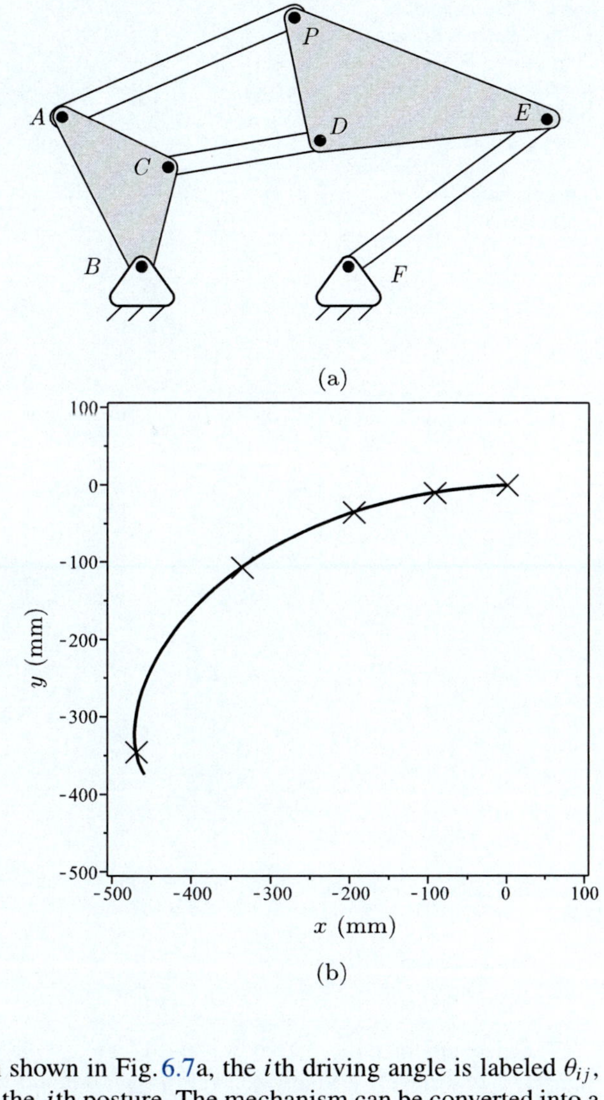

Fig. 6.6 The synthesized Stephenson-II linkage: **a** CAD model, **b** a segment of the coupler curve that includes the prescribed points

In the planar mechanism shown in Fig. 6.7a, the ith driving angle is labeled θ_{ij}, with $i = 1, 2, 3$, j denoting the jth posture. The mechanism can be converted into a 1-dof planar linkage if we enforce all three input angles θ_{ij} to be identical. This can be achieved by introducing two parallelogram linkages, $B_1B_2C_2G$ and $B_2B_3HC_2$, in the kinematic loops 1 and 2 of the mechanism, which leads to a linkage with 13 revolute joints, as shown in Fig. 6.7b. In the new linkage, we have two more ternary links, B_1C_1G and B_3C_3H, as used for the construction of the two parallelograms. The input angles are thus expressed as $\theta_{ij} = \theta_{0i} + \theta_j$, where θ_{0i} is the angle of the ith limb at the reference posture. As such, we have only one rotation variable, θ_j, for all three input links, the mechanism thus being endowed with only one dof. This statement can be verified via the Chebyshev-Grübler-Kutzbach formula.

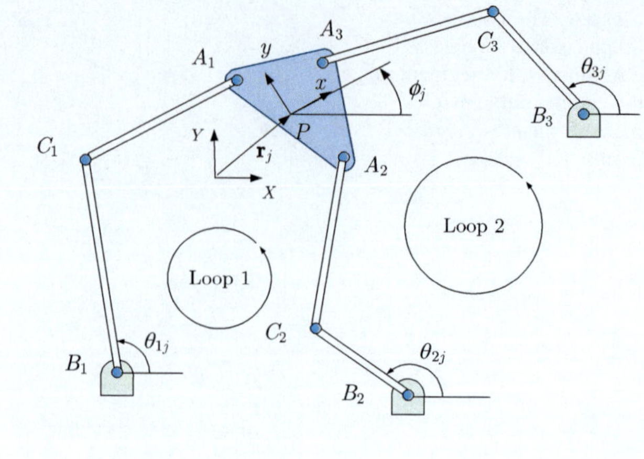

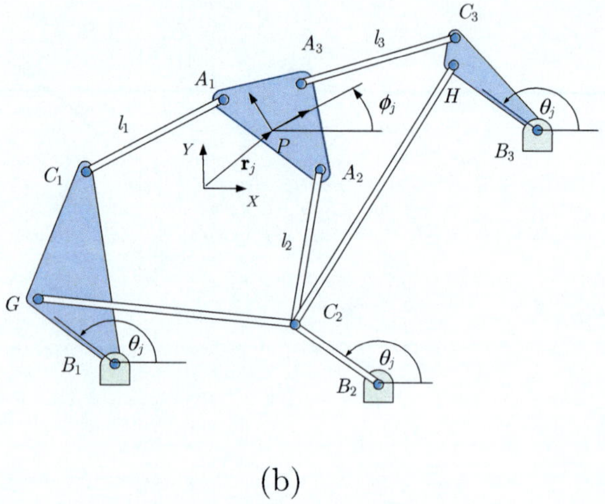

Fig. 6.7 Example of multiloop planar linkages, **a** a three-dof 3-RRR three-dof planar linkage, i.e., a planar three-dof robot; **b** a 10-bar, single-dof planar linkage, whereby the input links are constrained by two parallelograms

The mechanism in Fig. 6.7b is now used for exact motion synthesis. As the coupler link $A_1A_2A_3$ moves, it visits m given poses, defined by $\mathbf{r}_j$ and ϕ_j, $j = 0, \ldots, m$. All displacements are given with respect to the reference pose $\mathbf{r}_0 = \mathbf{0}$ and $\phi_0 = 0$. At this pose, the angle of rotation is $\theta_j \equiv \theta_0 = 0$.

It is noted that the mechanism in this procedure is obtained by constraining a 3-RRR planar parallel linkage. In this way, the degree of freedom is reduced from 3 to 1. Moreover, due to the presence of constraining parallelograms, the 10-bar planar mechanism thus obtained has a reduced design space when compared with its general 10-bar counterpart, e.g., the linkage proposed by Korendiy (2017).

6.6 Synthesis for *m* Prescribed Poses

Figure 6.8 shows one single limb of the 3-RRR mechanism of Fig. 6.7a. Without loss of generality, the mechanism under synthesis carries an arbitrary architecture, *not necessarily symmetric*.

Let dyad A_1C_1 have a length l_1, i.e.,

$$(\mathbf{a}_{1j} - \mathbf{c}_{1j})^T(\mathbf{a}_{1j} - \mathbf{c}_{1j}) = l_1^2, \quad j = 1, \ldots, m \tag{6.28}$$

$\mathbf{a}_{1j}$ and $\mathbf{c}_{1j}$ denoting the position vectors of joints A_1 and C_1 at the jth posture, l_1 being the length of link A_1C_1. The above relation is applicable to the other two *floating* dyads, A_2C_2 and A_3C_3, of Fig. 6.7b, of lengths l_2 and l_3, respectively. Thus, Eq. (6.28) is rewritten as

$$(\mathbf{a}_{ij} - \mathbf{c}_{ij})^T(\mathbf{a}_{ij} - \mathbf{c}_{ij}) = l_i^2, \quad i = 1, 2, 3, \quad j = 1, \ldots, m \tag{6.29}$$

with

$$\mathbf{a}_{ij} = \mathbf{r}_j + \mathbf{Q}_j\mathbf{a}_{0i} \tag{6.30a}$$
$$\mathbf{c}_{ij} = \mathbf{b}_i + \mathbf{R}_j\mathbf{c}_i' \tag{6.30b}$$
$$\mathbf{c}_i' = \mathbf{c}_{0i} - \mathbf{b}_i \tag{6.30c}$$

where $\mathbf{Q}_j = \mathbf{Q}(\phi_j)$ and $\mathbf{R}_j = \mathbf{R}(\theta_j)$ are 2×2 rotation matrices, while $\mathbf{a}_{0i}$ and $\mathbf{c}_{0i}$ are the position vectors of joints A_i and C_i at the reference posture. Moreover, the lengths of the three distal links A_iC_i are determined from the reference posture as

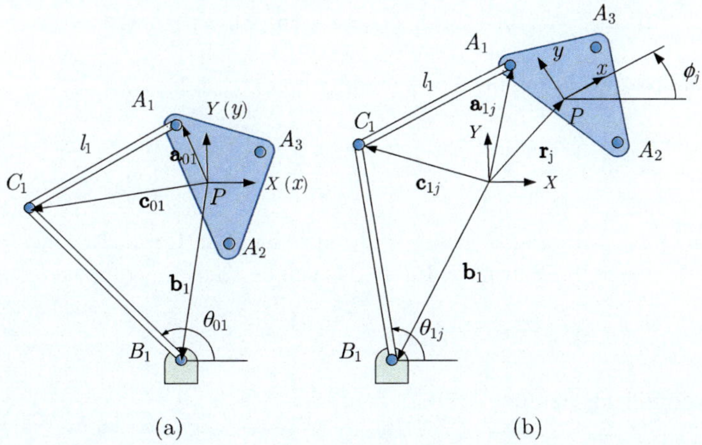

Fig. 6.8 A single limb showing **a** the reference and **b** the jth pose

$$l_i = ||\mathbf{a}_{0i} - \mathbf{c}_{0i}||, \quad i = 1, 2, 3 \tag{6.31}$$

Equation (6.29) represents a system of $3m$ equations. On the other hand, the unknowns of the system include both design parameters and motion variables. The design parameters are $\mathbf{a}_{0i}$, $\mathbf{c}_{0i}$ and $\mathbf{b}_i$, for $i = 1, 2, 3$, which amount to 18 scalar unknowns to be determined. The motion variable θ_j leads to m parameters, for $m + 1$ prescribed postures. The number m of postures for exact synthesis should thus satisfy:

$$3m = 18 + m \tag{6.32}$$

which leads to $m = 9$. This means that the linkage can visit exactly up to 10 prescribed postures, as described in Sect. 6.7

6.7 Solution of the 10-Posture Synthesis Equations

The system of synthesis equations (6.29) comprises both trigonometric and polynomial functions, in terms of design and motion variables, whose solution brings about numerical challenges. As the motion variables θ_j are auxiliary, we eliminate them to simplify the formulation, as described below.

First, Eq. (6.29) is expanded:

$$(\mathbf{r}_j + \mathbf{Q}_j\mathbf{a}_{0i} - \mathbf{b}_i - \mathbf{R}_j\mathbf{c}_i')^T(\mathbf{r}_j + \mathbf{Q}_j\mathbf{a}_{0i} - \mathbf{b}_i - \mathbf{R}_j\mathbf{c}_i') = l_i^2 \quad i = 1, 2, 3 \tag{6.33}$$

with $\mathbf{c}_i'$ defined in Eq. (6.30c). Moreover, given that $\mathbf{R}_j = \cos\theta_j\mathbf{1} + \sin\theta_j\mathbf{E}$, Eq. (6.33), after further expansion, becomes

$$\mathbf{d}_{ij}^T\mathbf{d}_{ij} - \mathbf{c}_i'^T\mathbf{c}_i' - 2\mathbf{d}_{ij}^T\mathbf{c}_i'\cos\theta_j - 2\mathbf{d}_{ij}^T\mathbf{E}\mathbf{c}_i'\sin\theta_j - l_i^2 = 0 \tag{6.34}$$

where $\mathbf{1}$, denotes, as usual, the 2×2 identity matrix, while

$$\mathbf{d}_{ij} = \mathbf{r}_j + \mathbf{e}_{ij}, \quad \text{and} \quad \mathbf{E} = \begin{bmatrix} 0 & -1 \\ 1 & 0 \end{bmatrix} \tag{6.35}$$

with $\mathbf{e}_{ij} \equiv \mathbf{Q}_j\mathbf{a}_{0i} - \mathbf{b}_i$ and $\mathbf{E}$ having been introduced in Eq. (1.1a), recalled above for quick reference. Furthermore, Eq. (6.34) can be recast in compact form as

$$A_i\cos\theta_j + B_i\sin\theta_j + C_i = 0, \quad i = 1, 2, 3 \tag{6.36a}$$

with the coefficients below

$$A_i = \mathbf{d}_{ij}^T \mathbf{c}_i' \tag{6.36b}$$

$$B_i = \mathbf{d}_{ij}^T \mathbf{E} \mathbf{c}_i' \tag{6.36c}$$

$$\begin{aligned} C_i &= (l_i^2 - \mathbf{d}_{ij}^T \mathbf{d}_{ij} - \mathbf{c}_i'^T \mathbf{c}_i')/2 \\ &= (l_i^2 - \mathbf{r}_j^T \mathbf{r}_j - \mathbf{c}_i'^T \mathbf{c}_i' - \mathbf{e}_{ij}^T \mathbf{e}_{ij})/2 - \mathbf{r}_j^T \mathbf{e}_{ij} \end{aligned} \tag{6.36d}$$

For the jth posture, Eq. (6.36a) yields a system of three linear equations in $\cos\theta_j$ and $\sin\theta_j$, for $j = 1, \ldots, m$, namely,

$$A_1 \cos\theta_j + B_1 \sin\theta_j + C_1 = 0 \tag{6.37a}$$

$$A_2 \cos\theta_j + B_2 \sin\theta_j + C_2 = 0 \tag{6.37b}$$

$$A_3 \cos\theta_j + B_3 \sin\theta_j + C_3 = 0 \tag{6.37c}$$

which are the constraint equations of the multiloop linkage under study. These equations can be used directly for synthesis. On the other hand, the equations can be further simplified, as described presently.

We notice that the three foregoing constraint equations include a quadratic term $\mathbf{r}_j^T \mathbf{r}_j$ in the expression for C_i in Eq. (6.36d). The system of equations can thus be simplified upon elimination of this quadratic term from two out of the three equations. Subtracting Eq. (6.37c) from Eqs. (6.37a) and (6.37b) yields, respectively,

$$A_1' \cos\theta_j + B_1' \sin\theta_j + C_1' = 0 \tag{6.38a}$$

$$A_2' \cos\theta_j + B_2' \sin\theta_j + C_2' = 0 \tag{6.38b}$$

where

$$A_i' = A_i - A_3, \quad B_i' = B_i - B_3, \quad C_i' = C_i - C_3, \quad i = 1, 2 \tag{6.39}$$

The *motion variable* θ_j is now eliminated. From Eqs. (6.37c) and (6.38a), we obtain

$$\cos\theta_j = \frac{B_1' C_3 - C_1' B_3}{A_1' B_3 - A_3 B_1'}; \quad \sin\theta_j = -\frac{A_1' C_3 - A_3 C_1'}{A_1' B_3 - A_3 B_1'} \tag{6.40}$$

Substituting the above expressions into the identity $\sin^2\theta_j + \cos^2\theta_j = 1$ yields

$$(B_1' C_3 - C_1' B_3)^2 + (A_1' C_3 - A_3 C_1')^2 - (A_1' B_3 - A_3 B_1')^2 = 0 \tag{6.41}$$

We can obtain likewise one similar equation from Eqs. (6.37c) and (6.38b), namely,

$$(B_2' C_3 - C_2' B_3)^2 + (A_2' C_3 - A_3 C_2')^2 - (A_2' B_3 - A_3 B_2')^2 = 0 \tag{6.42}$$

Equations (6.41) and (6.42) are applicable to the j-th pose. Proceeding similarly for all poses, $j = 1, \ldots, 9$, we finally obtain 18 equations containing design variables only. Upon assembling all these equations, a system of 18 polynomial equations in 18 unknowns is obtained. The system admits exact solutions to generate linkages capable of visiting all 10 prescribed postures. The same conclusion can be reached if equations (6.36a) are used for synthesis, with extra variables θ_j.

Once all joint coordinates at the reference pose are found from the synthesis equations, the coordinates of points G and H for constructing the two parallelograms are readily found. Indeed, taking the parallelograms of Fig. 6.7b, for example, the position vectors $\mathbf{g}$ and $\mathbf{h}$ are obtained as

$$\mathbf{g} = \mathbf{b}_1 + \mathbf{c}_2'; \quad \mathbf{h} = \mathbf{b}_3 + \mathbf{c}_2'; \tag{6.43}$$

6.8 A Synthesis Example

Below we include an example of exact synthesis and analysis for the 10-bar, 1-dof planar linkage. Prior to solving the synthesis equations, we *normalize* the displacements by means of a *characteristic length*. This is intended to reduce the effect of data disparate dimensions on round-off error amplification. For the problem at hand, we take the rms value of the translation Euclidian norms in the data as the characteristic length r, namely,

$$r = \sqrt{\frac{1}{m}\sum_{j=1}^{m} ||\mathbf{r}_j||^2} \tag{6.44}$$

the normalized displacement vectors thus becoming

$$\rho = \mathbf{r}_j / r \tag{6.45}$$

Linkage synthesis was conducted combining computer-algebra and number-crunching software,[5] with solver *fsolve* requiring an initial guess for all the unknows. In our implementation, these values are generated randomly from a predefined, bounded subset of the design space, upon which the solver returns a solution, *if one exists at all*. If the solution is feasible, it is further verified graphically and analyzed by means of computer algebra for order- and branch-defect verification. A solution is considered feasible if it is neither trivial, such as a zero link-length, nor contains two or three overlapping limbs.[6] Unfeasible solutions, i.e., linkages with order or branch defect, for example, will be discarded. The algorithm continues searching for a new solution until a defect-free linkage is obtained. The implementation of the linkage-synthesis algorithm is illustrated in Fig. 6.9.

---

[5] Matlab 17.0 and Maple 18. The former was used to find numerical solutions via the equation solver *fsolve*, the latter for symbolic calculation.

[6] As a matter of fact, in the realm of planar linkages, overlapping links can be readily accommodated, upon placing the links at stake in parallel planes.

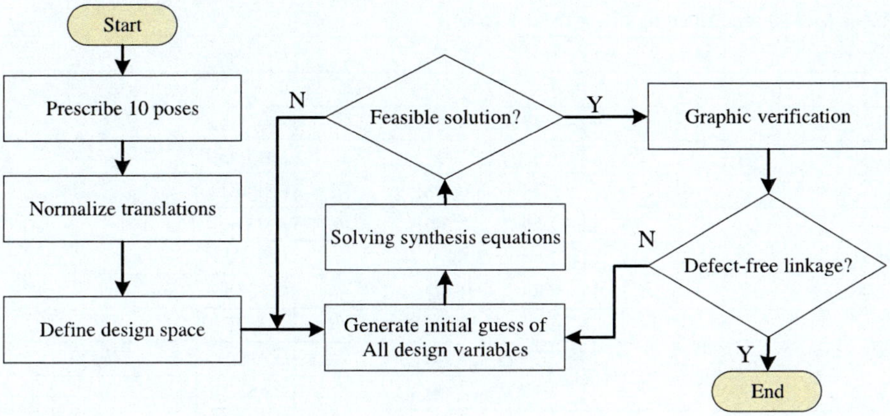

Fig. 6.9 Flowchart for 10-pose exact synthesis

In linkage synthesis, detection and elimination of linkage-defect is necessary to obtain a design capable of functioning as prescribed. Effective methods are needed to eliminate linkage designs with order and/or branch defect (Balli and Chand, 2002). In this book, we resort to *visual* means to verify the mechanism motion, which tests and verifies linkage feasibility to visit all postures in the prescribed order and free of branching defect. In doing so, we generate coupler curves using symbolic software to verify that all poses are located on one single branch. Once this is done, a CAD model is built to further verify the motion of the mechanism and ensure that all poses are visited in the prescribed order.

The coupler curve for defect-verification can be obtained from the intersection of the plots of Eqs. (6.41) and (6.42). However, their intersection brings about extraneous solutions, due to the presence of fractions with higher-order polynomials in their denominators—see Eq. (6.40). To avoid extraneous solutions, we resort to one more constraint, namely,

$$\det(\mathbf{M}) = 0 \tag{6.46}$$

with

$$\mathbf{M} = \begin{bmatrix} A_1' & B_1' & C_1' \\ A_2' & B_2' & C_2' \\ A_3 & B_3 & C_3 \end{bmatrix} \tag{6.47}$$

The surface represented by Eq. (6.46) is the *manifold* of all poses of the coupler link. Its intersection with the surface generated from Eq. (6.41) yields a 3D curve representing both the trajectory of point P and the coupler-link orientation. In this procedure, a software command[7] is utilized to generate the 3D curve from two equations.

[7] Maple's `intersection`.

Table 6.4 10 prescribed poses for the example

Pose	$\mathbf{r}_j^T$ (m)	ϕ_j (rad)
0	[0,0]	0
1	[−0.4054, 1.0531]	−0.0461
2	[−1.059, 2.0436]	−0.0759
3	[−2.2163, 3.0428]	−0.0814
4	[−3.5686, 3.5746]	−0.0430
5	[−4.6006, 3.62517]	0.0162
6	[−5.846, 3.2129]	0.1338
7	[−6.622, 2.5089]	0.2535
8	[−7.1395, 1.4513]	0.3956
9	[−7.2702, −0.0024]	0.5542

Table 6.5 Two synthesis results in terms of joint locations (m)

	i	$\mathbf{a}_{0i}^T$	$\mathbf{b}_i^T$	$\mathbf{c}_{0i}^T$
M1	1	[−1.5, 0]	[−7.062, 0.5149]	[−2.860, 3.2246]
	2	[2.0, 0]	[−2.1603, −3.2858]	[1.8291, −2.995]
	3	[0, 1.732]	[−1.7155, 2.4390]	[2.6189, 3.1953]
M2	1	[1.5774, 4.7255]	[−2.0036, 2.0849]	[−2.3191, 5.1655]
	2	[6.2113, 2.7950]	[−0.780, −2.6886]	[4.9965, 3.1993]
	3	[8.5986, −1.4432]	[−0.9712, −5.5318]	[8.0299, −2.0142]

The 10 poses for our example are given in Table 6.4. The characteristic length is calculated with the data value as $r = 5.30$ m, with which all displacements are normalized.

Following our synthesis procedure, solutions are obtained. Table 6.5 lists two solutions, their corresponding linkages being denoted M1 and M2. The synthesis error, defined by the rms value of the residuals of all synthesis equations, is equal to 6.703×10^{-5}m for M1, and 6.95×10^{-4}m for M2. The two linkages, M1 and M2, are displayed in Figs. 6.10 and 6.11, respectively, together with the 10 prescribed poses. In the two figures, solid squares stand for grounded joints, while small circles represent moving revolute joints.

For linkage M2, shown in Fig. 6.11, apparently the ground joint B_2 interferes with the ternary link B_3C_3H. Interference occurs only in the figure, as the designer need not locate all the links in the same plane!

While only two solutions are provided in this example, it is noted that the system of synthesis equations admits multiple solutions. The cardinality of the solution space can be obtained by means of the Bezout number, which provides an upper bound for the number of possible solutions. Notice that the system of constraint equations contains 18 octic equations with 18 variables, which implies a Bezout

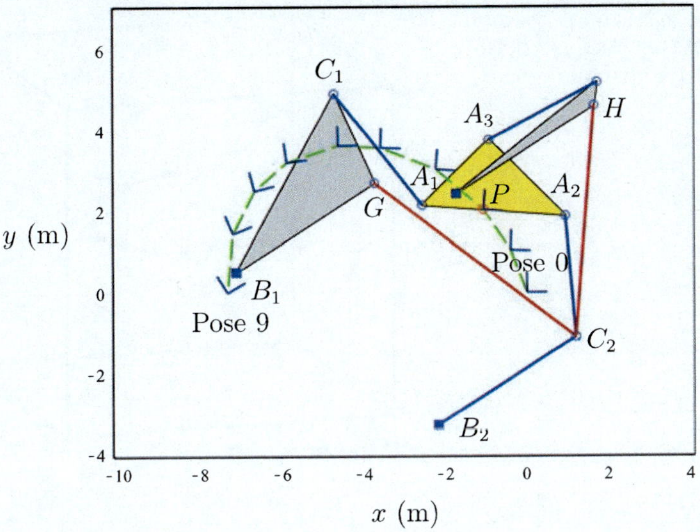

Fig. 6.10 Linkage M1 synthesized in the example

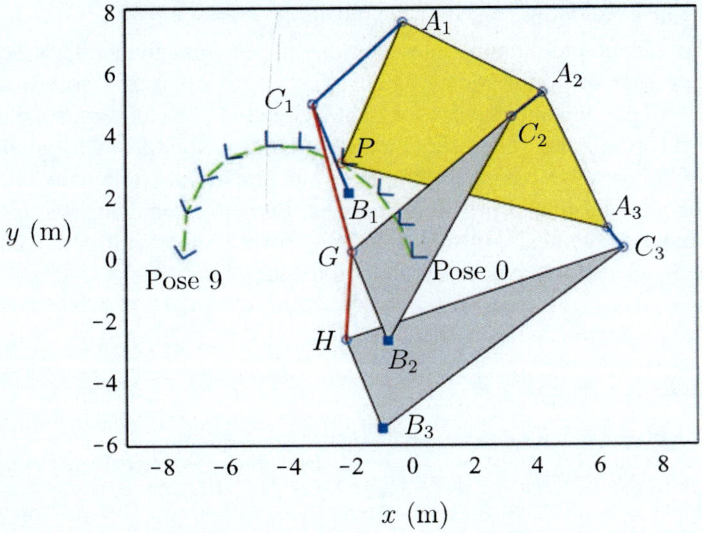

Fig. 6.11 Linkage M2 synthesized in the example

number of $8^{18} \doteq 1.8 \times 10^{16}$. Given such an extremely high upper bound, equation solving remains a challenging numerical problem.

The coupler curve of M1 is shown in Fig. 6.12. In this figure, two closed curves can be identified, which implies that the linkage has two branches. Moreover, in this

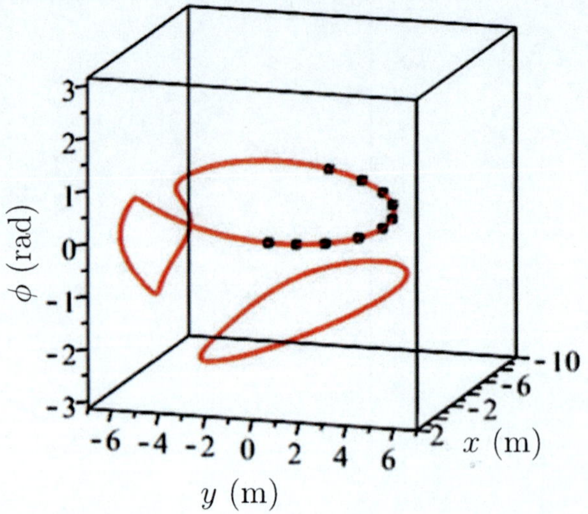

Fig. 6.12 Coupler-curve display of M1, all black dots indicating the prescribed poses

example, luckily, all prescribed points lie in one and the same branch. Had this not been the case, a new initial guess should be tried.

The above example demonstrates that the gist of linkage synthesis lies in how to formulate the constraint equations *effectively* and *concisely*. More examples of synthesis with this formulation are available (Bai et al., 2020). It must be noted that an effective formulation guarantees that the synthesis equations will yield the right results, while a concise formulation reduces the numerical complexity in equation-solving. As pointed out in previous chapters, numerical challenges remain in the realm of linkage synthesis. In this light, new formulations and new solutions are continuously needed to advance the state of the art.

References

Bai, S., Li, Z., & Li, R. (2020). Exact synthesis and input-output analysis of 1-dof planar linkages for visiting 10 poses. *Mechanism and Machine Theory, 143*, 103625.

Bai, S., Wang, D., & Dong, H. (2016). A unified formulation for dimensional synthesis of Stephenson linkages. *ASME Journal of Mechanisms and Robotics, 8*(4), 041009.

Balli, S. S., & Chand, S. (2002). Defects in link mechanisms and solution rectification. *Mechanism and Machine Theory, 37*(9), 851–876.

Cavic, M., Pencic, M., Rackov, M., Doric, J., & Lu, Z. (2019). Graphical synthesis of 6-bar dwell linkage mechanism. *FME Transactions, 47*, 226–233.

Dong, H., & Wang, D. (2007). New approach for optimum synthesis of six-bar dwell mechanisms by adaptive curve fitting. In *Proceedings of Twelfth World Congress in Mechanism and Machine Science* (pp. 17–21).

Kim, H., Hamid, S., & Soni, A. (1972). Synthesis of six-link mechanisms for point path generation. *Journal of Mechanisms, 6*(4), 447–461.

Korendiy, V. (2017). Structural and kinematic synthesis of the 1-dof eight-bar walking mechanism with revolute kinematic pairs. *Ukrainian Journal of Mechanical Engineering and Materials Science, 3*(2), 88–102.

McLarnan, C. W. (1963). Synthesis of six-link plane mechanisms by numerical analysis. *ASME Journal of Engineering for Industry, 85*(1), 5–10.

Norton, R. (2004). *Design of machinery: An introduction to the synthesis and analysis of mechanisms and machines*. McGraw-Hill Higher Education.

Plecnik, M., McCarthy, J., & Wampler, C. (2014). Kinematic synthesis of a Watt-I six-bar linkage for body guidance. In J. Lenarčič & O. Khatib (Eds)., *Advances in robot kinematics* (pp. 317–325). Springer.

Riva, R. (1981). Synthesis of Stephenson's six-bar linkages by an interactive technique. *Meccanica, 16*(3), 157–166.

Shen, Q., Lee, W.-T., Russell, K., & Sodhi, R. S. (2008). On motion generation of Watt I mechanisms for mechanical finger design. *Transactions of the CSME/de la SCGM, 32*(3–4), 411–421.

Soh, G. S., & McCarthy, J. M. (2008). The synthesis of six-bar linkages as constrained planar 3R chains. *Mechanism and Machine Theory, 43*(2), 160–170.

Appendix A
A Summary of Dual Algebra

The algebra of dual numbers is recalled here, with extensions to vector and matrix operations. Most materials in this chapter, except for A.3, are reproduced from a chapter in a *NATO Advanced Study Institute book*.[1] However, the material in the foregoing reference has been thoroughly revised herein.

A.1 Introduction

The aim of this Appendix is to outline the applications of dual algebra to kinematic analysis. To this end, the algebra of dual scalars, vectors, and matrices is first recalled. The applications refer to the computation of the parameters of both (i) the screw of a rigid body between *two finitely separated poses*, and (ii) the *instant screw*. However, the applications of dual numbers in kinematics go beyond that. Indeed, the well-known *Principle of Transference* (Bottema & Roth, 1978; Dimentberg, 1965; Rico Martinez & Duffy, 1993) has been found extremely useful in spatial kinematics, since it allows the derivation of spatial kinematic relations by simply *dualizing* the corresponding relations of spherical kinematics.

Dual numbers were first proposed by Clifford (1873), their applications to kinematics being attributed to both (Kotelnikov, 1895) and study (1903). A comprehensive analysis of dual numbers and their applications to the kinematic analysis of spatial linkages was conducted by Yang (1963) and Yang and Freudenstein (1964). Bottema and Roth (1978) include a study of theoretical kinematics using dual numbers. Thereafter, Agrawal (1987) reported on the application of dual quaternions to spatial kinematics, while Pradeep et al. (1989) used the *dual-matrix exponential* in

[1] Angeles, J., 1998, "The Application of Dual Algebra to Kinematic Analysis", in Angeles, J. and Zakhariev, E. (editors), *Computational Methods in Mechanical Systems*, Springer-Verlag, Heidelberg, Vol. 161, pp. 3–31.

J. Angeles and S. Bai, *Kinematics of Mechanical Systems*,
Mathematical Engineering, https://doi.org/10.1007/978-3-031-09544-3

the analysis of robotic manipulators. Shoham and Brodsky (1993, 1994) proposed a dual inertia operator for the dynamical analysis of mechanical systems. A comprehensive introduction to dual quaternions is due to McCarthy (1990), an abstract treatment by Chevallier (1991) following suit.

A.2 Definitions

A *dual number* $\hat{a}$ is defined as the sum of a *primal* part a, and a *dual* part a_0, namely

$$\hat{a} = a + \epsilon a_0 , \tag{A.1}$$

where ϵ is the dual unity, which verifies $\epsilon \neq 0$, $\epsilon^2 = 0$, a and a_0 being real numbers, the former denoting the *primal part* of $\hat{a}$, the latter its *dual part*. Actually, dual numbers with complex parts can be equally defined (Cheng & Thompson, 1996). For the purposes of the book, real numbers suffice.

If $a_0 = 0$, $\hat{a}$ is called a *real number*. Correspondingly, $\hat{a}$ is a *complex dual number* if $a = 0$ and $\hat{a}$ is called a *pure dual number*. If neither is zero, $\hat{a}$ is called a *proper dual number*.

Let $\hat{b} = b + \epsilon b_0$ be a second dual number. Equality, addition, multiplication, and division are defined, respectively, as

$$\hat{a} = \hat{b} \Leftrightarrow a = b, \quad a_0 = b_0 \tag{A.2a}$$

$$\hat{a} + \hat{b} = (a + b) + \epsilon(a_0 + b_0) \tag{A.2b}$$

$$\hat{a}\hat{b} = ab + \epsilon(ab_0 + a_0 b) \tag{A.2c}$$

$$\frac{\hat{a}}{\hat{b}} = \frac{a}{b} - \epsilon\left(\frac{ab_0 - a_0 b}{b^2}\right), \quad b \neq 0 \tag{A.2d}$$

From Eq. (A.2d), it is apparent that *the division by a pure dual number is not defined*. Hence, dual numbers do not form a *field* in the algebraic sense; they do form a *ring* (Simmons, 1963).

All formal operations involving dual numbers are identical to those of ordinary algebra, while taking into account that $\epsilon^2 = \epsilon^3 = \cdots = 0$. Therefore, the series expansion of the *analytic function* $f(\hat{x})$ of a dual argument $\hat{x}$ is given by

$$f(\hat{x}) = f(x + \epsilon x_0) = f(x) + \epsilon x_0 \frac{\mathrm{d}f(x)}{\mathrm{d}x} \tag{A.3}$$

As a direct consequence of Eq. (A.3), we have the expression below for the exponential of a dual number $\hat{x}$:

$$\mathrm{e}^{\hat{x}} = \mathrm{e}^{x} + \epsilon\, x_0 \mathrm{e}^{x} = \mathrm{e}^{x}(1 + \epsilon\, x_0) , \tag{A.4}$$

and hence, *the dual exponential cannot be a pure dual number*.

The *dual angle* $\hat{\theta}$ between two skew lines $\mathcal{L}_1$ and $\mathcal{L}_2$, introduced by Study (1903), is defined as

$$\hat{\theta} = \theta + \epsilon s \ , \tag{A.5}$$

where θ and s are, respectively, the twist angle and the distance between the two lines. The *dual trigonometric functions* of the dual angle $\hat{\theta}$ are derived directly from Eq. (A.3), namely

$$\cos\hat{\theta} = \cos\theta - \epsilon s \sin\theta, \quad \sin\hat{\theta} = \sin\theta + \epsilon s \cos\theta, \quad \tan\hat{\theta} = \tan\theta + \epsilon s \sec^2\theta \tag{A.6}$$

Moreover, all identities of ordinary trigonometry hold for dual angles. Likewise, the square root of a dual number can be readily found by a straightforward application of Eq. (A.3), namely

$$\sqrt{\hat{x}} = \sqrt{x} + \epsilon \frac{x_0}{2\sqrt{x}}, \quad x \neq 0 \tag{A.7}$$

A *dual vector* $\hat{\mathbf{a}}$ is defined as the sum of a primal vector part $\mathbf{a}$, and a dual vector part $\mathbf{a}_0$, namely

$$\hat{\mathbf{a}} = \mathbf{a} + \epsilon \mathbf{a}_0 \tag{A.8}$$

where both $\mathbf{a}$ and $\mathbf{a}_0$ are Cartesian, three-dimensional real vectors. Henceforth, all vectors are assumed to be of this kind. Further, let $\hat{\mathbf{a}}$ and $\hat{\mathbf{b}}$ be two dual vectors and $\hat{c}$ a dual scalar. The concepts of dual-vector equality, multiplication of a dual vector by a dual scalar, inner product, and vector product of two dual vectors are defined as follows:

$$\hat{\mathbf{a}} = \hat{\mathbf{b}} \quad \Leftrightarrow \quad \mathbf{a} = \mathbf{b} \ \text{ and } \ \mathbf{a}_0 = \mathbf{b}_0 \ ; \tag{A.9a}$$

$$\hat{c}\,\hat{\mathbf{a}} = c\,\mathbf{a} + \epsilon\,(c_0\mathbf{a} + c\,\mathbf{a}_0) \ ; \tag{A.9b}$$

$$\hat{\mathbf{a}} \cdot \hat{\mathbf{b}} = \mathbf{a} \cdot \mathbf{b} + \epsilon\,(\mathbf{a} \cdot \mathbf{b}_0 + \mathbf{a}_0 \cdot \mathbf{b}) \ ; \tag{A.9c}$$

$$\hat{\mathbf{a}} \times \hat{\mathbf{b}} = \mathbf{a} \times \mathbf{b} + \epsilon\,(\mathbf{a} \times \mathbf{b}_0 + \mathbf{a}_0 \times \mathbf{b}) \tag{A.9d}$$

In particular, when $\hat{\mathbf{b}} = \hat{\mathbf{a}}$, Eq. (A.9c) leads to the *Euclidean norm* of the dual vector $\hat{\mathbf{a}}$, i.e.,

$$\|\hat{\mathbf{a}}\|^2 = \|\mathbf{a}\|^2 + \epsilon\,2\mathbf{a} \cdot \mathbf{a}_0 \tag{A.9e}$$

Furthermore, the six *normalized Plücker coordinates* of a line $\mathcal{L}$ passing through a point P of position vector $\mathbf{p}$ and parallel to the unit vector $\mathbf{e}$ are given by the pair $(\mathbf{e}, \ \mathbf{p} \times \mathbf{e})$, the product $\mathbf{e}_0 \equiv \mathbf{p} \times \mathbf{e}$ denoting the *moment* of the line. The foregoing coordinates can be represented by a *dual unit vector* $\hat{\mathbf{e}}^*$, whose six real components in $\mathbf{e}$ and $\mathbf{e}_0$ are the Plücker coordinates of $\mathcal{L}$, namely

$$\hat{\mathbf{e}}^* = \mathbf{e} + \epsilon\,\mathbf{e}_0, \quad \text{with} \quad \|\mathbf{e}\| = 1 \ \text{ and } \ \mathbf{e} \cdot \mathbf{e}_0 = 0 \tag{A.10}$$

The reader is invited to verify the results summarized below.

Lemma A.2.1 *For* $\hat{\mathbf{e}}^* \equiv \mathbf{e} + \epsilon\,\mathbf{e}_0$ *and* $\hat{\mathbf{f}}^* \equiv \mathbf{f} + \epsilon\,\mathbf{f}_0$ *defined as two dual unit vectors representing lines* $\mathcal{L}$ *and* $\mathcal{M}$*, respectively, we have*

(*i*) *If* $\hat{\mathbf{e}}^* \times \hat{\mathbf{f}}^*$ *is a pure dual vector, then* $\mathcal{L}$ *and* $\mathcal{M}$ *are parallel;*
(*ii*) *if* $\hat{\mathbf{e}}^* \cdot \hat{\mathbf{f}}^*$ *is a pure dual number, then* $\mathcal{L}$ *and* $\mathcal{M}$ *are perpendicular;*
(*iii*) $\mathcal{L}$ *and* $\mathcal{M}$ *are coincident if and only if* $\hat{\mathbf{e}}^* \times \hat{\mathbf{f}}^* = \mathbf{0}$*; and*
(*iv*) $\mathcal{L}$ *and* $\mathcal{M}$ *intersect at right angles if and only if* $\hat{\mathbf{e}}^* \cdot \hat{\mathbf{f}}^* = 0$.

Dual matrices can be defined likewise, i.e., if $\mathbf{A}$ and $\mathbf{A}_0$ are two real $n \times n$ matrices, then the dual $n \times n$ matrix $\hat{\mathbf{A}}$ is defined as

$$\hat{\mathbf{A}} \equiv \mathbf{A} + \epsilon\,\mathbf{A}_0 \tag{A.11}$$

While we work mainly with 3×3 matrices in connection with dual Cartesian vectors, the above definition can be readily applied to any square matrices, which is the reason why n has been left arbitrary. Actually, rectangular dual matrices are also possible. Equality, multiplication by a dual scalar, and multiplication by a dual vector are defined as in the foregoing cases. Moreover, matrix multiplication is defined correspondingly, but then the order of multiplication must be respected. We thus have that, if $\hat{\mathbf{A}}$ and $\hat{\mathbf{B}}$ are two $n \times n$ dual matrices, with their primal and dual parts self-understood, then

$$\hat{\mathbf{A}}\hat{\mathbf{B}} = \mathbf{A}\mathbf{B} + \epsilon\,(\mathbf{A}\mathbf{B}_0 + \mathbf{A}_0\mathbf{B}) \tag{A.12}$$

Therefore, matrix $\hat{\mathbf{A}}$ is real if $\mathbf{A}_0 = \mathbf{O}$, where $\mathbf{O}$ denotes the $n \times n$ zero matrix; if $\mathbf{A} = \mathbf{O}$, then $\hat{\mathbf{A}}$ is called a *pure dual matrix*. Moreover, as we shall see below, *a square dual matrix admits an inverse if and only if its primal part is nonsingular*.

Now we can define the inverse of a dual matrix, if this is nonsingular. Indeed, it suffices to make $\hat{\mathbf{B}} = \hat{\mathbf{A}}^{-1}$ in Eq. (A.12) and the right-hand side of this equation equal to the $n \times n$ identity matrix, $\mathbf{1}$, thereby obtaining two matrix equations that allow us to find the primal and the dual parts of $\hat{\mathbf{A}}^{-1}$, namely

$$\mathbf{A}\mathbf{B} = \mathbf{1}, \quad \mathbf{A}\mathbf{B}_0 + \mathbf{A}_0\mathbf{B} = \mathbf{O}\,,$$

whence

$$\mathbf{B} = \mathbf{A}^{-1}, \quad \mathbf{B}_0 = -\mathbf{A}^{-1}\mathbf{A}_0\mathbf{A}^{-1}\,,$$

which are defined because $\mathbf{A}$ has been assumed invertible. Hence, for any nonsingular $n \times n$ dual matrix $\hat{\mathbf{A}}$,

$$\hat{\mathbf{A}}^{-1} = \mathbf{A}^{-1} - \epsilon\,\mathbf{A}^{-1}\mathbf{A}_0\mathbf{A}^{-1} \tag{A.13}$$

The striking similarity of the dual part of the foregoing expression with the time-derivative of the inverse of $\mathbf{A}(t)$, namely

$$\frac{\mathrm{d}}{\mathrm{d}t}[\mathbf{A}^{-1}(t)] = -\mathbf{A}^{-1}(t)\dot{\mathbf{A}}(t)\mathbf{A}^{-1}(t)$$

is to be highlighted.

In order to find an expression for the determinant of a $n \times n$ dual matrix, we need to recall the general expression for the dual function defined in Eq. (A.3). However, that expression has to be adapted to a dual-matrix argument, which leads to

$$f(\hat{\mathbf{A}}) = f(\mathbf{A}) + \epsilon\,\mathrm{tr}\left[\mathbf{A}_0\left(\frac{\mathrm{d}f}{\mathrm{d}\hat{\mathbf{A}}}\right)^T\right]\Bigg|_{\hat{\mathbf{A}}=\mathbf{A}} \tag{A.14}$$

In particular, when $f(\hat{\mathbf{A}}) = \det(\hat{\mathbf{A}})$, we have, recalling the formula for the derivative of the determinant with respect to its matrix argument (Angeles, 1982), for any $n \times n$ matrix $\mathbf{X}$,

$$\frac{\mathrm{d}}{\mathrm{d}\mathbf{X}}[\det(\mathbf{X})] = \det(\mathbf{X})\mathbf{X}^{-T} ,$$

where $\mathbf{X}^{-T}$ denotes the transpose of the inverse of $\mathbf{X}$ or, equivalently, the transpose of $\mathbf{X}^{-1}$. Therefore,

$$\mathrm{tr}\left[\mathbf{A}_0\left(\frac{\mathrm{d}f}{\mathrm{d}\hat{\mathbf{A}}}\right)^T\right]\Bigg|_{\hat{\mathbf{A}}=\mathbf{A}} = \det(\mathbf{A})\mathrm{tr}(\mathbf{A}_0\mathbf{A}^{-1}) ,$$

and hence,

$$\det(\hat{\mathbf{A}}) = \det(\mathbf{A})[1 + \epsilon\,\mathrm{tr}(\mathbf{A}_0\mathbf{A}^{-1})] \tag{A.15}$$

which reveals a significant result: *the invertibility of the dual part of a dual matrix does not play any role in the invertibility of the given dual matrix.*

Now we can define the eigenvalue problem for the dual matrix $\hat{\mathbf{A}}$ introduced above. Let $\hat{\lambda}$ and $\hat{\mathbf{e}}$ be a dual eigenvalue and the corresponding dual (unit) eigenvector of $\hat{\mathbf{A}}$, respectively. Then,

$$\hat{\mathbf{A}}\hat{\mathbf{e}} = \hat{\lambda}\hat{\mathbf{e}}, \quad \|\hat{\mathbf{e}}\| = 1 \tag{A.16a}$$

For the foregoing linear homogeneous equation to admit a non-trivial solution, we must have

$$\det(\hat{\lambda}\mathbf{1} - \hat{\mathbf{A}}) = 0 , \tag{A.16b}$$

which yields an nth-order dual polynomial in the dual number $\hat{\lambda}$, Eq. (A.16b) thus being the *dual characteristic equation* of $\hat{\mathbf{A}}$. Its n dual roots, real and complex, constitute the n *dual eigenvalues* of $\hat{\mathbf{A}}$. Notice that, associated with each dual eigenvalue $\hat{\lambda}_i$, a corresponding *dual (unit) eigenvector* $\hat{\mathbf{e}}_i^*$ is defined, for $i = 1, 2, \ldots, n$. Moreover, if we recall Eq. (A.4), we can write

$$\mathrm{e}^{\hat{\mathbf{A}}} = \mathrm{e}^{\mathbf{A}} + \epsilon\,\mathbf{A}_0\mathrm{e}^{\mathbf{A}} \tag{A.17}$$

Upon expansion, the foregoing expression can be cast in the form

$$\mathrm{e}^{\hat{\mathbf{A}}} = (\mathbf{1} + \epsilon\,\mathbf{A}_0)\mathrm{e}^{\mathbf{A}} \neq \mathrm{e}^{\mathbf{A}}(\mathbf{1} + \epsilon\,\mathbf{A}_0) \ , \tag{A.18}$$

the inequality arising because, in general, $\mathbf{A}$ and $\mathbf{A}_0$ do not commute. They do so only in the case in which they share the same set of eigenvectors. A special case in which this happens occurs when one matrix is an *analytic function* of the other. More formally, we have the following.

Lemma A.2.2 *If* $\mathbf{F}$ *is an analytic matrix function of matrix* $\mathbf{A}$*, then the two matrices*

(i) *share the same set of eigenvectors, and*
(ii) *commute under multiplication.*

Typical examples of analytic matrix functions are $\mathbf{F} = \mathbf{A}^N$ and $\mathbf{F} = \mathrm{e}^{\mathbf{A}}$, for an integer N.

A.3 Dual Algebra and Geometry

We include in this section the background on dual algebra required to represent the geometric relationships between (1) two lines, and (2) a point and a line.

A.3.1 Link-Dimension Determination Based on Two Skew Lines

The dual angle $\hat{\theta}$ between two skew lines $\mathcal{L}_1$ and $\mathcal{L}_2$, as shown in Fig. A.1, is defined as

$$\hat{\theta} = \theta + \epsilon s \tag{A.19}$$

where θ and s are, respectively, the twist angle and the distance between the two lines.

Let two skew lines $\mathcal{L}_1$ and $\mathcal{L}_2$ be represented by $\hat{\mathbf{a}} = \mathbf{a} + \epsilon\bar{\mathbf{a}}$ and $\hat{\mathbf{b}} = \mathbf{b} + \epsilon\bar{\mathbf{b}}$, respectively. The relations below then follow:

$$\hat{\mathbf{a}}^T\hat{\mathbf{b}} = \cos\hat{\theta}, \quad \hat{\mathbf{a}} \times \hat{\mathbf{b}} = \hat{\mathbf{n}}\sin\hat{\theta} \tag{A.20}$$

where $\hat{\theta}$ is the dual angle between lines $\mathcal{L}_1$ and $\mathcal{L}_2$, while $\hat{\mathbf{n}}$ is the dual unit vector of the common normal to the two lines, which is given by

$$\hat{\mathbf{n}} = \frac{1}{\sin\hat{\theta}}(\hat{\mathbf{a}} \times \hat{\mathbf{b}}) \tag{A.21}$$

That is

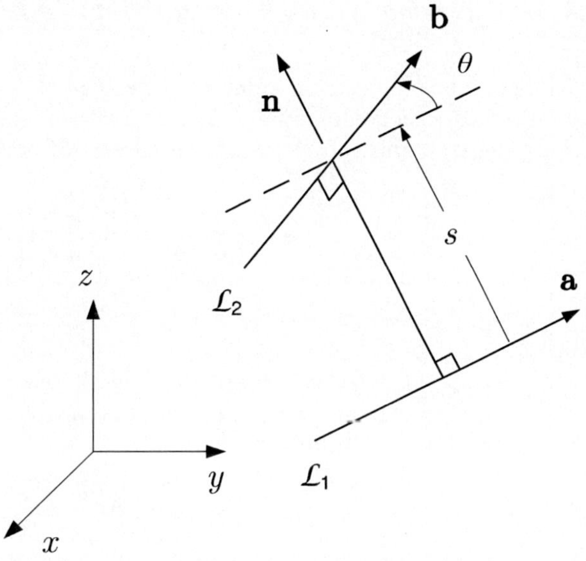

Fig. A.1 The dual angle between two lines

$$\hat{\mathbf{n}} = \frac{1}{\sin\theta + \epsilon s \cos\theta}(\mathbf{a} + \epsilon\bar{\mathbf{a}}) \times (\mathbf{b} + \epsilon\bar{\mathbf{b}}) \tag{A.22}$$

Expanding the right-hand side of the above equation yields, after simplification,

$$\hat{\mathbf{n}} = \frac{1}{\sin\theta}(\mathbf{a} \times \mathbf{b}) + \epsilon\left(\frac{1}{\sin\theta}(\mathbf{a} \times \bar{\mathbf{b}} + \bar{\mathbf{a}} \times \mathbf{b}) - \frac{s\cos\theta}{\sin^2\theta}(\mathbf{a} \times \mathbf{b})\right) \tag{A.23}$$

or, in compact form,

$$\hat{\mathbf{n}} = \mathbf{n} + \epsilon\bar{\mathbf{n}} \tag{A.24}$$

with

$$\mathbf{n} = \frac{1}{\sin\theta}(\mathbf{a} \times \mathbf{b}), \quad \bar{\mathbf{n}} = \frac{1}{\sin\theta}(\mathbf{a} \times \bar{\mathbf{b}} + \bar{\mathbf{a}} \times \mathbf{b}) - \frac{s\cos\theta}{\sin^2\theta}(\mathbf{a} \times \mathbf{b}) \tag{A.25}$$

The two dual vectors $\hat{\mathbf{a}}$ and $\hat{\mathbf{b}}$ are associated with two link dimensions, namely the twist angle θ and the *signed distance* s:

$$\mathbf{a}^T\mathbf{b} = \cos\theta \tag{A.26}$$

$$\mathbf{a}^T\bar{\mathbf{b}} + \bar{\mathbf{a}}^T\mathbf{b} = -s\sin\theta \tag{A.27}$$

A.3.2 Normal Vector From a Point to a Line

With reference to Fig. A.2, a line $\mathcal{L}$ is specified by a dual vector $\hat{\mathbf{a}}$. Point P' is the orthogonal projection of point P onto $\mathcal{L}$. We want to find the vector $\mathbf{n}$ shown in the same figure. To this end, two equations are established, namely

$$\mathbf{n} \cdot \mathbf{a} = 0 \tag{A.28}$$

$$(\mathbf{p} + \mathbf{n}) \times \mathbf{a} - \bar{\mathbf{a}} = \mathbf{0} \tag{A.29}$$

where $\mathbf{p}$ is the position vector of point P and $\bar{\mathbf{a}}$ is the *dual part* of the dual vector of line $\mathcal{L}$.

Moreover, Eq. (A.29) is rewritten in an alternative form, upon introduction of matrix $\mathbf{A}$, the skew-symmetric matrix of vector $\mathbf{a}$, best known as the *cross-product matrix (CPM)* of $\mathbf{a}$, thus obtaining

$$\mathbf{A}\mathbf{n} = -(\mathbf{A}\mathbf{p} + \bar{\mathbf{a}}) \tag{A.30}$$

which stands for three scalar equations, only two of which are independent, because *the cross-product matrix* is known to be of rank 2. A determined system of equations can be established from Eqs. (A.28) and (A.30) to find vector $\mathbf{n}$, as described presently.

First, a 4×3 system of equations is introduced

$$\mathbf{N}\mathbf{n} = \mathbf{w} \tag{A.31}$$

with

$$\mathbf{N} = \begin{bmatrix} \mathbf{a}^T \\ \mathbf{A} \end{bmatrix} \in \mathbb{R}^{4\times 3}, \quad \mathbf{w} = \begin{bmatrix} 0 \\ -(\mathbf{A}\mathbf{p} + \bar{\mathbf{a}}) \end{bmatrix} \in \mathbb{R}^4 \tag{A.32}$$

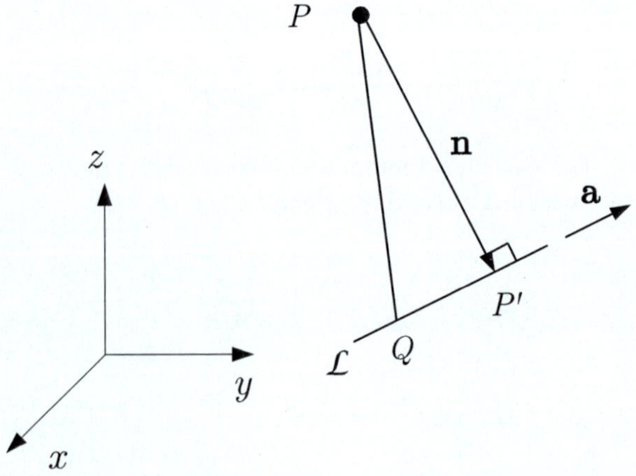

Fig. A.2 The distance from one point to a line

where matrix $\mathbf{N}$ is orthogonal. Indeed, let us consider the product below

$$\mathbf{N}^T\mathbf{N} = \mathbf{a}\mathbf{a}^T + \mathbf{A}^T\mathbf{A} \tag{A.33}$$

and

$$\mathbf{A}^T\mathbf{A} = -\mathbf{A}^2 = ||\mathbf{a}||^2\mathbf{1} - \mathbf{a}\mathbf{a}^T = \mathbf{1} - \mathbf{a}\mathbf{a}^T \tag{A.34}$$

Hence,

$$\mathbf{N}^T\mathbf{N} = \mathbf{1} \tag{A.35}$$

with $\mathbf{1}$ denoting the 3×3 identity matrix, thereby proving orthogonality. Therefore, multiplying both sides of Eq. (A.31) by $\mathbf{N}^T$ yields

$$\mathbf{n} = \mathbf{N}^T\mathbf{w} \tag{A.36}$$

The norm of vector $\mathbf{n}$ is then found to be

$$||\mathbf{n}|| = ||\mathbf{w}|| = ||\mathbf{A}\mathbf{p} + \bar{\mathbf{a}}|| \tag{A.37}$$

Alternatively, if we know a point Q on the line $\mathcal{L}$ of Fig. A.2, vector $\mathbf{n}$ is readily calculated:

$$\mathbf{n} = (\mathbf{p} - \mathbf{q}) - [(\mathbf{p} - \mathbf{q})^T\mathbf{a}]\mathbf{a} \tag{A.38}$$

where $\mathbf{q}$ is the position vector of point Q. Substituting $\mathbf{n}$, as given above, into Eq. (A.29), after simplification, leads to

$$\mathbf{n} \times \mathbf{a} = \bar{\mathbf{a}} - \mathbf{p} \times \mathbf{a} \tag{A.39}$$

which readily leads to

$$||\mathbf{n}|| = ||\bar{\mathbf{a}} - \mathbf{p} \times \mathbf{a}|| \tag{A.40}$$

i.e.,

$$\mathbf{n}^T\mathbf{n} = (\bar{\mathbf{a}} - \mathbf{p} \times \mathbf{a})^T(\bar{\mathbf{a}} - \mathbf{p} \times \mathbf{a}) \tag{A.41}$$

a restatement of Eq. (A.37).

A.4 Fundamentals of Rigid-Body Kinematics

We review in this section some basic facts from rigid-body kinematics. For the sake of conciseness, some proofs are not given, but the pertinent references are cited whenever necessary.

A.4.1 Finite Displacements

A *rigid body* is understood as a particular case of the *continuum*, with one special feature: under any given motion, *two arbitrary* points of the body remain equidistant. A rigid body is available through a *configuration* or *pose*, denoted as $\mathcal{B}$. Whenever a *reference configuration* is needed, this will be labeled $\mathcal{B}^0$. Moreover, the position vector of a point P of the body in configuration $\mathcal{B}$ will be denoted by $\mathbf{p}$, that in $\mathcal{B}^0$ being denoted, correspondingly,[2] as $\mathbf{p}^0$.

A rigid-body motion leaving a fixed point O of the body is called a *pure rotation*, represented by a *proper orthogonal matrix* $\mathbf{Q}$, i.e., $\mathbf{Q}$ verifies the two properties below

$$\mathbf{Q}\mathbf{Q}^T = \mathbf{1}, \quad \det(\mathbf{Q}) = +1 \tag{A.42}$$

According to *Euler's theorem* (Euler, 1776), a pure rotation leaves a set of points of the body *immutable*, this set lying on a line $\mathcal{L}$, which is termed the *axis of rotation*. If we draw the perpendicular from an arbitrary point P of the body to $\mathcal{L}$, and denote P' as its intersection with $\mathcal{L}$, the angle ϕ between $P'P^0$ and $P'P$, where, according to our convention, P^0 denotes the point P in the reference configuration $\mathcal{B}^0$ of the body, is called the *angle of rotation*. Notice that a direction must be specified along this line to define the sign of the angle, which is a real number. Furthermore, the direction of the line is specified by the unit vector $\mathbf{e}$. We term $\mathbf{e}$ and ϕ the *natural invariants* of $\mathbf{Q}$.

As a result of Euler's theorem, the rotation $\mathbf{Q}$ can be represented in terms of its natural invariants. This representation takes the form

$$\mathbf{Q} = \mathbf{e}\mathbf{e}^T + \cos\phi(\mathbf{1} - \mathbf{e}\mathbf{e}^T) + \sin\phi\mathbf{E}\ , \tag{A.43}$$

where $\mathbf{E}$ denotes the *cross-product matrix* of $\mathbf{e}$, i.e., for any three-dimensional vector $\mathbf{v}$,

$$\mathbf{e} \times \mathbf{v} = \mathbf{E}\mathbf{v} \tag{A.44}$$

As a result of the foregoing definition, $\mathbf{E}$ is skew-symmetric, i.e., $\mathbf{E} = -\mathbf{E}^T$ and, moreover, it has the properties below

$$\mathbf{E}^{2k+1} = (-1)^k\mathbf{E}, \quad \mathbf{E}^{2k} = (-1)^k(\mathbf{1} - \mathbf{e}\mathbf{e}^T), \quad \text{for} \quad k = 1, 2, \ldots$$

By virtue of the foregoing properties of the cross-product matrix $\mathbf{E}$ of $\mathbf{e}$, the rotation matrix $\mathbf{Q}$ can be written in the alternative form

$$\mathbf{Q} = \mathbf{1} + \sin\phi\mathbf{E} + (1 - \cos\phi)\mathbf{E}^2 \tag{A.45}$$

[2] This notation cannot possibly mislead the reader to indicate a "power" of vector $\mathbf{p}$, as the quoted concept is not pertinent to vector algebra.

Now, if we recall the *Cayley–Hamilton theorem* (Halmos, 1974), we can realize that the right-hand side of the foregoing equation is nothing but the exponential of $\phi\mathbf{E}$, i.e.,

$$\mathbf{Q} = e^{\phi\mathbf{E}} , \tag{A.46}$$

termed the *exponential form of the rotation matrix*. Now it is a simple matter to obtain the eigenvalues of the rotation matrix if we first notice that one eigenvalue of $\mathbf{E}$ is 0, the other eigenvalues being readily derived as $\pm\sqrt{-1}$, where $\sqrt{-1} \equiv i$ is the imaginary unit. Therefore, if $\mathbf{Q}$ is the exponential of $\phi\mathbf{E}$, then the eigenvalues of $\mathbf{Q}$ are the exponentials of the eigenvalues of $\phi\mathbf{E}$:

$$\lambda_1 = e^0 = 1, \quad \lambda_{2,3} = e^{\pm\sqrt{-1}\phi} = \cos\phi \pm \sqrt{-1}\sin\phi \tag{A.47}$$

Moreover, we recall below the *Cartesian decomposition* of a $n \times n$ matrix $\mathbf{A}$, namely

$$\mathbf{A} = \mathbf{A}_s + \mathbf{A}_{ss} , \tag{A.48a}$$

where $\mathbf{A}_s$ is symmetric and $\mathbf{A}_{ss}$ is skew-symmetric. These matrices are given by

$$\mathbf{A}_s \equiv \frac{1}{2}(\mathbf{A} + \mathbf{A}^T), \quad \mathbf{A}_{ss} \equiv \frac{1}{2}(\mathbf{A} - \mathbf{A}^T) \tag{A.48b}$$

Any 3×3 skew-symmetric matrix is fully defined by three scalars, which means that such a matrix is *isomorphic* to a three-dimensional vector. Indeed, let $\mathbf{S}$ be a 3×3 skew-symmetric matrix and $\mathbf{v}$ be an arbitrary three-dimensional vector. Then, one *unique* vector $\mathbf{s}$ can be defined such that

$$\mathbf{S}\mathbf{v} \equiv \mathbf{s} \times \mathbf{v} \tag{A.49}$$

When the above items are expressed in a given coordinate frame $\mathcal{F}$, the components of $\mathbf{S}$, indicated as $\{ s_{i,j} \}_{i,j=1}^3$, those of $\mathbf{s}$ as $\{ s_i \}_1^3$, bear the relations below

$$\mathbf{S} = \begin{bmatrix} 0 & -s_3 & s_2 \\ s_3 & 0 & -s_1 \\ -s_2 & s_1 & 0 \end{bmatrix}, \quad \mathbf{s} = \frac{1}{2}\begin{bmatrix} s_{32} - s_{23} \\ s_{13} - s_{31} \\ s_{21} - s_{12} \end{bmatrix} \tag{A.50}$$

Vector $\mathbf{s}$ is termed the *linear invariant*[3] of $\mathbf{S}$, the corresponding scalar invariant being $\text{tr}(\mathbf{S})$, which vanishes for skew-symmetric matrices, like $\mathbf{S}$. However, the scalar linear invariant of an arbitrary $n \times n$ matrix needn't vanish. In general, we define the *axial vector* of an arbitrary 3×3 matrix $\mathbf{A}$ in terms of the difference of its off-diagonal entries, as appearing in Eq. (A.50) for the entries of matrix $\mathbf{S}$. Apparently, the axial vector of any 3×3 matrix is identical to that of its skew-symmetric component; this

[3] A scalar quantity associated with a vector or a tensor {e.g., a matrix} that remains unchanged under a change of frame is termed as *invariant* of the array in questions.

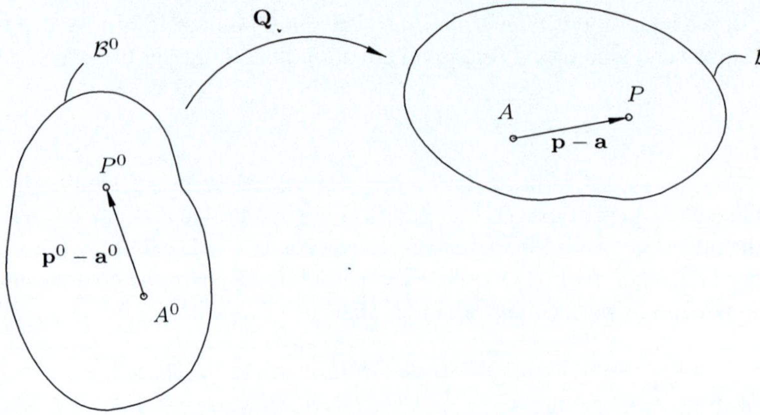

Fig. A.3 Displacements of two points of a rigid body in two finitely separated configurations

vector, represented as $\mathbf{a} \equiv \text{vect}(\mathbf{A})$, is the *vector linear invariant* of $\mathbf{A}$. The *scalar linear invariant* of the same matrix is its trace, $\text{tr}(\mathbf{A})$, which involves only its *diagonal entries*. With these definitions, notice that

$$\frac{1}{2}(\mathbf{A} - \mathbf{A}^T)\mathbf{v} = \mathbf{a} \times \mathbf{v}$$

Further, with reference to Fig. A.3, let A and P be two points of a rigid body, which is shown in its reference and its current configurations, $\mathcal{B}^0$ and $\mathcal{B}$, respectively.

We can regard vector $\mathbf{p} - \mathbf{a}$ as the image of $\mathbf{p}^0 - \mathbf{a}^0$ under the rotation $\mathbf{Q}$, namely

$$\mathbf{p} - \mathbf{a} = \mathbf{Q}(\mathbf{p}^0 - \mathbf{a}^0)\,, \tag{A.51}$$

and hence

$$\mathbf{p} = \mathbf{a} + \mathbf{Q}(\mathbf{p}^0 - \mathbf{a}^0) \tag{A.52}$$

Furthermore, the *displacement* $\mathbf{d}_A$ of A is defined as the difference $\mathbf{a} - \mathbf{a}^0$, with a similar definition for the displacement $\mathbf{d}_P$ of P. From the above equation, it is now apparent that a linear relation between the two displacements follows:

$$\mathbf{d}_P = \mathbf{d}_A + (\mathbf{Q} - \mathbf{1})(\mathbf{p}^0 - \mathbf{a}^0) \tag{A.53}$$

Therefore,

Theorem A.4.1 *The displacements of all the points of a rigid body have identical projections onto the axis of the concomitant rotation.*

The proof of the foregoing result follows upon dot-multiplying both sides of Eq. (A.53) by $\mathbf{e}$

$$\mathbf{e} \cdot \mathbf{d}_P = \mathbf{e} \cdot \mathbf{d}_A$$

From the previous result, it is apparent that $\|\mathbf{d}_P\|$ can attain infinitely large values, depending on $\|\mathbf{p}^0 - \mathbf{a}^0\|$, but, in general, $\mathbf{d}_P$ does not vanish. Hence, a minimum of $\|\mathbf{d}_P\|$ can be found, a result summarized in the Mozzi–Chasles theorem (Mozzi, 1763; Chasles, 1830. This theorem states that the points of $\mathcal{B}$ of minimum-norm displacement lie in a line $\mathcal{M}$ that is parallel to the axis of the concomitant rotation, represented by matrix **Q**, the minimum-norm displacement being a vector parallel to the same axis. If we recall that **e** and ϕ denote the natural invariants of **Q**, then the position vector $\mathbf{p}^*$ of the point P^* of $\mathcal{M}$ lying closest to the origin O is given by Angeles (1997)

$$\mathbf{p}^* = \frac{(\mathbf{Q} - \mathbf{1})^T(\mathbf{Q}\mathbf{a}^0 - \mathbf{a})}{2(1 - \cos\phi)}, \quad \text{for} \quad \phi \neq 0\,, \tag{A.54}$$

the special case in which $\phi = 0$ corresponding to a *pure translation*, whereby all points of $\mathcal{B}$ undergo identical displacements. In this case, then, the axis $\mathcal{M}$ is indeterminate, because all points of the body can be thought of as undergoing minimum-norm displacements.

Henceforth, line $\mathcal{M}$ will be termed the *Mozzi–Chasles axis*. Notice that the Plücker coordinates of the Mozzi–Chasles axis are given by **e** and $\mathbf{e}_0 \equiv \mathbf{p}^* \times \mathbf{e}$. We shall denote $\mathbf{d}^*$ the *minimum-norm displacement*, which can be represented in the form

$$\mathbf{d}^* = d^*\mathbf{e}, \quad \text{with} \quad d^* = \mathbf{d}_P \cdot \mathbf{e} \tag{A.55}$$

with $\mathbf{d}_P$ indicating the displacement undergone by an arbitrary point P.

Therefore, the body under study can be regarded as undergoing, from $\mathcal{B}^0$ to $\mathcal{B}$, a screw motion, as if the body were rigidly attached to the bolt of a screw of axis $\mathcal{M}$ and *pitch* p given by

$$p = \frac{d^*}{\phi} = \frac{\mathbf{e} \cdot \mathbf{d}_P}{\phi} \tag{A.56}$$

We list below further results.

Lemma A.4.1 *Let A and P be two points of a rigid body undergoing a general motion from a reference pose $\mathcal{B}^0$ to a current pose $\mathcal{B}$. Then, under the notation adopted above, the difference* $\mathbf{p} - \mathbf{Q}\mathbf{p}^0$ *remains constant and is denoted as* **d**, *i.e.,*

$$\mathbf{p} - \mathbf{Q}\mathbf{p}^0 = \mathbf{a} - \mathbf{Q}\mathbf{a}^0 = \mathbf{d} \tag{A.57}$$

Proof If we recall Eq. (A.52) and substitute the expression therein for **p** in the difference $\mathbf{p} - \mathbf{Q}\mathbf{p}^0$, we obtain

$$\mathbf{p} - \mathbf{Q}\mathbf{p}^0 = \mathbf{a} + \mathbf{Q}(\mathbf{p}^0 - \mathbf{a}^0) - \mathbf{Q}\mathbf{p}^0 = \mathbf{a} - \mathbf{Q}\mathbf{a}^0 = \mathbf{d}\,,$$

thereby completing the intended proof.

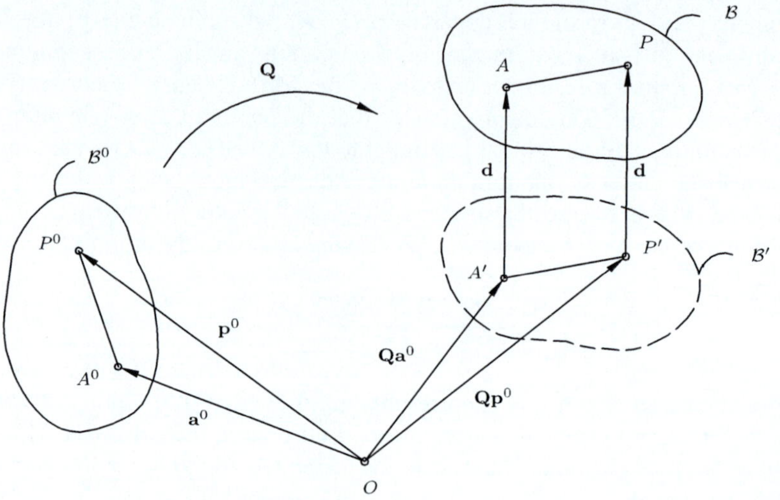

Fig. A.4 Geometric interpretation of Lemma 3.1

Furthermore, the kinematic interpretation of $\mathbf{d}$ follows directly from Eq. (A.57): $\mathbf{d}$ represents the displacement of the point of $\mathcal{B}$ that coincides with the origin in the reference pose $\mathcal{B}^0$. The geometric interpretation of the foregoing lemma is given in Fig. A.4. What this figure indicates is that the pose $\mathcal{B}$ can be attained from $\mathcal{B}^0$ in two stages: (a) first, the body is given a pure rotation $\mathbf{Q}$ about the origin O that takes the body to the intermediate pose $\mathcal{B}'$; (b) then, from $\mathcal{B}'$, the body is given a pure translation of displacement $\mathbf{d}$ that takes the body into $\mathcal{B}$.

Therefore, Eq. (A.54) for the position vector of the point of the Mozzi–Chasles axis lying closest to the origin can be expressed in terms of vector $\mathbf{d}$ as

$$\mathbf{p}^* = \frac{(\mathbf{1} - \mathbf{Q})^T\mathbf{d}}{2(1 - \cos\phi)}, \quad \text{for} \quad \phi \neq 0 \tag{A.58}$$

Notice that, in general, $\mathbf{d}$ is not of the minimum norm. Additionally, $\mathbf{d}$ is origin-dependent, and hence not an invariant of the motion under study. Now, if we choose the origin on the Mozzi–Chasles axis $\mathcal{M}$, then we have the layout of Fig. A.5, and vector $\mathbf{d}$ becomes a multiple of $\mathbf{e}$, in the sense $\mathbf{d} = d^*\mathbf{e}$.

We can now express the Plücker coordinates of a line $\mathcal{L}$ of a rigid body $\mathcal{B}$ in terms of those of the line in its reference configuration $\mathcal{L}^0$ (Bottema & Roth, (1978); Pradeep, et al. 1989), as shown in Fig. A.6. To this end, we let $\mathbf{f}$ be the unit vector parallel to $\mathcal{L}$, P being a point of $\mathcal{L}$, and arrange the Plücker coordinates of $\mathcal{L}^0$ and $\mathcal{L}$ into the six-dimensional arrays $\boldsymbol{\lambda}^0$ and $\boldsymbol{\lambda}$, respectively, both defined as

$$\boldsymbol{\lambda}^0 \equiv \begin{bmatrix} \mathbf{f}^0 \\ \mathbf{p}^0 \times \mathbf{f}^0 \end{bmatrix}, \quad \boldsymbol{\lambda} \equiv \begin{bmatrix} \mathbf{f} \\ \mathbf{p} \times \mathbf{f} \end{bmatrix} \tag{A.59}$$

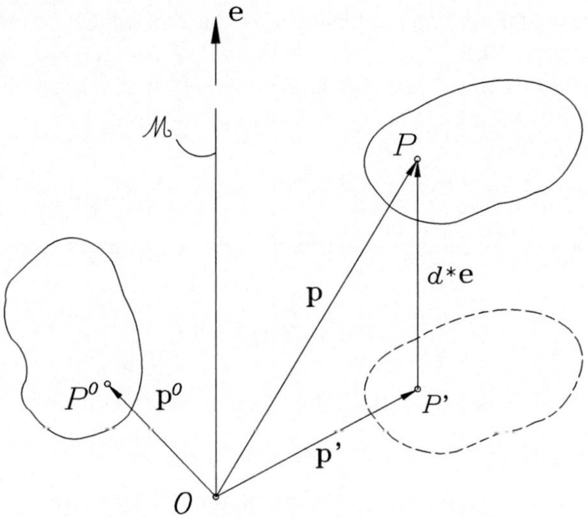

Fig. A.5 Rigid-body displacement with origin on the Mozzi–Chasles axis

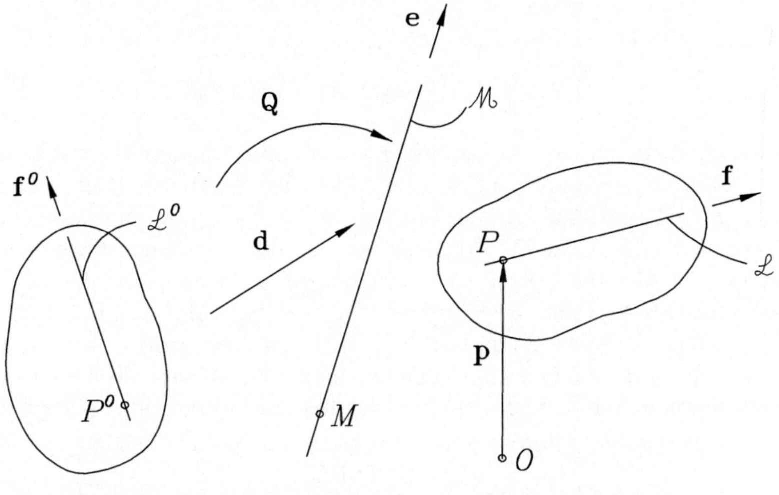

Fig. A.6 The reference and the current configurations of a body and one of its lines

We thus have

$$\mathbf{f} = \mathbf{Q}\mathbf{f}^0, \quad \mathbf{p} = \mathbf{Q}\mathbf{p}^0 + \mathbf{d}\ ,$$

and hence,

$$\mathbf{p} \times \mathbf{f} = (\mathbf{Q}\mathbf{p}^0 + \mathbf{d}) \times \mathbf{Q}\mathbf{f}^0 = (\mathbf{Q}\mathbf{p}^0) \times \mathbf{Q}\mathbf{f} + \mathbf{d} \times \mathbf{Q}\mathbf{f}^0$$

Now, the first term of the rightmost-hand side of the above equation can be simplified upon noticing that *the cross product of two rotated vectors is identical to the rotated cross product*. Furthermore, the second term of the same side can be expressed in terms of $\mathbf{D}$, the cross-product matrix of $\mathbf{d}$, thereby obtaining

$$\mathbf{p} \times \mathbf{f} = \mathbf{Q}(\mathbf{p}^0 \times \mathbf{f}^0) + \mathbf{D}\mathbf{Q}\mathbf{f}^0$$

Upon substitution of the foregoing expressions for $\mathbf{f}$ and $\mathbf{p} \times \mathbf{f}$ into Eq. (A.59), we obtain

$$\boldsymbol{\lambda} = \begin{bmatrix} \mathbf{Q}\mathbf{f}^0 \\ \mathbf{D}\mathbf{Q}\mathbf{f}^0 + \mathbf{Q}(\mathbf{p}^0 \times \mathbf{f}^0) \end{bmatrix},$$

which can be readily cast in the form of a linear transformation of $\boldsymbol{\lambda}^0$, i.e.,

$$\begin{bmatrix} \mathbf{f} \\ \mathbf{p} \times \mathbf{f} \end{bmatrix} = \begin{bmatrix} \mathbf{Q} & \mathbf{O} \\ \mathbf{D}\mathbf{Q} & \mathbf{Q} \end{bmatrix} \begin{bmatrix} \mathbf{f}^0 \\ \mathbf{p}^0 \times \mathbf{f}^0 \end{bmatrix}, \tag{A.60a}$$

$\mathbf{O}$ denoting the 3×3 zero matrix.

As the reader can readily verify, the inverse relation of Eq. (A.60a) takes the form

$$\begin{bmatrix} \mathbf{f}^0 \\ \mathbf{p}^0 \times \mathbf{f}^0 \end{bmatrix} = \begin{bmatrix} \mathbf{Q}^T & \mathbf{O} \\ -\mathbf{Q}^T\mathbf{D} & \mathbf{Q}^T \end{bmatrix} \begin{bmatrix} \mathbf{f} \\ \mathbf{p} \times \mathbf{f} \end{bmatrix} \tag{A.60b}$$

By inspection of the above equations, and recalling the dual-unit-vector representation of a line, as given in Eq. (A.10), we can realize that the dual unit vector of $\mathcal{L}$ can be expressed as the image of the dual unit vector of $\mathcal{L}^0$ upon a linear transformation given by a dual matrix, $\hat{\mathbf{Q}}$. Moreover, the dual matrix of interest can be readily derived from the real matrix of Eq. (A.60a). Indeed, it can be realized, from Sect. 2, that the difference between the primal and the dual parts of a dual quantity is that the units of the dual part are those of the primal part times units of length. Hence, the primal part of the dual matrix sought is bound to be $\mathbf{Q}$, which is dimensionless, the corresponding dual part being $\mathbf{DQ}$, which has units of length. A plausible form of the matrix sought is, then,

$$\hat{\mathbf{Q}} = \mathbf{Q} + \epsilon\, \mathbf{D}\mathbf{Q} \tag{A.61}$$

The correctness of the above expression can be readily verified. Indeed, let $\hat{\mathbf{f}}^* = \mathbf{f} + \epsilon\, \mathbf{p} \times \mathbf{f}$ and $\hat{\mathbf{f}}^{0*} = \mathbf{f}^0 + \epsilon\, \mathbf{p}^0 \times \mathbf{f}^0$ be the dual unit vectors of $\mathcal{L}$ and $\mathcal{L}^0$, respectively. Then upon performing the product $\hat{\mathbf{Q}}\hat{\mathbf{f}}^{0*}$, we notice that the product is rightfully $\hat{\mathbf{f}}^*$, i.e., $\hat{\mathbf{f}}^* = \hat{\mathbf{Q}}\hat{\mathbf{f}}^{0*}$. In the derivations below, we will need expressions for the vector and scalar linear invariants of the product of two matrices, one of which is skew-symmetric. A derivation of these expressions is available (Angeles, 1997). For quick reference, we recall these relations below.

Theorem A.4.2 *Let both* $\mathbf{A}$ *and* $\mathbf{S}$ *be* 3×3 *matrices, the former arbitrary, the latter skew-symmetric. Then,*

$$\mathrm{vect}(\mathbf{SA}) = \frac{1}{2}[\mathrm{tr}(\mathbf{A})\mathbf{1} - \mathbf{A}]\mathbf{s}\ , \tag{A.62}$$

where $\mathbf{s} \equiv \mathrm{vect}(\mathbf{S})$.

Now, as a direct consequence of the above result, we have the following.

Corollary A.4.1 *If* **A** *in Theorem A.4.2 is skew-symmetric, its trace vanishes, the axial vector of the product* **SA** *thus reducing to*

$$\mathrm{vect}(\mathbf{SA}) = -\frac{1}{2}\mathbf{As} = -\frac{1}{2}\mathbf{a} \times \mathbf{s}\ , \tag{A.63}$$

where $\mathbf{a} \equiv \mathrm{vect}(\mathbf{A})$.

Moreover, we have the following theorem.

Theorem A.4.3 *Let* **A**, **S**, *and* **s** *be defined as in Theorem A.4.2. Then,*

$$\mathrm{tr}(\mathbf{SA}) = -2\mathbf{s} \cdot [\mathrm{vect}(\mathbf{A})] \tag{A.64}$$

Furthermore, we prove now that $\hat{\mathbf{Q}}$ is proper orthogonal. Indeed, orthogonality can be proven by performing the product $\hat{\mathbf{Q}}\hat{\mathbf{Q}}^T$ and noticing that this product yields, rightfully, the 3×3 identity matrix, i.e., $\hat{\mathbf{Q}}\hat{\mathbf{Q}}^T = \mathbf{1}$. Proper orthogonality is proven, in turn, upon application of formula (A.15) to matrix $\hat{\mathbf{Q}}$, as given by Eq. (A.61), namely

$$\det(\hat{\mathbf{Q}}) = \det(\mathbf{Q})[1 + \epsilon\,\mathrm{tr}(\mathbf{DQQ}^{-1})] = \det(\mathbf{Q})[1 + \epsilon\,\mathrm{tr}(\mathbf{D})] = 1\ ,$$

thus completing the proof.

The exponential form of the dual rotation matrix can be obtained if we notice that the exponential of a pure dual number $\hat{x} = \epsilon x_0$ reduces to[4]

$$\mathrm{e}^{\epsilon x_0} = 1 + \epsilon x_0 \tag{A.65}$$

On the other hand, we can write

$$\hat{\mathbf{Q}} = (\mathbf{1} + \epsilon\mathbf{D})\mathbf{Q} \tag{A.66}$$

In analogy with Eq. (A.65), the above expression takes the form

$$\hat{\mathbf{Q}} = \mathrm{e}^{\epsilon\mathbf{D}}\mathbf{Q}$$

Furthermore, if we recall the exponential form of **Q**, as given in Eq. (A.46), the foregoing expression simplifies to

[4] The result follows upon the series expansion of the exponential of ϵx_0.

$$\hat{\mathbf{Q}} = e^{\epsilon \mathbf{D}} e^{\phi \mathbf{E}} \tag{A.67}$$

However, since **D** and **E** are unrelated, *they do not share the same set of eigenvectors, and hence, they do not commute under multiplication*, the foregoing expression thus not being further reducible into one single exponential. Nevertheless, if the origin is placed on the Mozzi–Chasles axis, as depicted in Fig. A.5, then the dual rotation matrix becomes

$$\hat{\mathbf{Q}} = \mathbf{Q} + \epsilon\, d^* \mathbf{E}\mathbf{Q}\,, \tag{A.68}$$

where $d^*\mathbf{E}$ is, apparently, the cross-product matrix of vector $d^*\mathbf{e}$. Furthermore, the exponential form of the dual rotation matrix, Eq. (A.67), then simplifies to $\hat{\mathbf{Q}} = e^{(\phi+\epsilon\, d^*)\mathbf{E}}$ or, if we let $\hat{\phi} = \phi + \epsilon\, d^*$, $\hat{\mathbf{Q}}$ can be expressed in a more intuitive form

$$\hat{\mathbf{Q}} = e^{\hat{\phi}\mathbf{E}} \tag{A.69}$$

A.4.2 Velocity Analysis

Upon differentiation of both sides of Eq. (A.51) with respect to time, we obtain

$$\dot{\mathbf{p}} - \dot{\mathbf{a}} = \dot{\mathbf{Q}}(\mathbf{p}^0 - \mathbf{a}^0)\,,$$

and, if we solve for $(\mathbf{p}^0 - \mathbf{a}^0)$ from the aforementioned equation, we obtain

$$\dot{\mathbf{p}} - \dot{\mathbf{a}} = \dot{\mathbf{Q}}\mathbf{Q}^T(\mathbf{p} - \mathbf{a})\,, \tag{A.70}$$

where $\dot{\mathbf{Q}}\mathbf{Q}^T$ is defined as the *angular-velocity matrix* of the motion under study, represented as Ω, namely

$$\Omega \equiv \dot{\mathbf{Q}}\mathbf{Q}^T \tag{A.71a}$$

It can be readily proven that the foregoing matrix is skew-symmetric, i.e.,

$$\Omega^T = -\Omega \tag{A.71b}$$

Moreover, the axial vector of Ω is the angular-velocity vector ω:

$$\omega = \text{vect}(\Omega) \tag{A.71c}$$

We can now write Eq. (A.70) in the form

$$\dot{\mathbf{p}} = \dot{\mathbf{a}} + \Omega(\mathbf{p} - \mathbf{a}) = \dot{\mathbf{a}} + \omega \times (\mathbf{p} - \mathbf{a}) \tag{A.72}$$

Hence,

$$\dot{\mathbf{p}} - \boldsymbol{\omega} \times \mathbf{p} = \dot{\mathbf{a}} - \boldsymbol{\omega} \times \mathbf{a} \equiv \mathbf{v}^0 = \text{const} \tag{A.73}$$

Therefore, the difference $\dot{\mathbf{p}} - \boldsymbol{\omega} \times \mathbf{p}$ is the same for all points of a rigid body. The kinematic interpretation of this quantity is straightforward: If we rewrite $\mathbf{v}^0$ in the form $\mathbf{v}^0 = \dot{\mathbf{p}} + \boldsymbol{\omega} \times (-\mathbf{p})$, then we can readily realize that, $-\mathbf{p}$ being the vector directed from point P of the rigid body to the origin O, $\mathbf{v}^0$ represents the velocity of the point of the body that coincides instantaneously with the origin. Furthermore, we express $\mathbf{d}$, as given by Eq. (A.57), in terms of the position vector $\mathbf{p}$ of an arbitrary point P, thus obtaining

$$\mathbf{d} = \mathbf{p} - \mathbf{Q}\mathbf{p}^0 \tag{A.74}$$

Differentiation of the two sides of the above expression with respect to time leads to

$$\dot{\mathbf{d}} = \dot{\mathbf{p}} - \dot{\mathbf{Q}}\mathbf{p}^0 \, ,$$

which can be readily expressed in terms of the current value of the position vector of P, upon solving for $\mathbf{p}^0$ from Eq. (A.74), namely

$$\dot{\mathbf{d}} = \dot{\mathbf{p}} - \boldsymbol{\Omega}(\mathbf{p} - \mathbf{d}) \quad \text{or} \quad \dot{\mathbf{d}} - \boldsymbol{\omega} \times \mathbf{d} = \dot{\mathbf{p}} - \boldsymbol{\omega} \times \mathbf{p} \tag{A.75}$$

Hence, the difference $\dot{\mathbf{d}} - \boldsymbol{\omega} \times \mathbf{d}$ is identical to the difference $\dot{\mathbf{p}} - \boldsymbol{\omega} \times \mathbf{p}$, i.e.,

$$\dot{\mathbf{d}} - \boldsymbol{\omega} \times \mathbf{d} = \mathbf{v}^0 \tag{A.76}$$

Furthermore, upon dot-multiplying the two sides of Eq. (A.73) by $\boldsymbol{\omega}$, we obtain an interesting result, namely

$$\boldsymbol{\omega} \cdot \dot{\mathbf{p}} = \boldsymbol{\omega} \cdot \dot{\mathbf{a}} \tag{A.77}$$

Therefore, we have the following.

Theorem A.4.4 *The velocities of all points of a rigid body have the same projection onto the angular-velocity vector of the motion under study.*

Similar to the Mozzi–Chasles theorem, we have now the following.

Theorem A.4.5 *Given a rigid body $\mathcal{B}$ under general motion, a set of its points, on one specific line $\mathcal{L}$, undergoes the identical minimum-magnitude velocity $\mathbf{v}^*$ parallel to the angular velocity $\boldsymbol{\omega}$.*

The Plücker coordinates of line $\mathcal{L}$, grouped in the six-dimensional array $\boldsymbol{\lambda}$, are given as

$$\boldsymbol{\lambda} \equiv \begin{bmatrix} \mathbf{f} \\ \boldsymbol{\pi} \times \mathbf{f} \end{bmatrix}, \quad \mathbf{f} \equiv \frac{\boldsymbol{\omega}}{\|\boldsymbol{\omega}\|}, \quad \boldsymbol{\pi} \equiv \frac{\boldsymbol{\omega} \times \mathbf{v}^0}{\|\boldsymbol{\omega}\|^2} \, , \tag{A.78}$$

where $\mathbf{v}^0$ was previously introduced as the velocity of the point of $\mathcal{B}$ that coincides instantaneously with the origin. Line $\mathcal{L}$ is termed the *instant screw axis*—ISA, for brevity.

Thus, the instantaneous motion of $\mathcal{B}$ is defined by a screw of axis $\mathcal{L}$ and pitch p', given by

$$p' = \frac{\dot{\mathbf{p}} \cdot \boldsymbol{\omega}}{\|\boldsymbol{\omega}\|^2}, \tag{A.79}$$

where $\dot{\mathbf{p}}$ is the velocity of an arbitrary point P of $\mathcal{B}$, the product $\dot{\mathbf{p}} \cdot \boldsymbol{\omega}$ being constant by virtue of Theorem A.4.4. A detailed proof of the foregoing results is available in Angeles (1997).

A.4.3 *The Linear Invariants of the Dual Rotation Matrix*

We start by recalling the *linear invariants* of the real rotation matrix (Angeles, 1997), namely

$$\mathbf{q} \equiv \text{vect}(\mathbf{Q}) = (\sin\phi)\mathbf{e}, \quad q_0 \equiv \frac{\text{tr}(\mathbf{Q}) - 1}{2} = \cos\phi \tag{A.80a}$$

Notice that the linear invariants of any 3×3 matrix can be obtained, respectively, from simple differences of its off-diagonal entries and the sum of its diagonal entries. Once the foregoing invariants are calculated, the natural invariants can be obtained uniquely as indicated below. First, the sign of $\mathbf{e}$ can be changed without affecting $\mathbf{q}$ if the sign of ϕ is changed accordingly, which means that the sign of ϕ—or that of $\mathbf{e}$, for that matter—is undefined. In order to define this sign uniquely, we will adopt a positive sign for $\sin\phi$, which means that ϕ is assumed, henceforth, to lie in the interval $0 \le \phi \le \pi$.

We can thus obtain the inverse relations of Eq. (A.80a) in the form

$$\mathbf{e} = \frac{\mathbf{q}}{\|\mathbf{q}\|}, \quad \phi = \arctan\left(\frac{\|\mathbf{q}\|}{q_0}\right), \quad \mathbf{q} \neq \mathbf{0}, \tag{A.80b}$$

the case $\mathbf{q} = \mathbf{0}$ being handled separately. Indeed, $\mathbf{q}$ vanishes under two cases: (a) $\phi = 0$, in which case the body undergoes a pure translation and the axis of rotation is obviously undefined *in position*[5]; and (b) $\phi = \pi$, in which case $\mathbf{Q}$ is symmetric and takes the form

$$\text{For } \phi = \pi: \quad \mathbf{Q} = -\mathbf{1} + 2\mathbf{e}\mathbf{e}^T, \tag{A.80c}$$

the natural invariants thus becoming apparent. Moreover, they can be readily extracted from $\mathbf{Q}$.

Similar to the linear invariants of the real rotation matrix, in the dual case we have

$$\hat{\mathbf{q}} \equiv \text{vect}(\hat{\mathbf{Q}}), \quad \hat{q}_0 \equiv \frac{\text{tr}(\hat{\mathbf{Q}}) - 1}{2} \tag{A.81}$$

[5] It does not exist!

Expressions for the foregoing quantities in terms of the motion parameters are derived below; in the sequel, we also derive expressions for the *dual natural invariants* in terms of the same parameters. We start by expanding the vector linear invariant:

$$\operatorname{vect}(\hat{\mathbf{Q}}) = \operatorname{vect}(\mathbf{Q} + \epsilon\, \mathbf{DQ}) = \operatorname{vect}(\mathbf{Q}) + \epsilon \operatorname{vect}(\mathbf{DQ}) \tag{A.82a}$$

But, by virtue of Eq. (A.43),

$$\operatorname{vect}(\mathbf{Q}) = (\sin\phi)\mathbf{e} \tag{A.82b}$$

Furthermore, the second term of the rightmost-hand side of Eq. (A.82a) can be readily calculated if we recall Theorem A.4.2, with $\mathbf{d} \equiv \operatorname{vect}(\mathbf{D})$:

$$\operatorname{vect}(\mathbf{DQ}) = \frac{1}{2}[\operatorname{tr}(\mathbf{Q})\mathbf{1} - \mathbf{Q}]\mathbf{d} \tag{A.82c}$$

As well, from expression (A.43), we have

$$\operatorname{tr}(\mathbf{Q})\mathbf{1} - \mathbf{Q} = (1 + \cos\phi)\mathbf{1} - \sin\phi\mathbf{E} - (1 - \cos\phi)\mathbf{e}\mathbf{e}^T$$

Upon substitution of the foregoing relation into Eq. (A.82c), the desired expression for $\operatorname{vect}(\mathbf{DQ})$ is obtained:

$$\operatorname{vect}(\mathbf{DQ}) = \frac{1}{2}[(1 + \cos\phi)\mathbf{d} - \sin\phi\mathbf{e} \times \mathbf{d} - (1 - \cos\phi)(\mathbf{e}\cdot\mathbf{d})\mathbf{e}]\ , \tag{A.82d}$$

and hence,

$$\hat{\mathbf{q}} = (\sin\phi)\mathbf{e} + \epsilon\frac{1}{2}[(\cos\phi)(\mathbf{e}\cdot\mathbf{d})\mathbf{e} + (1 + \cos\phi)\mathbf{d} + (\sin\phi)\mathbf{d} \times \mathbf{e} \quad - (\mathbf{e}\cdot\mathbf{d})\mathbf{e}] \tag{A.82e}$$

On the other hand, the position vector $\mathbf{p}^*$ of the Mozzi–Chasles axis, given by Eq. (A.58), can be expressed as

$$\mathbf{p}^* = \frac{1}{2}\frac{\sin\phi}{1 - \cos\phi}\mathbf{e} \times \mathbf{d} + \frac{1}{2}\mathbf{d} - \frac{1}{2}(\mathbf{e}\cdot\mathbf{d})\mathbf{e}\ , \tag{A.83a}$$

and hence,

$$\mathbf{p}^* \times \mathbf{e} = \frac{1}{2}\frac{\sin\phi}{1 - \cos\phi}\mathbf{d} - \frac{1}{2}\frac{\sin\phi}{1 - \cos\phi}(\mathbf{e}\cdot\mathbf{d})\mathbf{e} + \frac{1}{2}\mathbf{d} \times \mathbf{e} \tag{A.83b}$$

Moreover, let us recall the identity[6]

[6] This is just an alternative form of the well-known identify $\cos^2\phi + \sin^2\phi = 1$.

$$\frac{1+\cos\phi}{\sin\phi} = \frac{\sin\phi}{1-\cos\phi}, \tag{A.83c}$$

which allows us to rewrite Eq. (A.83b) in the form

$$\mathbf{p}^* \times \mathbf{e} = \frac{1}{2}\frac{1+\cos\phi}{\sin\phi}\mathbf{d} - \frac{1}{2}\frac{1+\cos\phi}{\sin\phi}(\mathbf{e}\cdot\mathbf{d})\mathbf{e} + \frac{1}{2}\mathbf{d}\times\mathbf{e} \tag{A.83d}$$

Therefore,

$$(\sin\phi)\mathbf{p}^* \times \mathbf{e} = \frac{1}{2}[(1+\cos\phi)\mathbf{d} - (1+\cos\phi)(\mathbf{e}\cdot\mathbf{d})\mathbf{e} + (\sin\phi)\mathbf{d}\times\mathbf{e}] ,$$

$\hat{\mathbf{q}}$ then taking the form

$$\hat{\mathbf{q}} = (\sin\phi)\mathbf{e} + \epsilon\,[(\cos\phi)(\mathbf{e}\cdot\mathbf{d})\mathbf{e} + (\sin\phi)\mathbf{p}^* \times \mathbf{e}] \tag{A.84}$$

If we now recall Eqs.(A.55) and (A.56), $\mathbf{d}\cdot\mathbf{e} \equiv d^* = p\phi$, while $\mathbf{p}^* \times \mathbf{e}$ is the moment of the associated Mozzi–Chasles axis, parallel to vector $\mathbf{e}_0$, and hence, Eq. (A.84) becomes

$$\hat{\mathbf{q}} = (\sin\phi)\mathbf{e} + \epsilon\,[(\cos\phi)\,p\phi\mathbf{e} + (\sin\phi)\mathbf{e}_0] \tag{A.85}$$

Moreover, $\hat{\mathbf{q}}$ can be further simplified to

$$\hat{\mathbf{q}} = \hat{\mathbf{e}}^* \sin\hat{\phi}, \quad \hat{\phi} \equiv \phi\,(1+\epsilon\,p) , \tag{A.86}$$

where $\hat{\mathbf{e}}^*$ is the dual unit vector representing the Mozzi–Chasles axis, i.e., $\hat{\mathbf{e}}^* = \mathbf{e} + \epsilon\,\mathbf{e}_0$.

Now, such as in the real case, we can calculate the dual natural invariants of the motion under study in terms of the foregoing dual linear invariants. We do this by mimicking Eqs. (A.80b), namely

$$\hat{\mathbf{e}}^* = \frac{\hat{\mathbf{q}}}{\|\hat{\mathbf{q}}\|}, \quad \hat{\phi} = \arctan\left(\frac{\|\hat{\mathbf{q}}\|}{\hat{q}_0}\right), \quad \|\hat{\mathbf{q}}\| \neq 0 , \tag{A.87}$$

where $\|\hat{\mathbf{q}}\|$ is calculated from Eq. (A.9e), which gives $\|\hat{\mathbf{q}}\|^2$, the square root of the latter then following from Eq. (A.7), thus obtaining

$$\|\hat{\mathbf{q}}\| = \sin\hat{\phi} = \sin\phi + \epsilon\,(\mathbf{e}\cdot\mathbf{d})\cos\phi , \tag{A.88}$$

and hence, upon simplification,

$$\hat{\mathbf{e}}^* = \mathbf{e} + \epsilon\mathbf{p}^* \times \mathbf{e} = \mathbf{e} + \epsilon\,\mathbf{e}_0 , \tag{A.89}$$

which is rightfully the dual unit vector of the Mozzi–Chasles axis. Furthermore,

$$\mathrm{tr}(\hat{\mathbf{Q}}) = \mathrm{tr}(\mathbf{Q}) + \epsilon\,\mathrm{tr}(\mathbf{DQ}) \;, \tag{A.90a}$$

where, from Theorem A.4.3, $\mathrm{tr}(\mathbf{DQ})$ turns out to be

$$\mathrm{tr}(\mathbf{DQ}) = -2[\mathrm{vect}(\mathbf{Q})] \cdot \mathbf{d} = -2\sin\phi(\mathbf{e} \cdot \mathbf{d}) \;, \tag{A.90b}$$

and hence,

$$\mathrm{tr}(\hat{\mathbf{Q}}) = 1 + 2\cos\phi - \epsilon\,2(\sin\phi)\mathbf{e} \cdot \mathbf{d} \tag{A.90c}$$

Moreover, from the second of Eqs. (A.81),

$$\hat{q}_0 \equiv \cos\hat{\phi} = \cos\phi - \epsilon\,(\sin\phi)(\mathbf{e} \cdot \mathbf{d}) \;,$$

which, by virtue of Eqs. (A.55), leads to

$$\hat{q}_0 = \cos\phi - \epsilon\,(\sin\phi)d^*, \quad \hat{\phi} = \phi + \epsilon\,d^* = \phi(1 + \epsilon\,p) \tag{A.90d}$$

In summary, the dual angle of the dual rotation under study comprises both the angle of rotation of $\mathbf{Q}$ in its primal part and, in its dual part, the axial component of the displacement of all points of the moving body onto the Mozzi–Chasles axis. Upon comparison of the dual angle between two lines, as given in Eq. (A.5), with the dual angle of rotation $\hat{\phi}$, it is then apparent that the primal part of the latter plays the role of the angle between two lines, while the corresponding dual part plays the role of the distance s between those lines. It is noteworthy that a pure rotation has a dual angle of rotation that is real, while a pure translation has an angle of rotation that is a pure dual number.

Example 1: Determination of the screw parameters of a rigid-body motion.

We take here an example from the literature (Angeles, 1997): The cube of Fig. A.7 is displaced from configuration $A^0B^0 \ldots H^0$ into configuration $AB \ldots H$. Find the Plücker coordinates of the Mozzi–Chasles axis of the motion undergone by the cube.

Solution: We start by constructing $\hat{\mathbf{Q}}$: $\hat{\mathbf{Q}} \equiv [\,\hat{\mathbf{i}}^* \;\; \hat{\mathbf{j}}^* \;\; \hat{\mathbf{k}}^*\,]$, where $\hat{\mathbf{i}}^*$, $\hat{\mathbf{j}}^*$, and $\hat{\mathbf{k}}^*$ are the dual unit vectors of lines AB, AD, and AE, respectively. These lines are, in turn, the images of line segments A^0B^0, A^0D^0, and A^0E^0 under the rigid-body motion at hand. The dual unit vectors of the foregoing segments are denoted as $\hat{\mathbf{i}}^{0*}$, $\hat{\mathbf{j}}^{0*}$, and $\hat{\mathbf{k}}^{0*}$, respectively. They are parallel to the X-, Y-, and Z-axes of the figure. We thus have

Fig. A.7 Motion of a cube

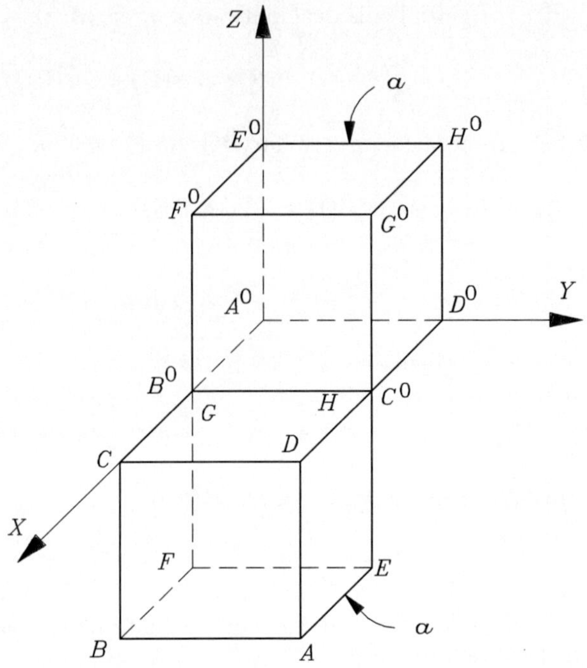

$$\hat{\mathbf{i}}^* = -\mathbf{j}^0 + \epsilon\,\mathbf{a} \times (-\mathbf{j}^0), \quad \hat{\mathbf{j}}^* = \mathbf{k}^0 + \epsilon\,\mathbf{a} \times \mathbf{k}^0, \quad \hat{\mathbf{k}}^* = -\mathbf{i}^0 + \epsilon\,\mathbf{a} \times (-\mathbf{i}^0)\,,$$

where $\mathbf{a}$ is the position vector of A, namely

$$\mathbf{a} = \begin{bmatrix} 2 & 1 & -1 \end{bmatrix}^T a$$

Hence,

$$\begin{aligned} \hat{\mathbf{i}}^* &= -\mathbf{j}^0 + \epsilon\, a(-\mathbf{i}^0 - 2\mathbf{k}^0) \\ \hat{\mathbf{j}}^* &= \mathbf{k}^0 + \epsilon\, a(\mathbf{i}^0 - 2\mathbf{j}^0) \\ \hat{\mathbf{k}}^* &= -\mathbf{i}^0 + \epsilon\, a(\mathbf{j}^0 + \mathbf{k}^0) \end{aligned}$$

Therefore,

$$\hat{\mathbf{Q}} = \begin{bmatrix} -\epsilon\, a & +\epsilon\, a & -1 \\ -1 & -\epsilon\, 2a & +\epsilon\, a \\ -\epsilon\, 2a & 1 & +\epsilon\, a \end{bmatrix},$$

its vector and scalar linear invariants then following

$$\text{vect}(\hat{\mathbf{Q}}) = \frac{1}{2}\begin{bmatrix} 1 - \epsilon\, a \\ -1 + \epsilon\, 2a \\ -1 - \epsilon\, a \end{bmatrix}, \quad \text{tr}(\hat{\mathbf{Q}}) = -\, \epsilon\, (2a)$$

Hence,

$$\|\text{vect}(\hat{\mathbf{Q}})\|^2 = \|\frac{1}{2}\begin{bmatrix} 1 \\ -1 \\ -1 \end{bmatrix}\|^2 + \epsilon\, 2\frac{1}{2}\begin{bmatrix} 1 & -1 & -1 \end{bmatrix}\begin{bmatrix} -1 \\ 2 \\ -1 \end{bmatrix}\frac{a}{2} = \frac{3}{4} - \epsilon\, a$$

Thus,

$$\|\text{vect}(\hat{\mathbf{Q}})\| = \frac{\sqrt{3}}{2} + \epsilon\, \frac{-a}{\sqrt{3}} = \frac{\sqrt{3}}{2} - \epsilon\, \frac{\sqrt{3}}{3} a$$

Therefore, the unit dual vector representing the Mozzi–Chasles axis of the motion at hand, $\hat{\mathbf{e}}^*$, is given by $\hat{\mathbf{e}}^* = \text{vect}(\hat{\mathbf{Q}})/\|\text{vect}(\hat{\mathbf{Q}})\|$, i.e.,

$$\hat{\mathbf{e}}^* = \frac{1}{\sqrt{3}/2}\frac{1}{2}\begin{bmatrix} 1 \\ -1 \\ -1 \end{bmatrix} - \epsilon\, \frac{a}{3/4}\left(\frac{1}{2}\begin{bmatrix} 1 \\ -1 \\ -1 \end{bmatrix}\frac{-\sqrt{3}}{3} - \frac{1}{2}\begin{bmatrix} -1 \\ 2 \\ -1 \end{bmatrix}\frac{\sqrt{3}}{2}\right)$$

After various stages of simplification, the foregoing expression reduces to

$$\hat{\mathbf{e}}^* = \frac{\sqrt{3}}{3}\begin{bmatrix} 1 \\ -1 \\ -1 \end{bmatrix} + \epsilon\, \frac{\sqrt{3}}{9}\begin{bmatrix} -1 \\ 4 \\ -5 \end{bmatrix} a$$

Thus, the Mozzi–Chasles axis is parallel to the unit vector $\mathbf{e}$, which is given by the primal part of $\hat{\mathbf{e}}^*$. The dual part of the same dual unit vector, in turn, represents the moment of the Mozzi–Chasles axis, from which the position vector $\mathbf{p}^*$ of P^*, its point closest to the origin, is readily found:

$$\mathbf{p}^* = \mathbf{e} \times \mathbf{e}_0 = \frac{a}{3}\begin{bmatrix} 3 & 2 & 1 \end{bmatrix}^T$$

A.4.4 The Dual Euler–Rodrigues Parameters of a Rigid-Body Motion

We first recall the definition of the Euler–Rodrigues parameters of a pure rotation, which are isomorphic to the *quaternion* of the rotation (Hamilton, 1844). These are most naturally introduced as the linear invariants of the square root of the rotation at hand, and represented, upon paralleling the definition of the linear invariants, as

$$\mathbf{r} \equiv \operatorname{vect}(\sqrt{\mathbf{Q}}), \quad r_0 \equiv \frac{\operatorname{tr}(\sqrt{\mathbf{Q}}) - 1}{2}, \tag{A.91}$$

the *proper orthogonal* square root of $\mathbf{Q}$ being given as Angeles (1997)[7]

$$\sqrt{\mathbf{Q}} = \mathbf{1} + \sin\left(\frac{\phi}{2}\right)\mathbf{E} + \left[1 - \cos\left(\frac{\phi}{2}\right)\right]\mathbf{E}^2 \tag{A.92}$$

and $\mathbf{E}$ defined in Eq. (A.44). The *dual Euler–Rodrigues parameters* of a rigid-body motion are thus defined as

$$\hat{\mathbf{r}} \equiv \operatorname{vect}(\sqrt{\hat{\mathbf{Q}}}), \quad \hat{r}_0 \equiv \frac{\operatorname{tr}(\sqrt{\hat{\mathbf{Q}}}) - 1}{2} \tag{A.93}$$

Below, we derive an expression for $\sqrt{\hat{\mathbf{Q}}}$. Prior to this, we introduce a relation that will prove useful.

Lemma A.4.2 *Let* $\mathbf{a}$ *and* $\mathbf{b}$ *be arbitrary three-dimensional vectors, and* $\mathbf{c} \equiv \mathbf{a} \times \mathbf{b}$. *The cross-product matrix* $\mathbf{C}$ *of* $\mathbf{c}$ *is given by*

$$\mathbf{C} = \mathbf{b}\mathbf{a}^T - \mathbf{a}\mathbf{b}^T \tag{A.94}$$

Proof This follows by noticing that, for any three-dimensional vector $\mathbf{u}$,

$$\mathbf{c} \times \mathbf{u} = (\mathbf{a} \times \mathbf{b}) \times \mathbf{u} = \mathbf{b}(\mathbf{a}^T\mathbf{u}) - \mathbf{a}(\mathbf{b}^T\mathbf{u}),$$

which readily leads to

$$\mathbf{C}\mathbf{u} = (\mathbf{b}\mathbf{a}^T - \mathbf{a}\mathbf{b}^T)\mathbf{u},$$

thereby completing the proof.

Now we proceed to determine $\sqrt{\hat{\mathbf{Q}}}$. To this end, we regard the motion at hand, from a reference configuration $\mathcal{B}^0$ to a current configuration $\mathcal{B}$, as consisting of a rotation $\mathbf{Q}$ about the origin O followed by a translation $\mathbf{d}$. Then, this motion is decomposed into two parts, as shown in Fig. A.8: First, the body is rotated about the origin O by a rotation $\sqrt{\mathbf{Q}}$ and a translation $\mathbf{d}_s$, as yet to be determined; then, from the configuration $\mathcal{B}'$ thus attained, the body is given a new rotation $\sqrt{\mathbf{Q}}$ about O as well, followed by the same translation $\mathbf{d}_s$.

It is apparent that, from the general expression for the dual rotation matrix, Eq. (A.66), $\sqrt{\hat{\mathbf{Q}}}$ can be represented as

[7] The reader is invited to verify the validity of this expression upon simply squaring the right-hand side of Eq.(A.92), then recalling the Cayley–Hamilton theorem.

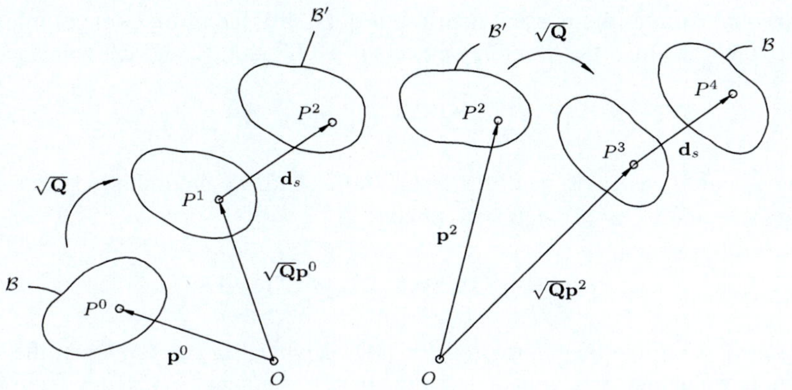

Fig. A.8 Decomposition of the motion of a rigid body

$$\sqrt{\hat{\mathbf{Q}}} = (\mathbf{1} + \epsilon\,\mathbf{D}_s)\sqrt{\mathbf{Q}}\,, \tag{A.95}$$

the calculation of $\sqrt{\hat{\mathbf{Q}}}$ thus reducing to that of the skew-symmetric matrix $\mathbf{D}_s$, which is the cross-product matrix of $\mathbf{d}_s$. This matrix is calculated below in terms of $\sqrt{\mathbf{Q}}$ and $\mathbf{D}$. We have[8]

$$\mathbf{p}^2 \equiv \sqrt{\mathbf{Q}}\mathbf{p}^0 + \mathbf{d}_s\,, \tag{A.96}$$

and hence,

$$\mathbf{p}^4 = \sqrt{\mathbf{Q}}\mathbf{p}^2 + \mathbf{d}_s = \mathbf{Q}\mathbf{p}^0 + (\mathbf{1} + \sqrt{\mathbf{Q}})\mathbf{d}_s \tag{A.97}$$

Moreover, $\mathbf{p}^4$ is the position vector of point P in $\mathcal{B}$, which can be attained by a rotation $\mathbf{Q}$ about O, followed by a translation $\mathbf{d}$, i.e.,

$$\mathbf{p}^4 = \mathbf{Q}\mathbf{p}^0 + \mathbf{d} \tag{A.98}$$

Upon comparing the right-hand sides of Eqs. (A.97) and (A.98), we obtain

$$(\mathbf{1} + \sqrt{\mathbf{Q}})\mathbf{d}_s = \mathbf{d}\,,$$

and hence,

$$\mathbf{d}_s = (\mathbf{1} + \sqrt{\mathbf{Q}})^{-1}\mathbf{d} \tag{A.99}$$

An expression for the above inverse can be derived if we realize that this inverse is an *analytic function* of $\sqrt{\mathbf{Q}}$, which is, in turn, an *analytic function* of $\mathbf{Q}$. We can thus conclude that by virtue of the Cayley–Hamilton theorem, invoked when deriving the

[8] The reader is reminded that "powers of vectors" are meaningless, and hence, superscripts 2 and 0 in Eq. (A.96) cannot possibly be interpreted as "powers" of their associated vectors.

exponential form of the rotation matrix in Eq. (A.46), the inverse sought must be a linear combination of the first three powers of $\mathbf{E}$: $\mathbf{E}^0 \equiv \mathbf{1}$, $\mathbf{E}$, and $\mathbf{E}^2$, namely

$$(\mathbf{1}+\sqrt{\mathbf{Q}})^{-1} = \alpha\mathbf{1} + \beta\mathbf{E} + \gamma\mathbf{E}^2\ , \tag{A.100}$$

with α, β, and γ as yet to be determined. To this end, we premultiply both sides of the above equation by $(\mathbf{1}+\sqrt{\mathbf{Q}})$, to obtain

$$(\mathbf{1}+\sqrt{\mathbf{Q}})(\alpha\mathbf{1} + \beta\mathbf{E} + \gamma\mathbf{E}^2) = \mathbf{1}$$

If we now substitute the expression for $\sqrt{\mathbf{Q}}$ displayed in Eq. (A.92) in the above equation, we obtain three equations for the three unknowns α, β, and γ, from which it is a simple matter to solve for these unknowns, namely

$$\alpha = \frac{1}{2}, \quad \beta = -\frac{\sin(\phi/2)}{2[1+\cos(\phi/2)]}, \quad \gamma = 0\ , \tag{A.101}$$

the inverse sought thus taking the form

$$(\mathbf{1}+\sqrt{\mathbf{Q}})^{-1} = \frac{1}{2}\left[\mathbf{1} - \frac{\sin(\phi/2)}{1+\cos(\phi/2)}\mathbf{E}\right] \tag{A.102}$$

Therefore, Eq. (A.99) yields

$$\mathbf{d}_s = (\mathbf{1}+\sqrt{\mathbf{Q}})^{-1}\mathbf{d} = \frac{1}{2}\left[\mathbf{1} - \frac{\sin(\phi/2)}{1+\cos(\phi/2)}\mathbf{E}\right]\mathbf{d}\ ,$$

i.e.,

$$\mathbf{d}_s = \frac{1}{2}\left[\mathbf{d} - \frac{\sin(\phi/2)}{1+\cos(\phi/2)}\mathbf{e}\times\mathbf{d}\right] \tag{A.103}$$

Hence, $\mathbf{d}_s$ is the sum of two vectors, $\mathbf{D}_s$ thus reducing to the sum of the corresponding cross-product matrices. Moreover, the cross-product matrix of the first term of the right-hand side of the foregoing equation is apparently proportional to $\mathbf{D}$, and that of the second term being proportional to the cross-product matrix of $\mathbf{e}\times\mathbf{d}$. The latter can be readily obtained by application of Lemma A.4.2, which leads to

$$\mathbf{D}_s = \frac{1}{2}\left[\mathbf{D} - \frac{\sin(\phi/2)}{1+\cos(\phi/2)}(\mathbf{d}\mathbf{e}^T - \mathbf{e}\mathbf{d}^T)\right] \tag{A.104}$$

Hence, $\sqrt{\hat{\mathbf{Q}}}$, defined in Eq. (A.95), takes the form

$$\sqrt{\hat{\mathbf{Q}}} = \left\{1 + \epsilon\frac{1}{2}\left[\mathbf{D} - \frac{\sin(\phi/2)}{1+\cos(\phi/2)}(\mathbf{d}\mathbf{e}^T - \mathbf{e}\mathbf{d}^T)\right]\right\}\sqrt{\mathbf{Q}} \tag{A.105}$$

Moreover, the linear invariants of $\sqrt{\hat{\mathbf{Q}}}$ are readily found from the same Eq. (A.98):

$$\operatorname{vect}\left(\sqrt{\hat{\mathbf{Q}}}\right) = \operatorname{vect}\left(\sqrt{\mathbf{Q}}\right) + \epsilon \operatorname{vect}\left(\mathbf{D}_s\sqrt{\mathbf{Q}}\right) \tag{A.106a}$$

and

$$\operatorname{tr}\left(\sqrt{\hat{\mathbf{Q}}}\right) = \operatorname{tr}\left(\sqrt{\mathbf{Q}}\right) + \epsilon \operatorname{tr}\left(\mathbf{D}_s\sqrt{\mathbf{Q}}\right) \tag{A.106b}$$

An expression for $\operatorname{vect}\left(\sqrt{\mathbf{Q}}\right)$, appearing in the first term of $\operatorname{vect}\left(\sqrt{\hat{\mathbf{Q}}}\right)$, can be obtained from Eq. (A.92), namely

$$\operatorname{vect}\left(\sqrt{\mathbf{Q}}\right) = \sin\left(\frac{\phi}{2}\right)\operatorname{vect}(\mathbf{E}) = \sin\left(\frac{\phi}{2}\right)\mathbf{e}\,, \tag{A.107}$$

with the two symmetric terms being filtered out because the vector of each vanishes. Moreover, an expression for the second term of the right-hand side of the same equation, (A.96), is obtained by application of Theorem A.4.2:

$$\operatorname{vect}\left(\mathbf{D}_s\sqrt{\mathbf{Q}}\right) = \frac{1}{2}\left[\operatorname{tr}\left(\sqrt{\mathbf{Q}}\right)\mathbf{1} - \sqrt{\mathbf{Q}}\right]\mathbf{d}_s\,,$$

which can be further expanded, without intermediate lengthy derivations, if we realize that the above expression is the counterpart of that appearing in Eq. (A.82c); the latter is expanded in Eq. (A.82d). Thus, all we need now is to mimic Eq. (A.82d), if with ϕ and $\mathbf{d}$ substituted by their counterparts $\phi/2$ and $\mathbf{d}_s$, respectively, i.e.,

$$\begin{aligned}\operatorname{vect}(\mathbf{D}_s\sqrt{\mathbf{Q}}) = \frac{1}{2}\Bigg\{&\left[1 + \cos\left(\frac{\phi}{2}\right)\right]\mathbf{d}_s - \sin\left(\frac{\phi}{2}\right)\mathbf{e}\times\mathbf{d}_s \\ &-\left[1 - \cos\left(\frac{\phi}{2}\right)\right](\mathbf{e}\cdot\mathbf{d}_s)\mathbf{e}\Bigg\}\end{aligned} \tag{A.108}$$

If we now simplify the above expression, and substitute the simplified expression into Eq. (A.106a), along with Eq. (A.107), we obtain the desired expression for $\hat{\mathbf{r}}$, as defined in Eq. (A.93), and hence,

$$\hat{\mathbf{r}} = \sin\left(\frac{\phi}{2}\right)\mathbf{e} + \epsilon\left[\cos\left(\frac{\phi}{2}\right)p_s\frac{\phi}{2}\mathbf{e} + \sin\left(\frac{\phi}{2}\right)\mathbf{e}_0\right], \tag{A.109}$$

where p_s is the pitch associated with the motion represented by $\sqrt{\hat{\mathbf{Q}}}$, namely

$$p_s \equiv \frac{2}{\phi}\mathbf{d}_s\cdot\mathbf{e} \tag{A.110}$$

where we have recalled the expression for $\mathbf{d}_s$ displayed in Eq. (A.103). Similar to Eq. (A.86), then, the dual vector of the Euler–Rodrigues parameters is given by

$$\hat{\mathbf{r}} = \hat{\mathbf{e}}^* \sin\left(\frac{\hat{\phi}}{2}\right), \quad \text{with} \quad \hat{\phi} \equiv \phi + \epsilon d_s^* \quad \& \quad d_s^* \equiv \mathbf{d}_s \cdot \mathbf{e} \tag{A.111}$$

The scalar of the Euler–Rodrigues parameters under study, $\hat{r}_0$, is now found in terms of the trace of $\sqrt{\hat{\mathbf{Q}}}$, which is displayed in Eq. (A.106b). In that equation,

$$\mathrm{tr}(\sqrt{\mathbf{Q}}) = 1 + 2\cos\left(\frac{\phi}{2}\right),$$

the dual part of the right-hand side of Eq. (A.106b) being calculated by application of Theorem A.4.3:

$$\mathrm{tr}(\mathbf{D}_s\sqrt{\mathbf{Q}}) = -2\mathbf{d}_s \cdot \mathrm{vect}(\sqrt{\mathbf{Q}}) = -2\mathbf{d}_s \cdot \mathbf{e}\sin\left(\frac{\phi}{2}\right)$$

or, in terms of the corresponding pitch p_s,

$$\mathrm{tr}(\mathbf{D}_s\sqrt{\mathbf{Q}}) = -2p_s \sin\left(\frac{\phi}{2}\right)$$

Therefore,

$$\mathrm{tr}(\sqrt{\hat{\mathbf{Q}}}) = 1 + 2\cos\left(\frac{\phi}{2}\right) - \epsilon\, 2p_s \sin\left(\frac{\phi}{2}\right),$$

and hence,

$$\hat{r}_0 = \cos\left(\frac{\phi}{2}\right) - \epsilon\, p_s \sin\left(\frac{\phi}{2}\right), \tag{A.112}$$

which is the counterpart of the second of Eqs. (A.80a). The set $(\hat{\mathbf{r}}, \hat{r}_0)$ constitutes the *dual quaternion* of the motion under study (McCarthy, 1990).

A.5 The Dual Angular Velocity

Similar to the angular-velocity matrix $\mathbf{\Omega}$ introduced in Eq. (A.71a), the *dual angular-velocity matrix* $\hat{\mathbf{\Omega}}$ is defined as

$$\hat{\mathbf{\Omega}} \equiv \dot{\hat{\mathbf{Q}}}\hat{\mathbf{Q}}^T \tag{A.113}$$

Next, we differentiate with respect to time the expression for $\hat{\mathbf{Q}}$ in Eq. (A.66), which yields

$$\dot{\hat{\mathbf{Q}}} = (\mathbf{1} + \epsilon\,\mathbf{D})\dot{\mathbf{Q}} + \epsilon\dot{\mathbf{D}}\mathbf{Q}$$

Upon substitution of the above expression for $\dot{\hat{\mathbf{Q}}}$ and of the expression for $\hat{\mathbf{Q}}$ of Eq. (A.66) into Eq. (A.113), we obtain

$$\hat{\mathbf{\Omega}} = \mathbf{\Omega} + \epsilon\,(\mathbf{D\Omega} - \mathbf{\Omega D} + \dot{\mathbf{D}}) \tag{A.114}$$

The *dual angular-velocity vector* $\hat{\omega}$ of the motion under study is then obtained as the axial vector of the foregoing expression, namely

$$\hat{\omega} = \text{vect}(\hat{\mathbf{\Omega}}) = \omega + \epsilon\,[\text{vect}(\mathbf{D\Omega} - \mathbf{\Omega D}) + \dot{\mathbf{d}}]\,, \tag{A.115}$$

with $\dot{\mathbf{d}}$ being the time-derivative of vector $\mathbf{d}$, introduced in Eq. (A.57). Thus, in order to determine $\hat{\omega}$, all we need is the axial vector of the difference $\mathbf{D\Omega} - \mathbf{\Omega D}$. An expression for this difference can be obtained in various manners, one of which is outlined below. First, we notice that this difference is skew-symmetric, and hence,

$$\text{vect}(\mathbf{D\Omega} - \mathbf{\Omega D}) = 2\,\text{vect}(\mathbf{D\Omega})$$

Further, the vector of $\mathbf{D\Omega}$ is computed by means of Corollary A.4.1, Eq. (A.63), upon substituting $\mathbf{A}$ by $\mathbf{\Omega}$ in that expression. Thus,

$$\text{vect}(\mathbf{D\Omega}) = -\frac{1}{2}\omega \times \mathbf{d} \tag{A.116}$$

Therefore,

$$\hat{\omega} = \omega + \epsilon\,(\dot{\mathbf{d}} - \omega \times \mathbf{d})\,, \tag{A.117}$$

and, if we recall Eq. (A.76), the foregoing expression takes the alternative form

$$\hat{\omega} = \omega + \epsilon\,\mathbf{v}^0 \tag{A.118}$$

In consequence, the dual angular velocity is the *dual representation* of the *twist* $\mathbf{t}$—introduced in Sect. 1.2—of $\mathcal{B}$, defined as the six-dimensional array

$$\mathbf{t} \equiv \begin{bmatrix} \omega \\ \mathbf{v}^0 \end{bmatrix} \tag{A.119}$$

with $\mathbf{v}^0$ defined in Eq. (A.73).

We can therefore find the angular-velocity vector and the moment of the ISA about the given origin—i.e., the *instant-screw parameters* of the motion at hand—if we are given enough information to allow us to compute $\hat{\omega}$. The information required to determine the screw parameters of the motion under study is included in the position and the velocity vectors of three noncollinear points of a rigid body (Angeles, 1997).

However, notice that the dual rotation matrix was obtained in Example 1 in terms of the dual unit vectors representing three mutually orthogonal lines. As well, by virtue of Lemma A.2.1, the three lines of Example 1 were chosen concurrently and mutually orthogonal.

Now, in order to find the instant-screw parameters of interest, we need the time-derivatives of the dual unit vectors representing three concurrent, mutually orthogonal lines, but all we have at our disposal is the position and velocity vectors of three noncollinear points. Nevertheless, once we know three noncollinear points of a rigid body, say A, B, and C, along with their velocities, it is possible to find the position and velocity vectors of three pairs of points defining a triad of concurrent, mutually orthogonal lines, an issue that falls beyond the scope of this book. Rather than discussing the problem at hand in its fullest generality, we limit ourselves to the special case in which the position vector $\mathbf{p}$ of a point P of the rigid body under study can be determined so that the three lines PA, PB, and PC are mutually orthogonal. Further, we let the position vectors of the three given points be, correspondingly, $\mathbf{a}$, $\mathbf{b}$, and $\mathbf{c}$. Thus, point P of the body in this case forms a rectangular trihedron with vertex at P and edges PA, PB, and PC. We can thus express $\mathbf{p}$ as a nonlinear function of the three position vectors $\mathbf{a}$, $\mathbf{b}$, and $\mathbf{c}$:

$$\mathbf{p} = \mathbf{p}(\mathbf{a},\ \mathbf{b},\ \mathbf{c}) \tag{A.120}$$

Moreover, the velocity of point P, $\dot{\mathbf{p}}$, can be calculated now as a linear combination of the velocities of the three given points, by straightforward differentiation of the foregoing expression, namely

$$\dot{\mathbf{p}} = \mathbf{P}_a\dot{\mathbf{a}} + \mathbf{P}_b\dot{\mathbf{b}} + \mathbf{P}_c\dot{\mathbf{c}}\ , \tag{A.121}$$

where $\mathbf{P}_a$, $\mathbf{P}_b$, and $\mathbf{P}_c$ denote the partial derivatives of $\mathbf{p}$ with respect to $\mathbf{a}$, $\mathbf{b}$, and $\mathbf{c}$, respectively. Once the position and the velocity vectors of point P are known, it is possible to determine the time-rates of change of the dual unit vectors representing the three lines PA, PB and PC, as described below.

Let $\hat{\mathbf{e}}^*$ denote the dual unit vector representing the line determined by points A and P, its primary and dual parts, $\mathbf{e}$ and $\mathbf{e}_0$, being given by

$$\mathbf{e} = \frac{\mathbf{a}-\mathbf{p}}{\|\mathbf{a}-\mathbf{p}\|}, \quad \mathbf{e}_0 = \mathbf{p} \times \frac{\mathbf{a}-\mathbf{p}}{\|\mathbf{a}-\mathbf{p}\|} \tag{A.122}$$

Straightforward differentiation of the foregoing expressions with respect to time leads to

$$\dot{\mathbf{e}} = \frac{1}{\|\mathbf{a}-\mathbf{p}\|}\left(\dot{\mathbf{a}} - \dot{\mathbf{p}} - \mathbf{e}\frac{\mathrm{d}}{\mathrm{d}t}\|\mathbf{a}-\mathbf{p}\|\right),$$
$$\dot{\mathbf{e}}_0 = \dot{\mathbf{p}} \times \frac{\mathbf{a}-\mathbf{p}}{\|\mathbf{a}-\mathbf{p}\|} + \mathbf{p} \times \frac{1}{\|\mathbf{a}-\mathbf{p}\|}\left(\dot{\mathbf{a}} - \dot{\mathbf{p}} - \mathbf{e}\frac{\mathrm{d}}{\mathrm{d}t}\|\mathbf{a}-\mathbf{p}\|\right)$$

Upon simplification, we obtain the desired expression for $\dot{\hat{\mathbf{e}}}^*$, namely

$$\dot{\hat{\mathbf{e}}}^* = \frac{1}{\|\mathbf{a}-\mathbf{p}\|}\left[\dot{\mathbf{a}} - \dot{\mathbf{p}} + \epsilon\,(\mathbf{p}\dot{\mathbf{a}} + \dot{\mathbf{p}} \times \mathbf{a})\right] \tag{A.123}$$

Therefore, knowing the velocity of two points of a line, we can determine the time-rate of change of the dual unit vector representing the line. The foregoing idea is best illustrated with the aid of an example.

Example 2: Determination of the ISA of a rigid-body motion

For comparison purposes, we take an example from the literature (Angeles, 1997): The three vertices of the equilateral triangular plate of Fig. A.9, which lie in the X-Y plane, $\{ P_i \}_1^3$, have the position vectors $\{ \mathbf{p}_i \}_1^3$. Moreover, the origin of the coordinate frame $X,\ Y,\ Z$ lies at the centroid C of the triangle, and the velocities of the foregoing points, $\{ \dot{\mathbf{p}}_i \}_1^3$, are given in this coordinate frame as

$$\dot{\mathbf{p}}_1 = \frac{4-\sqrt{2}}{4}\begin{bmatrix}0\\0\\1\end{bmatrix}, \quad \dot{\mathbf{p}}_2 = \frac{4-\sqrt{3}}{4}\begin{bmatrix}0\\0\\1\end{bmatrix}, \quad \dot{\mathbf{p}}_3 = \frac{4+\sqrt{2}}{4}\begin{bmatrix}0\\0\\1\end{bmatrix}$$

With the above information, compute the instant-screw parameters of the motion under study.

Solution: Since the centroid C of the triangle coincides with that of the three given points, we have $\mathbf{c} = \mathbf{0}$, where $\mathbf{c}$ is the position vector of C. Moreover,

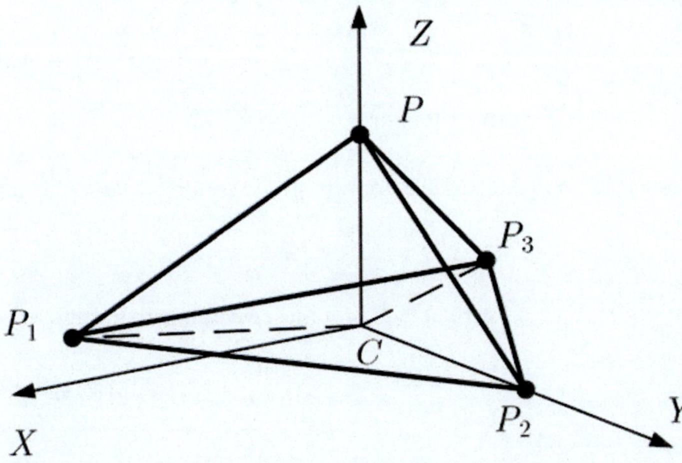

Fig. A.9 A rigid triangular plate undergoing a motion given by the velocity of its vertices

$$\mathbf{p}_1 = \begin{bmatrix} 1/2 \\ -\sqrt{3}/6 \\ 0 \end{bmatrix}, \quad \mathbf{p}_2 = \begin{bmatrix} 0 \\ \sqrt{3}/3 \\ 0 \end{bmatrix}, \quad \mathbf{p}_3 = \begin{bmatrix} -1/2 \\ -\sqrt{3}/6 \\ 0 \end{bmatrix}$$

First and foremost, we have to verify the compatibility of the data. To do this, we calculate the component of the relative velocities of two given points onto the line that they define. It can be readily shown that the data are compatible, and hence, the motion is possible. Next, we obtain the position vector of the point P that, along with $\{P_i\}_1^3$, forms an orthogonal, regular trihedron, i.e., all four sides bear the same length, unity in our case. It is not difficult to realize that the position vector of point P can be expressed as[9]

$$\mathbf{p} = \mathbf{c} + \frac{\sqrt{2}}{3}(\mathbf{p}_2 - \mathbf{p}_1) \times (\mathbf{p}_3 - \mathbf{p}_1) = [0, 0, \sqrt{(6)}/6]^T$$

and hence,

$$\dot{\mathbf{p}} = \dot{\mathbf{c}} + \frac{\sqrt{2}}{3}[(\mathbf{p}_3 - \mathbf{p}_2) \times \dot{\mathbf{p}}_1 + (\mathbf{p}_1 - \mathbf{p}_3) \times \dot{\mathbf{p}}_2 + (\mathbf{p}_2 - \mathbf{p}_1) \times \dot{\mathbf{p}}_3],$$

with the numerical values of $\mathbf{p}$ and $\dot{\mathbf{p}}$ given as follows:

$$\mathbf{p} = \frac{\sqrt{6}}{6}\begin{bmatrix} 0 \\ 0 \\ 1 \end{bmatrix}, \quad \dot{\mathbf{p}} = \frac{1}{12}\begin{bmatrix} 2\sqrt{3} \\ \sqrt{6} \\ 12 - \sqrt{3} \end{bmatrix}$$

Now, let $\hat{\mathbf{e}}_i^*$ denote the dual unit vector representing the line that passes through P and P_i, i.e.,

$$\hat{\mathbf{e}}_i^* = \frac{1}{\|\mathbf{p}_i - \mathbf{p}\|}[\mathbf{p}_i - \mathbf{p} + \epsilon \mathbf{p} \times \mathbf{p}_i], \quad i = 1, 2, 3$$

where

$$\|\mathbf{p}_i - \mathbf{p}\| = \frac{\sqrt{2}}{2}, \quad i = 1, 2, 3$$

Next, the three foregoing dual unit vectors are stored column-wise in the dual rotation matrix $\hat{\mathbf{Q}}$, i.e.,

$$\hat{\mathbf{Q}} = [\hat{\mathbf{e}}_1^* \;\; \hat{\mathbf{e}}_2^* \;\; \hat{\mathbf{e}}_3^*]$$

Upon substitution of the numerical values of these vectors into the above expression, we obtain

[9] Although $\mathbf{c} = \mathbf{0}$ in this case, $\dot{\mathbf{c}} \neq \mathbf{0}$, and hence, $\mathbf{c}$ must be written explicitly in the expression for $\mathbf{p}$.

$$\hat{\mathbf{Q}} = \frac{\sqrt{12}}{12}\begin{bmatrix} 6+\epsilon 2 & -\epsilon 2\sqrt{2} & -6+\epsilon\sqrt{2} \\ -2\sqrt{3}+\epsilon\sqrt{6} & 4\sqrt{3} & -2\sqrt{3}-\epsilon\sqrt{6} \\ -2\sqrt{6} & -2\sqrt{6} & -2\sqrt{6} \end{bmatrix}$$

Likewise, the time-derivative of $\hat{\mathbf{Q}}$ is computed as

$$\dot{\hat{\mathbf{Q}}} = \frac{\sqrt{2}}{24}\left(\begin{bmatrix} -4\sqrt{3} & -4\sqrt{3} & -4\sqrt{3} \\ -2\sqrt{6} & -2\sqrt{6} & -2\sqrt{6} \\ -6\sqrt{2}+2\sqrt{3} & -4\sqrt{3} & 6\sqrt{2}+2\sqrt{3} \end{bmatrix}\right.$$

$$\left. + \epsilon\begin{bmatrix} -1+4\sqrt{3} & 2-8\sqrt{3} & -1+4\sqrt{3} \\ 12-\sqrt{3} & 0 & -12+\sqrt{3} \\ -(2+\sqrt{6}) & 4 & -2+\sqrt{6} \end{bmatrix}\right)$$

Therefore,

$$\hat{\mathbf{\Omega}} = \dot{\hat{\mathbf{Q}}}\hat{\mathbf{Q}}^T = \frac{1}{12}\begin{bmatrix} 0 & -\epsilon(12-\sqrt{3}) & 6\sqrt{2} \\ +\epsilon(12-\sqrt{3}) & 0 & 6 \\ -6\sqrt{2} & -6 & 0 \end{bmatrix},$$

which, as expected, is a dual skew-symmetric matrix. Hence,

$$\hat{\omega} = \text{vect}(\hat{\mathbf{\Omega}}) = \frac{1}{2}\begin{bmatrix} -1 \\ \sqrt{2} \\ 0 \end{bmatrix} + \epsilon\frac{12-\sqrt{3}}{12}\begin{bmatrix} 0 \\ 0 \\ 1 \end{bmatrix},$$

from which we can readily identify

$$\boldsymbol{\omega} = \frac{1}{2}\begin{bmatrix} -1 \\ \sqrt{2} \\ 0 \end{bmatrix}, \quad \mathbf{v}^0 = \frac{12-\sqrt{3}}{12}\begin{bmatrix} 0 \\ 0 \\ 1 \end{bmatrix}$$

Furthermore, the position vector $\boldsymbol{\pi}^*$ of the point P^* of the ISA lying closest to the origin can be obtained from $\mathbf{v}^0$. Indeed, let $\mathbf{v}^*$ be the velocity of P^*, which thus allows us to write

$$\mathbf{v}^0 = \mathbf{v}^* + \boldsymbol{\omega} \times (-\mathbf{p}^*) = \mathbf{v}^* + \mathbf{p}^* \times \boldsymbol{\omega}$$

Upon cross-multiplying the two sides of the foregoing expression by $\boldsymbol{\omega}$, we obtain

$$\mathbf{v}^0 \times \boldsymbol{\omega} = \mathbf{v}^* \times \boldsymbol{\omega} + (\mathbf{p}^* \times \boldsymbol{\omega}) \times \boldsymbol{\omega},$$

whose first term of the right-hand side vanishes because $\mathbf{v}^*$ and $\boldsymbol{\omega}$ are parallel. Therefore,

$$\mathbf{v}^0 \times \boldsymbol{\omega} = (\mathbf{p}^* \times \boldsymbol{\omega}) \times \boldsymbol{\omega} = (\mathbf{p}^* \cdot \boldsymbol{\omega})\boldsymbol{\omega} - \|\boldsymbol{\omega}\|^2\mathbf{p}^*$$

The first term of the rightmost-hand side of the foregoing equation vanishes because, $\mathbf{p}^*$ being the position vector of the point of the ISA that lies closest to the origin, and the ISA being parallel to $\boldsymbol{\omega}$, these two vectors are mutually orthogonal. Now $\mathbf{p}^*$ follows readily from the above expression, which yields

$$\mathbf{p}^* = -\frac{\mathbf{v}^0 \times \boldsymbol{\omega}}{\|\boldsymbol{\omega}\|^2}$$

The quantities involved in the foregoing expression are now evaluated:

$$-\mathbf{v}^0 \times \boldsymbol{\omega} = \boldsymbol{\omega} \times \mathbf{v}^0 = \frac{12 - \sqrt{3}}{24} \begin{bmatrix} \sqrt{2} \\ 1 \\ 0 \end{bmatrix}, \quad \|\boldsymbol{\omega}\|^2 = \frac{3}{4}$$

Finally, $\mathbf{p}^* = \{[(12 - \sqrt{3})]/18\}[\sqrt{2} \;\; 1 \;\; 0]^T$, which coincides with the results reported elsewhere (Angeles, 1997), obtained by an alternative method.

A.6 Conclusions

We revisited dual algebra in the context of kinematic analysis, which led to a straightforward introduction of dual quaternions. In the process, we showed that the parameters of both the finite screw and the instant screw of a rigid-body motion can be computed from the sum of the diagonal and the difference of the off-diagonal entries of the dual rotation and, correspondingly, the dual angular-velocity matrices.

References

Agrawal, O. P. (1987). Hamilton operators and dual-number-quaternions in spatial kinematics. *Mechanism and Machine Theory, 22*(6), 569–575.

Angeles, J. (1982). *Spatial kinematic chains. Analysis, synthesis, optimization.* Springer.

Angeles, J. (1997). *Robotic mechanical systems. Theory, methods, and algorithms.* Springer.

Bottema, O., & Roth, B. (1978). *Theoretical kinematics.* North-Holland Publishing Company.

Brodsky, V., & Shoham, M. (1994). The dual inertia operator and its application to robot dynamics. *ASME Journal of Mechanical Design, 116*, 1089–1095.

Chasles, M. (1830). Notes sur les propriétés générales de deux corps semblables entr'eux et placés d'une manière quelconque dans l'espace, et sur le déplacement fini ou infiniment petit d'un corps solide libre. *Bulletin des Sciences Mathematiques Ferrusaac, 14*, 321–32.

Cheng, H. H., & Thompson, S. (1996). Dual polynomials and complex dual numbers for analysis of spatial mechanisms. In *Proceedings of ASME Design Engineering Technical Conference and Computers in Engineering Conference*. Irvine.

Chevallier, D. P. (1991). Lie algebras, modules, dual quaternions and algebraic methods in kinematics. *Mechanism and Machine Theory, 26*(6), 613–627.

Clifford, W. (1873). Preliminary sketch of biquaternions. *The Proceedings of the London Mathematical Society, 4*, 381–395.
Dimentberg, F. M. (1965). *The screw calculus and its applications in mechanics*. Izdat.
Euler, L. (1776). Nova methodus motum corporum rigidorum degerminandi. *Novii Comentarii Academiæ Scientiarum Petropolitanæ, 20*, 208–238.
Halmos, P. (1974). *Finite-dimensional vector spaces*. Springer.
Hamilton, W. R. (1844). On quaternions: Or on a new system of imaginaries in algebra. *Philosophical Magazine, 25*(3), 489–495.
Kotelnikov, A. P. (1895). Screw calculus and some applications to geometry and mechanics. *Annals of the Imperial University of Kazan, 24.*
McCarthy, J. M. (1990). *An introduction to theoretical kinematics*. The MIT Press.
Mozzi, G. (1763). *Discorso Matematico Sopra il Rotamento Momentaneo dei Corpi*. Stamperia di Donato Campo, Naples.
Pradeep, A. K., Yoder, P. J., & Mukundan, R. (1989). On the use of dual-matrix exponentials in robotic kinematics. *The International Journal of Robotics Research, 8*(5), 57–66.
Rico Martínez, J. M., & Duffy, J. (1993). The principle of transference: History, statement and proof. *Mechanism and Machine Theory, 28*(1), 165–177.
Shoham, M., & Brodsky, V. (1993). Analysis of mechanisms by the dual inertia operator. In J. Angeles, G. Hommel, & P. Kovács (Eds.), *Computational kinematics* (pp. 129–138). Kluwer Academic Publishers.
Simmons, G. F. (1963). *Introduction to topology and modern analysis*. McGraw-Hill Book Co.
Study, & E. (1903). *Geometrie der Dynamen*. Teubner.
Yang, A. T. (1963). *Application of Quaternion Algebra and Dual Numbers to the Analysis of Spatial Mechanisms*. Ph.D. thesis, Columbia University. No. 64-2803 (University Microfilm, Ann Arbor, Michigan).
Yang, A. T., & Freudenstein, F. (1964). Application of dual-number quaternion algebra to the analysis of spatial mechanisms. *Journal of Applied Mechanics, 31*(2), 300–308.

Appendix B
Coupler Curves

The problem of *coupler-curve generation*, a.k.a. *path synthesis*, consists in finding a four-bar linkage whose coupler link carries a point that traces a prescribed curve. We summarize here coupler curves of planar and spherical linkages, largely based on earlier works reported by the authors (Bai & Angeles, 2015; Wu, et al. 2021; Bai, 2021).

B.1 Planar Four-Bar Linkages

A coupler curve is traced by one point on the coupler link of a four-bar linkage. The algebraic coupler curve of a planar four-bar linkage is a sextic bivariate polynomial (Cayley, 1875), namely

$$f(x, y) = \sum_{i,j=0}^{6} P_{ij} x^i y^j = 0, \quad i + j \leq 6 \tag{B.1}$$

where P_{ij} is a constant coefficient, whose arguments are the linkage dimensions. It is also known that the planar four-bar-linkage coupler curve is a *special form* of a sextic bivariate polynomial, *with only 15 coefficients*, out of a total of 28 in the general sextic (Blechschmidt & Uicker, 1986).

With reference to Fig. B.1, whose coupler curve is Γ, as traced by a landmark point P of the coupler link, the linkage is fully defined by nine parameters, as stored in the vector array $\mathbf{x}$ given by

$$\mathbf{x} = [m, h, b_1, b_2, d_1, d_2, l_2, l_3, l_4]^T, \tag{B.2}$$

with link lengths l_2, l_3, l_4, and (b_1, b_2) denoting the coordinates of point B, and (d_1, d_2) those of point D. The length l_1 of the base link is calculated from the coor-

J. Angeles and S. Bai, *Kinematics of Mechanical Systems*,
Mathematical Engineering, https://doi.org/10.1007/978-3-031-09544-3

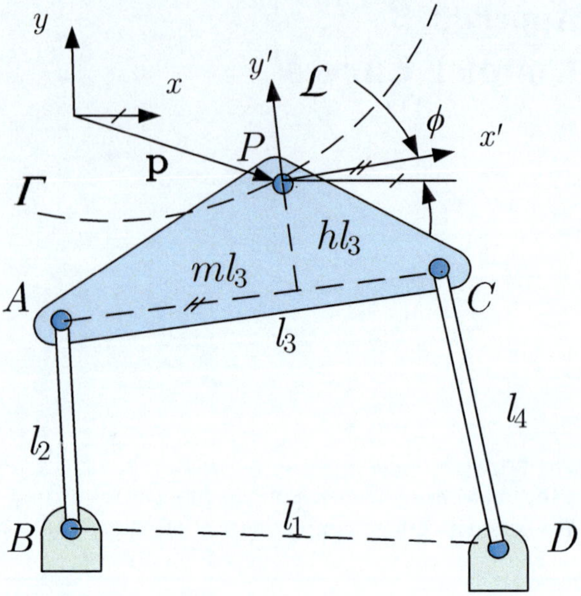

Fig. B.1 A generic four-bar linkage with nine design parameters, including link lengths l_2–l_4, and the coordinates of point P in the coupler link

dinates of points B and D. All nine pertinent geometric parameters of the problem at hand are included in Fig. B.1.

We formulate the coupler-curve equation based on *Burmester theory*. In this regard, a local coordinate frame $\mathcal{L}$, of axes x' and y', fixed to the coupler link with its origin at point P, is defined. Under the usual rigid-body assumption, the geometric constraint on the RR dyad AB is readily derived for any point P on the coupler curve:

$$\|(\mathbf{p}-\mathbf{b})+\mathbf{Q}\mathbf{a}_0\|^2 = l_2^2 \tag{B.3}$$

where $\mathbf{b}$ is the position vector of point B, while $\mathbf{a}_0 = [-ml_3, -hl_3]^T$ is that of point A of the coupler link in $\mathcal{L}$ at its reference pose. Moreover, $\mathbf{Q}$ denotes the rotation matrix of the local frame through an angle ϕ, i.e.,

$$\mathbf{Q} = \begin{bmatrix} \cos\phi & -\sin\phi \\ \sin\phi & \cos\phi \end{bmatrix} \tag{B.4}$$

Upon expansion of Eq. (B.3) and simplification of the expression thus resulting, we obtain

$$(\mathbf{p}-\mathbf{b})^T\mathbf{Q}\mathbf{a}_0 + \frac{1}{2}\left[(\mathbf{p}-\mathbf{b})^T(\mathbf{p}-\mathbf{b}) + \mathbf{a}_0^T\mathbf{a}_0 - l_2^2\right] = 0 \tag{B.5}$$

Similarly, the constraint equation for the RR dyad CD is obtained as

$$\|(\mathbf{p}-\mathbf{d})+\mathbf{Q}\mathbf{c}_0\|^2 = l_4^2$$

which, upon expansion of its right-hand side, readily leads to

$$(\mathbf{p}-\mathbf{d})^T\mathbf{Q}\mathbf{c}_0+\frac{1}{2}\left[(\mathbf{p}-\mathbf{d})^T(\mathbf{p}-\mathbf{d})+\mathbf{c}_0^T\mathbf{c}_0-l_4^2\right]=0 \tag{B.6}$$

with $\mathbf{d}$ denoting the position vector of point D, while $\mathbf{c}_0=[(1-m)l_3,\,-hl_3]^T$ is the position vector of point C of the coupler link at its reference pose. Further, we write $\mathbf{Q}$ in the form $\mathbf{Q}=c\mathbf{1}+s\mathbf{E}$, in which, as usual, $\mathbf{1}$ is the 2×2 identity matrix, while $\mathbf{E}$ is the rotation matrix through an angle of 90°. Moreover, we let $s\equiv\sin\phi$ and $c\equiv\cos\phi$. Upon expansion and simplification, Eq. (B.3) yields

$$A_1c+B_1s+C_1=0 \tag{B.7a}$$

with coefficients

$$A_1=\mathbf{p}^T\mathbf{a}_0-\mathbf{b}^T\mathbf{a}_0 \tag{B.7b}$$
$$B_1=\mathbf{p}^T\mathbf{E}\mathbf{a}_0-\mathbf{b}^T\mathbf{E}\mathbf{a}_0 \tag{B.7c}$$
$$C_1=\frac{1}{2}\left[(\mathbf{p}-\mathbf{b})^T(\mathbf{p}-\mathbf{b})+\mathbf{a}_0^T\mathbf{a}_0-l_2^2\right] \tag{B.7d}$$

Likewise, the synthesis equation for dyad CD is obtained as

$$A_2c+B_2s+C_2=0 \tag{B.7e}$$

with

$$A_2=\mathbf{r}^T\mathbf{c}_0-\mathbf{d}^T\mathbf{c}_0 \tag{B.7f}$$
$$B_2=\mathbf{r}^T\mathbf{E}\mathbf{c}_0-\mathbf{d}^T\mathbf{E}\mathbf{c}_0 \tag{B.7g}$$
$$C_2=\frac{1}{2}\left[(\mathbf{r}-\mathbf{d})^T(\mathbf{r}-\mathbf{d})+\mathbf{c}_0^T\mathbf{c}_0-l_4^2\right] \tag{B.7h}$$

Equations (B.7a) and (B.7e) yield expressions for c and s, namely

$$c=\frac{B_1C_2-C_1B_2}{A_1B_2-A_2B_1};\quad s=-\frac{A_1C_2-A_2C_1}{A_1B_2-A_2B_1} \tag{B.7i}$$

Finally, the substitution of the above expressions into the identity $s^2+c^2=1$ yields the coupler-curve equation, namely

$$\begin{aligned}&A_1^2C_2^2-A_1^2B_2^2+2\,A_1A_2B_1B_2-2\,A_1A_2C_1C_2-A_2^2B_1^2\\&+A_2^2C_1^2+B_1^2C_2^2-2\,B_1B_2C_1C_2+B_2^2C_1^2=0\end{aligned} \tag{B.8}$$

Equation (B.8) describes the coupler curve of the planar four-bar linkage, applicable to any point P on the coupler link. This equation is a special case of a sixth-order bivariate polynomial, which admits the general expression

$$\begin{aligned} f(x,y) =& K_1(x^2+y^2)^3 + (K_2x+K_3y)(x^2+y^2)^2 + (K_4x^2+K_5xy+K_6y^2)(x^2+y^2) \\ &+ K_7x^3 + K_8x^2y + K_9xy^2 + K_{10}y^3 + K_{11}x^2 + K_{12}xy + K_{13}y^2 \\ &+ K_{14}x + K_{15}y + K_{16} = 0 \end{aligned} \tag{B.9}$$

where coefficients K_i, $i = 1, \ldots, 16$, are functions of the linkage parameters. In the above equation, $K_1 = -{l_3}^2/4 < 0$. By introducing $k_i = K_{i+1}/K_1$, $i = 1, \ldots, 15$, Eq. (B.9) becomes

$$\begin{aligned} f(x,y) =& (x^2+y^2)^3 + (k_1x+k_2y)(x^2+y^2)^2 + (k_3x^2+k_4xy+k_5y^2)(x^2+y^2) \\ &+ k_6x^3 + k_7x^2y + k_8xy^2 + k_9y^3 + k_{10}x^2 + k_{11}xy + k_{12}y^2 \\ &+ k_{13}x + k_{14}y + k_{15} = 0 \end{aligned} \tag{B.10}$$

which is a sextic polynomial bivariate equation. In total, we have 15 coefficients $\{k_i\}_1^{15}$, all functions of nine linkage parameters.

By means of symbolic calculations, the expressions for the 15 parameters are found, as shown below only for $k_1, \ldots, k_5$, all other parameters being available online[10]:

$$k_1(\mathbf{x}) = -2hb_2 + 2hd_2 + 2mb_1 - 2md_1 - 4b_1 - 2d_1 \tag{B.11}$$

$$k_2(\mathbf{x}) = 2hb_1 - 2hd_1 + 2mb_2 - 2md_2 - 4b_2 - 2d_2 \tag{B.12}$$

$$\begin{aligned} k_3(\mathbf{x}) =& -8h^2r^2 + h^2b_1^2 - 2h^2b_1d_1 + h^2b_2^2 - 2h^2b_2d_2 + h^2d_1^2 + h^2d_2^2 - 8m^2r^2 \\ &+ m^2b_1^2 - 2m^2b_1d_1 + m^2b_2^2 - 2m^2b_2d_2 + m^2d_1^2 + m^2d_2^2 \\ &+ 4hb_1b_2 - 6hb_1d_2 + 6hb_2d_1 - 4hd_1d_2 + 8mr^2 - 6mb_1^2 + 2mb_1d_1 \\ &- 2mb_2^2 + 2mb_2d_2 + 4md_1^2 + 2ml_2^2 - 2ml_4^2 + 6b_1^2 + 8b_1d_1 + 2b_2^2 \\ &+ d_1^2 + d_2^2 - 2l_2^2 \end{aligned} \tag{B.13}$$

$$\begin{aligned} k_4(\mathbf{x}) =& -4hb_1^2 + 4hb_2^2 + 4hd_1^2 - 4hd_2^2 - 8mb_1b_2 + 8d_1md_2 + 8b_1b_2 \\ &+ 8b_1d_2 + 8b_2d_1 \end{aligned} \tag{B.14}$$

$$\begin{aligned} k_5(\mathbf{x}) =& 2ml_2^2 - 2l_2^2 - 2ml_4^2 - 2m^2b_2d_2 - 2h^2b_2d_2 - 2m^2b_1d_1 + 2mb_1d_1 \\ &- 6hb_1d_2 - 4hb_1b_2 + 6hb_2d_1 - 8h^2r^2 - 8m^2r^2 + 8mr^2 - 2h^2b_1d_1 \\ &+ 2mb_2d_2 + m^2b_1^2 + h^2d_1^2 + m^2b_2^2 + m^2d_1^2 + 4md_2^2 + h^2d_2^2 + m^2d_2^2 \\ &+ 8b_2d_2 + h^2b_2^2 - 2mb_1^2 - 6mb_2^2 + 4hd_1d_2 + h^2b_1^2 + 2b_1^2 + 6b_2^2 + d_1^2 + d_2^2 \end{aligned} \tag{B.15}$$

where $r = l_3/2$ is used for simplification. While $k_6, \ldots, k_{15}$ are not displayed for brevity, it is noted that the highest degree of these coefficients is eight.

[10] https://drive.google.com/file/d/1MOL3qUCqkSfF2DNaa7aU_jLKSKFFXbVK/view?usp=sharing.

Further analysis on the coefficient expressions reveals their properties:

1 Coefficients k_1, k_2, and k_4 are independent of r, l_2, and l_4;
2 Coefficients k_3 and k_5 are linear in l_2^2, l_4^2, and r^2; moreover, the coefficients of l_2^2 in k_3 and k_5 are identical. The same feature holds for the terms in l_4^2 and in r^2;
3 Coefficient pairs (k_6, k_8) and (k_7, k_9) have the same features as the pair (k_3, k_5);
4 Coefficients $k_{10}, \ldots, k_{15}$ contain only terms in l_2^4, $l_2^2 l_4^2$, l_4^4, l_2^2, l_4^2.

Given any algebraic equation of a four-bar linkage coupler curve with known coefficients k_i^*, for $i = 1, \ldots, 15$, a system of coupler-curve coefficient equations with all nine linkage parameters can be established:

$$k_i(\mathbf{x}) - k_i^* = 0; \quad i = 1, \ldots, 15 \tag{B.16}$$

B.1.1 Special Cases

Some special cases are noteworthy. These are cases whereby the origin coincides with a pivoting point, either B or D, and the ground link BD is aligned with either the X- or the Y-axis. In such cases, the expressions for k_3 and k_5 are identical. The same holds for the coefficient pairs (k_6, k_8) and (k_7, k_9). Furthermore, in these special cases, if $h = 0$, which means that point P lies along the line defined by AC, up to six coefficients vanish, thereby leading to a polynomial with only nine coefficients.

B.2 Spherical Four-Bar Linkages

We recall the spherical four-bar linkage in Fig. B.2, reproduced from Chap. 5, with its four linkage dimensions denoted as $\{\alpha_j\}_1^4$. The two grounded revolute joints are labeled B and D, the points at which their axes intersect the unit sphere. The two moving revolute joints are labeled A and C, the points at which their axes intersect the same sphere as well. The axes of the rotation of one dyad are thus defined by the segments $\overline{OB}$ and $\overline{OA}$; the position vectors of points A and B, $\mathbf{a}$ and $\mathbf{b}$, respectively, are both of unit magnitude, i.e.,

$$\|\mathbf{b}\| = 1, \quad \|\mathbf{a}\| = 1 \tag{B.17}$$

Likewise, the position vectors of points C and D are denoted by the unit vectors $\mathbf{c}$ and $\mathbf{d}$, respectively. Moreover, a coupler point is defined on the coupler link, namely point P, with a unit position vector $\mathbf{p}$. The other dimensions of the coupler link are denoted as ρ_1 and ρ_2, in addition to α_3. The problem now consists in finding an algebraic equation that describes the trajectory of the coupler point P, bearing all the foregoing dimensions in its parameters.

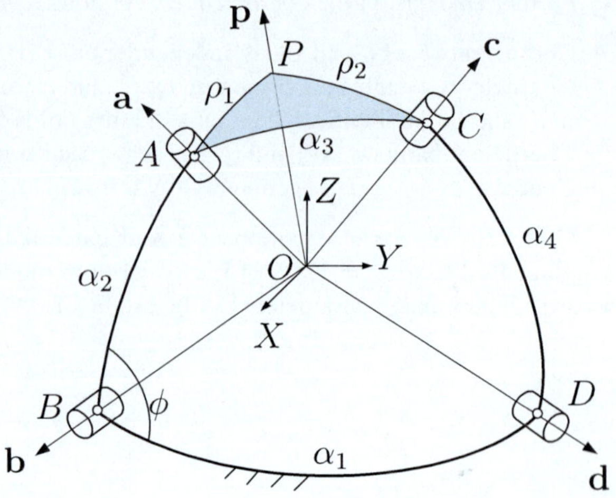

Fig. B.2 Illustration of a spherical four-bar linkage

B.2.1 Parametrized Coordinates

Prior to formulating an algebraic equation of the coupler curve, we introduce an alternative set of coordinates for points on the coupler link, for more than three points. The alternative coordinates are called *parametrized coordinates*, applicable to any link with more than three distinct points: for example, a ternary link, either in planar or spherical mechanisms. With parametrized coordinates, the position vector of a third point on the link can be obtained as a linear combination of three vectors, namely the position vectors of the other two points and their cross product. A similar formulation was earlier introduced and used for planar linkage synthesis (Bai, et al. 2016), whereby a unified formulation of linkage synthesis applicable to all three Stephenson linkages was developed. In this Appendix, we extend the parametrized coordinates to spherical mechanisms.

We take an arbitrary spherical triangle ABC, as shown in Fig. B.3, to describe the parametrized coordinates, thereby forming a basis of $\mathbb{R}^3$. With reference to Fig. B.3, the position vector of point C, namely the unit vector $\mathbf{n}_3$, is expressed in terms of the coordinates of points A and B, which are given as

$$\mathbf{n}_3 = \lambda\mathbf{n}_1 + \mu\mathbf{n}_2 + \nu\mathbf{n}_1 \times \mathbf{n}_2 \tag{B.18}$$

where $\mathbf{n}_1$ and $\mathbf{n}_2$ are unit vectors parallel to $\overrightarrow{OA}$ and $\overrightarrow{OB}$, respectively. Moreover, λ, μ, and ν are dimensionless parameters, namely the *parametrized coordinates* of point C, which can be uniquely determined from Eq. (B.18), as described presently.

Upon dot-multiplying both sides of Eq. (B.18) by vectors $\mathbf{n}_1^T$, $\mathbf{n}_2^T$, and $\mathbf{n}_3^T$ separately, we obtain, respectively

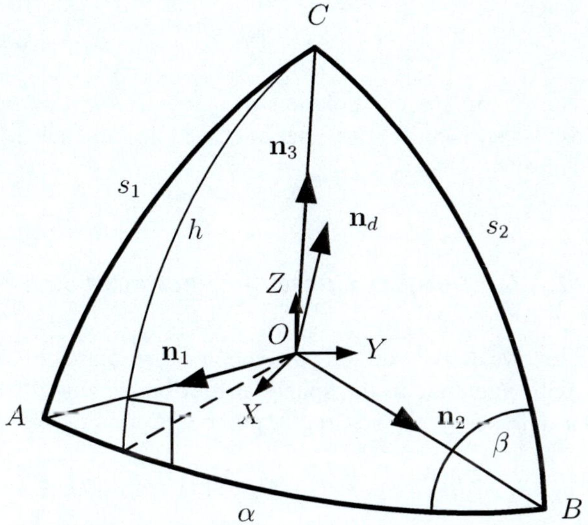

Fig. B.3 Illustration of parametrized coordinates for a spherical triangle with side lengths s_1, s_2, and α

$$\cos s_1 = \lambda + \mu \cos \alpha \tag{B.19a}$$
$$\cos s_2 = \lambda \cos \alpha + \mu \tag{B.19b}$$
$$1 = \lambda \cos s_1 + \mu \cos s_2 + \nu p \tag{B.19c}$$

and hence,

$$\lambda = \frac{\cos s_1 - \cos s_2 \cos \alpha}{\sin^2 \alpha} \tag{B.20a}$$
$$\mu = \frac{\cos s_2 - \cos s_1 \cos \alpha}{\sin^2 \alpha} \tag{B.20b}$$
$$\nu = \frac{1 - \mu \cos s_1 - \nu \cos s_2}{p} \tag{B.20c}$$

where $p = (\mathbf{n}_1 \times \mathbf{n}_2) \cdot \mathbf{n}_3$. From spherical trigonometry, it can be shown that $p = \sin \alpha \sin h$, with $h = \sin^{-1}(\sin \beta \sin s_1)$ denoting the *height* of triangle ABC. As parameters λ, μ, and ν are determined from the triangle dimensions, an important property of the parametrized coordinates is obtained:

The parametrized coordinates are dependent only on the relative location of the three points on the same body, but not on the body attitude.[11]

Equation (B.18) can be expressed in compact form as

$$\mathbf{n}_3 = \mathbf{N}\mathbf{n}_3' \tag{B.21}$$

[11] This result should come as no surprise, given that the pose, and hence, the orientation of a rigid body are fully defined by the positions of three noncollinear body points.

where

$$\mathbf{N} = [\mathbf{n}_1, \mathbf{n}_2, \mathbf{n}_d], \quad \mathbf{n}_3' = [\lambda, \mu, \nu]^T \tag{B.22}$$

As we can see, the columns of matrix $\mathbf{N}$ form an orthogonal basis for $\mathbb{R}^3$, while vector $\mathbf{n}_3'$ includes three parametrized coordinates. Moreover, the foregoing vector is frame-independent.

B.2.2 Coupler Curve of Spherical Four-Bar Linkages

Now we derive explicitly the coupler-curve equation for spherical four-bar linkages. With reference to the spherical four-bar linkage of Fig. 5.10, the position vector $\mathbf{p} = [x, y, z]^T$ of the coupler point P is parallel to OP and has a unit norm, i.e.,

$$g(x, y, z) = ||\mathbf{p}||^2 - 1 = 0. \tag{B.23}$$

Furthermore, With reference to the coupler link APC of Fig. B.2, the angle between $\overline{OA}$ and $\overline{OP}$ is given by

$$\mathbf{p}^T\mathbf{a} = \cos\rho_1 \tag{B.24}$$

which is constant; moreover,

$$\mathbf{a} = \mathbf{R}\mathbf{a}_0 \tag{B.25}$$

with $\mathbf{a}_0$ denoting the initial—reference—position of point A, i.e., a unit vector. Moreover, $\mathbf{R}$ represents the rotation matrix of link BA about $\overline{OB}$. Using the natural invariants of the rigid-body rotation, the general expression for matrix $\mathbf{R}$, representing a rotation about an axis parallel to the unit vector $\mathbf{e}$ about an angle φ, takes the form (Angeles 2014)

$$\mathbf{R} = \mathbf{e}\mathbf{e}^T + s\mathbf{E} + c(\mathbf{1} - \mathbf{e}\mathbf{e}^T), \quad c \equiv \cos\varphi, \quad s \equiv \sin\varphi \tag{B.26}$$

where $\mathbf{E}$ denotes the cross-product matrix of vector $\mathbf{e}$. For the problem at hand, link AB rotates about an axis parallel to $\overline{OB}$. Upon replacing $\mathbf{e}$ in Eq. (B.26) with $\mathbf{b}$, we can obtain the rotation matrix $\mathbf{R}$, Eq. (B.25) thus becoming

$$\mathbf{a} = \underbrace{[\mathbf{b}\mathbf{b}^T + s\mathbf{B} + c(\mathbf{1} - \mathbf{b}\mathbf{b}^T)]}_{\mathbf{R}}\mathbf{a}_0 \tag{B.27}$$

where $\mathbf{B} \equiv \mathrm{CPM}(\mathbf{b})$. Finally, Eq. (B.24) is expressed in the form

$$\mathbf{p}^T[\mathbf{b}\mathbf{b}^T + s\mathbf{B} + c(\mathbf{1} - \mathbf{b}\mathbf{b}^T)]\mathbf{a}_0 = \cos\rho_1 \tag{B.28}$$

Now we move to dyad DC. Using Eq. (B.21), we can express the unit vector $\mathbf{c}$ in terms of vectors $\mathbf{p}$ and $\mathbf{a}$ as

$$\mathbf{c} = \mathbf{M}\mathbf{c}' \tag{B.29}$$

where

$$\mathbf{M} = [\mathbf{p},\ \mathbf{a},\ \mathbf{p} \times \mathbf{a}] = [\mathbf{p},\ \mathbf{R}\mathbf{a}_0,\ \mathbf{P}\mathbf{R}\mathbf{a}_0], \quad \mathbf{c}' = [\lambda_c,\ \mu_c,\ \nu_c]^T \tag{B.30}$$

with $\mathbf{P} \equiv \mathrm{CPM}(\mathbf{p})$.

For any given spherical linkage, $\mathbf{c}'$ can be found as

$$\mathbf{c}' = \mathbf{M}_0^{-1}\mathbf{c}_0, \quad \mathbf{M}_0 = [\mathbf{p}_0,\ \mathbf{a}_0,\ \mathbf{p}_0 \times \mathbf{a}_0] \tag{B.31}$$

$\mathbf{c}_0$ being the unit vector parallel to the reference location of segment $\overline{OC}$.

Furthermore, the angle between $\overline{OC}$ and $\overline{OD}$ remains constant during a rotation. One more relation is thus obtained upon imposing this geometric constraint:

$$\mathbf{d}^T\mathbf{c} = \cos\alpha_4 \tag{B.32}$$

or

$$\mathbf{d}^T\mathbf{M}\mathbf{c}' = \cos\alpha_4 \tag{B.33}$$

Expansion of Eq. (B.33) leads to

$$\lambda_c\mathbf{d}^T\mathbf{p} + \mu_c\mathbf{d}^T\mathbf{R}\mathbf{a}_0 + \nu_c\mathbf{d}^T\mathbf{P}\mathbf{R}\mathbf{a}_0 = \cos\alpha_4 \tag{B.34}$$

Substitution of Eq. (B.26) into Eq. (B.34) yields, after simplification,

$$\lambda_c\mathbf{d}^T\mathbf{p} + \mathbf{d}^T(\mu_c\mathbf{1} + \nu_c\mathbf{P})[\mathbf{e}\mathbf{e}^T + s\mathbf{E} + c(\mathbf{1} - \mathbf{e}\mathbf{e}^T)]\mathbf{a}_0 = \cos\alpha_4 \tag{B.35}$$

Further, upon expansion, Eqs. (B.28) and (B.35) are rewritten as

$$A_1c + B_1s + C_1 = 0 \tag{B.36}$$

$$A_2c + B_2s + C_2 = 0 \tag{B.37}$$

with

$$A_1 = \mathbf{p}^T(\mathbf{1} - \mathbf{b}\mathbf{b}^T)\mathbf{a}_0 \tag{B.38a}$$

$$B_1 = \mathbf{p}^T\mathbf{B}\mathbf{a}_0 \tag{B.38b}$$

$$C_1 = \mathbf{p}^T\mathbf{b}\mathbf{b}^T\mathbf{a}_0 - \cos\rho_1 \tag{B.38c}$$

$$A_2 = \mathbf{d}^T(\mu\mathbf{1} + \nu\mathbf{P})(\mathbf{1} - \mathbf{b}\mathbf{b}^T)\mathbf{a}_0 \tag{B.38d}$$

$$B_2 = \mathbf{d}^T(\mu\mathbf{1} + \nu\mathbf{P})\mathbf{B}\mathbf{a}_0 \tag{B.38e}$$

$$C_2 = \mathbf{d}^T(\mu\mathbf{1} + \nu\mathbf{P})\mathbf{b}\mathbf{b}^T\mathbf{a}_0 + \lambda\mathbf{d}^T\mathbf{p} - \cos\alpha_4 \tag{B.38f}$$

Equations (B.36) and (B.37) thus yield

$$c = \frac{B_1C_2 - C_1B_2}{A_1B_2 - A_2B_1}; \quad s = -\frac{A_1C_2 - A_2C_1}{A_1B_2 - A_2B_1} \tag{B.39}$$

Finally, substitution of the above expressions for c and s into the identity $s^2 + c^2 = 1$ yields

$$\begin{aligned} f(x, y, z) =& A_1^2C_2^2 - 2\,A_1C_2A_2C_1 + A_2^2C_1^2 + B_1C_2A_1B_2 - B_1^2C_2A_2 \\ & - C_1B_2^2A_1 + C_1B_2A_2B_1 - A_1^2B_2^2 + 2\,A_1B_2A_2B_1 - A_2^2B_1^2 = 0 \end{aligned} \tag{B.40}$$

which is the *coupler-curve equation for the spherical four-bar linkage ABCD*. This is a trivariate quartic polynomial, namely

$$f(x, y, z) = \sum_{i,j,k=0}^{4} K_{ijk}x^iy^jz^k = 0, \quad i + j + k \le 4 \tag{B.41}$$

whose coefficients K_{ijk} are all functions of all the linkage dimensions. The above quartic equation defines a conic surface, its intersection with the unit sphere of Eq. (B.23) generating the coupler curve.

Equation (B.41) can attain alternative forms, depending on the specific coordinates adopted. For example, if the coupler-point position is expressed in *spherical coordinates*, namely *longitude* ϑ_P and *latitude* φ_P, we have

$$\mathbf{p} = \begin{bmatrix} \cos\varphi_P\cos\vartheta_P \\ \cos\varphi_P\sin\vartheta_P \\ \sin\varphi_P \end{bmatrix} \tag{B.42}$$

Substituting Eq. (B.42) into Eq. (B.40) yields the equation of the coupler curve in spherical coordinates.

Alternatively, we can substitute the trigonometric functions in Eq. (B.42) with their tan-half-angle identities, namely

$$\cos\vartheta_P = \frac{1-u^2}{1+u^2}; \quad \sin\vartheta_P = \frac{2u}{1+u^2}; \quad u = \tan(\vartheta_P/2) \tag{B.43a}$$

$$\cos\varphi_P = \frac{1-v^2}{1+v^2}; \quad \sin\varphi_P = \frac{2v}{1+v^2}; \quad v = \tan(\varphi_P/2) \tag{B.43b}$$

which leads to a bivariate polynomial, similar to Eq. (B.1). As shown in Sect. 5.2.3, this is a bivariate 16th-degree polynomial.

In addition to the above transformation, the coupler curve can be projected into a plane through a *gnomonic projection*[12] to obtain a *projected curve*. This means that the coupler curve is projected from the center O of the sphere onto a plane tangent to the sphere at a selected point. Next, we project the coupler curve onto the plane $x = 1$, using the coordinate transformation given by Hall et al. (2006), namely

$$x = 1/\rho; \quad y = X/\rho; \quad z = Y/\rho \tag{B.44a}$$

with

$$\rho = \sqrt{U}; \quad U = 1 + X^2 + Y^2 \tag{B.44b}$$

where X and Y are the coordinates on the projective plane.

B.3 Conclusions

In this Appendix we described in detail the derivation of coupler curves for planar and spherical four-bar linkages. The derivation leads to the implicit functions of the coupler curves, namely a sextic bivariate polynomial for planar linkages and a quartic trivariate polynomial for spherical four-bar linkages. The latter is to be complemented with the equation of the unit sphere. The geometric and algebraic properties of both curves are analyzed with regard to the synthesis of planar and spherical linkages.

The derivation of the coupler curve of spatial RCCC linkages is included in Chap. 5, as given in implicit form in Eqs. (5.82a) and (5.82b).

References

Angeles, J. (2014). *Fundamentals of robotic mechanical systems: Theory, methods, and algorithms* (4th ed.). Springer.

Bai, S. (2021). Algebraic coupler curve of spherical four-bar linkages and its applications. *Mechanism and Machine Theory, 158*, 104218.

Bai, S., & Angeles, J. (2015). Coupler-curve synthesis of four-bar linkages via a novel formulation. *Mechanism and Machine Theory, 94*, 177–187.

Bai, S., Wang, D., & Dong, H. (2016). A unified formulation for dimensional synthesis of Stephenson linkages. *ASME Journal of Mechanisms and Robotics,8*(4), 041009.

Blechschmidt, J. L., & Uicker, J. J. (1986). Linkage synthesis using algebraic curves. *ASME Journal of Mechanical Design, 108*(4), 543–548.

Cayley, A. (1875). On three-bar motion. *Proceedings of The London Mathematical Society, s1–7*(1), 136–166.

[12] https://en.wikipedia.org/wiki/Gnomonic_projection.

Hall, E., Rosselot, D., Aull, M., & Balapa, M. (2006). A comparison of two- and three-dimensional imaging. In *Proceedings of SPIE Intelligent Robots and Computer Vision XXIV: Algorithms, Techniques, and Active Vision* (pp. 109–122).
Wu, R., Li, R., & Bai, S. (2021). A fully analytical method for coupler-curve synthesis of planar four-bar linkages. *Mechanism and Machine Theory, 155*, 104070.

Index

J. Angeles and S. Bai, *Kinematics of Mechanical Systems*, Mathematical Engineering, https://doi.org/10.1007/978-3-031-09544-3